Canadian Medicinal Crops

Ginseng in a commercial lath house. The shadow of the wood lath slats that provide the required shade are evident in the photograph. The distinctive red fruits are also apparent. Ginseng is Canada's most valuable medicinal plant, worth nearly 100 million dollars annually.

A Publication of the
National Research Council of Canada
Monograph Publishing Program

Canadian Medicinal Crops

Ernest Small
Principal Research Scientist

and **Paul M. Catling**
Research Scientist

Eastern Cereal and Oilseed Research Centre
Biological Resources Program
Research Branch
Agriculture and Agri-food Canada
Ottawa, Ontario, Canada K1A OC6

NRC·CNRC
NRC Research Press
Ottawa 1999

Printed in Canada on acid-free paper. 

ISBN 0-660-17534-7
NRC No. 42252

Canadian Cataloguing in Publication Data

Small, Ernest, 1940-

Canadian medicinal crops

Includes bibliographical references.
Issued by the National Research Council of Canada.
ISBN 0-660-17534-7

1. Medicinal plants — Canada. 2. Materia medica, Vegetable — Canada. 3. Botanical drug industry — Canada.
I. Catling, Paul M. II. National Research Council Canada. III. Title.

SB294.C3S52 1999 581'6'34'0971 C99-980039-6

Enquiries: Monograph Publishing Program, NRC Research Press, National Research Council of Canada, Ottawa, Ontario K1A 0R6, Canada

Correct citation for this publication: Small, E., and Catling, P.M. 1999. *Canadian Medicinal Crops*. NRC Research Press, Ottawa, Ontario, Canada. 240 p.

CONTENTS

CAUTIONARY NOTE AND DISCLAIMER

Medical self diagnosis and self medication, as noted frequently in this work, are potentially hazardous and are not recommended. Folkloric information is often flawed, and is given here for historical perspective only. The authors are professional botanists, and can not provide medical advice. References to the medicinal uses of plants are not intended to replace the medical advice of trained health-care professionals. This work presents extensive information gathered from the literature, and some error and omission is inevitably incorporated into compilations of this type. Original sources of information should be checked in critical studies. Liability arising directly or indirectly from the use of any of the information provided here is specifically disclaimed.

ACKNOWLEDGMENTS

Much of the text for the individual chapters on medicinal plants is adapted from our series "Poorly Known Economic Plants of Canada" in the *Bulletin of the Canadian Botanical Association*. We thank the editors of this newsletter, J.F. Gerrath and D. Lauzer, for their editorial assistance. The following were coauthors of the articles on the plants mentioned in parenthesis: J. Cayouette (*Hierochloe*), L. Druehl (*Laminariales*), and E. Haber (*Panax*). "A regional review of medicinal plant research in Canada" (Appendix 1) is solely authored by C. Simmons, and we thank her for a very useful contribution. We are also grateful to J. Cayouette and W.J. Cody for constructive criticism of portions of the manuscript. We particularly thank B. Brookes and S. Porebski for extensive help with searching for information, preparation of maps and drawings, layout, and proofreading. Additional valuable technical help was provided by G. Mitrow. B. Brookes prepared or modified several drawings, and several others were prepared by B. Flahey. Most of the illustrations have been adapted from 19th century copyright-free drawings and paintings in books that are generally unavailable to the public; although the artists have passed away, their work lives on. Finally, we thank the anonymous reviewers for their very constructive comments.

EXECUTIVE SUMMARY

The explosive interest in herbal products that provide medicinal or health benefits has made information on this topic vitally important to the public and crucial to farmers, merchants, economists, teachers, the pharmaceutical industry, and the medical arts professions. Canada has the potential to capitalize on tremendous global agribusiness opportunities and is in an excellent position to take advantage of the rapidly expanding market for "nutraceutical crops." Many of these are native to Canada and grow well here. This comprehensive reference guide to important indigenous medicinal plants meets the need for an overview of available information.

Chapters feature species that either are now, or have the potential to be, commercially profitable, such as ginseng, echinacea, Pacific yew, goldenseal, cascara, witch hazel, and kelp. The reader can quickly find details on a particular topic by examining the categories of information, which include: names (scientific, English and French), description and classification, medicinal uses, non-medicinal uses, toxicity, chemistry, importance, ecology, agricultural and commercial aspects, human interest information, selected key literature, and web links. All species are attractively illustrated and distribution maps are included. Additional chapters address: the business of growing medicinal plants; the regulatory and legal framework in Canada for producing and marketing; hazards associated with medicinal plants; and medicinal plant research in Canada. Also provided are: a guide to resources helpful to the farmer and marketer interested in growing medicinal crops in Canada; a guide to experts who can be consulted on various aspects of medicinal plants; a glossary of medicinal and pharmacological terms; a general list of books, review articles and research articles related to Canadian medicinal plants; and an extensive set of general World Wide Web links.

This work should help Canadians share in the economic and health benefits from the extraordinary current growth of the medicinal plant industry.

PREFACE

The need for this book became apparent to us with an ever-increasing number of requests for information on medicinal plants from growers, processors, marketers, and others in the agricultural community. There is no modern comprehensive analysis of native medicinal plants of Canada that possess commercial value, and this work is intended as a first step toward analysing the most promising possibilities. As many as a thousand of Canada's appproximately 3200 native species have been used historically for one or more medicinal purposes by Canada's native peoples. Although a very large number of Canadian plants have medicinal properties and deserve scientific study for the possibility that they may have commercial value, at present only a few dozen species are sold in significant quantities. This work is dedicated to these especially important indigenous medicinal plants, as well as to several that are not important at present but have characteristics that suggest that they have exceptional commercial potential as crops. Contrary to expectation, for most of these valuable species information is widely scattered, often difficult to obtain, and often very technical. Our goal has been to present this information in a form that is easily understood and readily retrieved. Medicinal plants are economically important for both the agricultural and agri-business communities, and so we have emphasized information of value to these sectors. Medicinal plants are also of interest to a wide variety of professions and the general public. Consequently our work is intended to serve as wide a variety of needs as possible. The illustrations we include, and the category of "Myths, Legends, Tales, Folklore and Interesting Facts," respectively relate to our basic human need to see objects, and to learn what is entertaining about them. We have particularly emphasized internet resources, with many links to various aspects of medicinal plants. This work is available as a Web product, and also as a book, since both formats are in demand today.

We will be expanding and improving the content periodically, adding chapters on additional native species of Canada with commercial value. Therefore, we would be very grateful to readers who notice errors and bring these to our attention, and we would also appreciate both criticism and suggestions. Our addresses:

Dr. Ernest Small
ECORC, CEF
Research Branch
Agriculture and Agri-Food Canada
Ottawa, ON K1A OC6
Tel. (613) 759–1370
Fax: (613) 759–1599
E-mail: smalle@em.agr.ca

Dr. Paul Catling
ECORC, CEF
Research Branch
Agriculture and Agri-Food Canada
Ottawa, ON K1A OC6
Tel. (613) 759–1373
Fax: (613) 759–1599
E-mail: catlingp@em.agr.ca

ORGANIZATION AND RATIONALE

The information in this work is encyclopedic, and naturally not all of the material will be of interest to everyone. In the Introduction, we have briefly reviewed a number of topics that are of general interest, and we acquaint the reader with important developments concerning medicinal plants. In later chapters we deal with a number of topics that are more narrowly focused to the needs of those directly concerned with specific aspects of the medicinal plant industry.

The major part of this work is a series of detailed accounts of important native Canadian medicinal plants. The information is organized in categories, which for the most part are self-evident. Under **Morphology**, a basic description of the plant is provided. Since excellent illustrations are presented, the descriptive information is limited to complementary details. Although the material here is not intended to replace identification keys, it should be helpful in verifying the identification of the species. Under **Classification and Geography**, information is given on the most authoritative delimitation of species, and a distribution map is presented, along with complementary notes. The section **Ecology** deals with habitat, pollination, seed dispersal, and a variety of environmental relationships. The ecological information is generally critical to successful cultivation of the species. **Medicinal Uses** includes a listing of older applications, followed by a more detailed account of current usage. The section on **Toxicity** points out possible harmful effects and indicates where caution is required. Under **Chemistry**, a summary of chemical constituents that are important medicinally is provided (detailed information may be found in the articles listed in the selected references). **Non-medicinal Uses** includes interesting minor uses, but is focussed on major alternative uses that need to be considered in developing a medicinal crop. Information helpful to developing and marketing the featured species as a medicinal crop is in **Agricultural and Commercial Aspects**. The section **Myths, Legends, Tales, Folklore, and Interesting Facts** provided us with an opportunity to share some of the amusing and entertaining facts and stories that we encountered in preparing this work. **Selected References** are articles, books, or reviews that give comprehensive or key information, or provide guides to additional information. The **World Wide Web Links** are, like the selected references, key sources of considerable useful information, and also numerous other useful links. For many, Web links will be more easily accessed, and more easily understood, than the scientific literature listed.

A chapter on medicinal cautions includes guidelines for using and marketing medicinal herbs, and summarizes information regarding legal requirements. The intent is to promote safe and responsible manufacture, sale, and use of herbal medicines.

A chapter on business aspects provides advice and background information of practical value to those wishing to participate in the medicinal crop industry. This is followed by chapters on information sources, including a guide to experts willing to provide advice, organizations focussed on medicinal plants, guides to a variety of catalogues, publications, magazines, newsletters, and Web sites that can be helpful, and a review of recent medicinal plant research in various regions of Canada.

A glossary of pharmacological and medical terms relevant to medicinal plants concludes this work. This is quite extensive, reflecting the many ways that medicinal plants have been and continue to be used for health purposes. Readers may find the glossary useful as a general guide to medical conditions, since we have tried to explain terms in comprehensible language. Also, the glossary should be a valuable companion to an ever-increasing literature on medicinal plants that employs medical terms.

We hope that readers will find that the information provided is organized in a way that makes it readily accesible, that it is easy to understand, and that it satisfies current needs.

INTRODUCTION

Medicinal plants from ancient times to the present

Plants have been used in treating human diseases for thousands of years. Some 60,000 years ago, it appears that Neanderthal man valued herbs as medicinal agents; this conclusion is based on a grave in Iran in which pollen grains of eight medicinal plants were found (Solecki and Shanidar 1975). One of these allegedly ancient medicinal herbs, yarrow, is discussed in this work as a modern medicinal plant. Since prehistoric times, shamans or medicine men and women of Eurasia and the Americas acquired a tremendous knowledge of medicinal plants. All of the native plant species discussed in detail in this work were used by native people in traditional medicine. The fact that hundreds of additional species were also used by First Nations Canadians (Arnason et al. 1981) suggests that many of these also have important pharmacological constituents that could be valuable in modern medicine. Up until the 18th century, the professions of doctor and botanist were closely linked. Indeed, the first modern botanic gardens, which were founded in 16th century Italy, in Pisa, Padova and Florence, were medicinal plant gardens attached to medical faculties or schools.

The use of medicinal plants is not just a custom of the distant past. Perhaps 90% of the world's population still relies completely on raw herbs and unrefined extracts as medicines (Duke 1985). A 1997 survey showed that 23% of Canadians have used herbal medicines (http://www.herbsociety.ca/times.html). In addition, as much as 25% of modern pharmaceutical drugs contain plant ingredients (Duke 1993).

The number of medicinal plants

There is a huge number of medicinal plants. In the US, almost 1800 medicinal plant species are commercially available (Muller and Clauson 1998: http://www.medscape.com/SCP/DBT/1998/v10.n05/d3287.mulL/d3287.mull-01.html). It has been estimated that about 13,000 species of plants have been employed for at least a century as traditional medicines by various cultures around the world (Tyler 1993a). A list of over 20,000 medicinal plants has been published (see details in Deans and Svoboda 1990), and very likely a much larger number of the world's flowering plant species have been used medicinally. Sometimes the figure of 70,000 medicinal plant species is cited, but this includes many algae, fungi, and microrganisms that are not really plants as the word is understood by botanists. In any event, there is no other category of plants useful to man (with the possible exception of ornamental plants) that includes so many species, and the question naturally arises why such a staggering number of plants have useful medicinal properties.

Medicinal chemicals

The medicinal qualities of plants are of course due to chemicals. Plants synthesize many compounds called primary metabolites that are critical to their existence. These include proteins, fats, and carbohydrates that serve a variety of purposes indispensable for sustenance and reproduction, not only for the plants themselves, but also for animals that feed on them. Plants also synthesize a dazzling array of additional components, called secondary metabolites, whose function has been debated. Many secondary metabolites are "antibiotic" in a broad sense, protecting the plants against fungi, bacteria, animals, and even other plants. Every plant species contains chemicals that can affect some animals or microorganisms negatively, strongly supporting the interpretation that secondary metabolites play a vital role in combatting diseases and herbivores. "Plants have been a rich source of medicines because they produce a host of bioactive molecules, most of which probably evolved as chemical defenses against predation or infection" (Cox and Balick 1994). Many animals too have evolved chemical defences, but on the whole the plant kingdom appears to far surpass the animal kingdom in this respect. It may be that sedentary animals, i.e., those that spend most of their lives attached to a given substrate just like plants (e.g., reef-building coral species, barnacles), have also evolved extensive chemical defences to protect themselves against predators, just like plants. However, most sedentary animals (the majority of which are found in the seas of the world) are very difficult to collect or grow. For all these reasons, there are far more plants with medicinal uses than animals.

There are major questions regarding just how *beneficial* medicinal herbs can be, and just how *harmful* they can be. As pointed out above, plants contain a very wide variety of secondary compounds. It is clear that some of these compounds, at least in a pure state and at some dosage, are medicinal or toxic. However, it does not necessarily follow that the same compounds present in the herb are as toxic or medicinal as when extracted from the herb, since synergistic (interactive) effects of the chemical components of the herb are possible. An intriguing question about the blend of useful and toxic components of medicinal herbs was posed by Duke (1985, p. 101; repeated on pages 366 and 414): "Can the homeostatic human body selectively take the one it needs?"

Most animals, including humans, have adapted over millions of years to a regular diet of plants. Consequently, the human system is adapted to a regular intake of plant constituents. Essential dietary constituents of plants are reasonably well understood, but the possible therapeutic role of most components of plants is not. Almost certainly humans have been unconsciously ingesting and benefitting from medicinal plant components for hundreds of thousands of years. By contrast, humans are generally not naturally adapted to consuming the powerful modern purified drugs that have become the mainstay of Western Medicine, and deadly adverse reactions (as noted later) are far more common than with herbal medicines. Traditional medicinal usage of herbs by humans, however imperfect and "unscientific" by modern standards, is the result of countless trial-and-error tests that people have conducted, and so traditional usage points the way to natural therapeutic usage. As we later stress, however, "natural" does not necessarily mean "safe." Some herbal products are extremely effective but so dangerous that they should only be used in the hands of skilled medical professionals. Others, however, are sufficiently safe that they can by used by laypeople to help prevent or alleviate minor health problems. Sometimes the herbal drugs are preferable, but as we stress throughout this work, qualified medical personnel should always be consulted.

Herbal medicines vs. pharmaceuticals

Two classes of use of medicinal preparations are commonly recognized, often under the titles herbal and pharmaceutical. Pharmaceuticals, discussed below, are refined or synthesized drugs. The World Health Organization has defined medicinal herbals as follows (WHO 1996):

"Finished, labelled medicinal products that contain as active ingredients aerial or underground parts of plants, or other plant material, or combinations thereof, whether in the crude state or as plant preparations. Plant material includes juices, gums, fatty oils, essential oils, and any other substances of this nature. Herbal medicines may contain excipients [inert additives such as starch used to improve adhesive quality in order to prepare pills or tablets] in addition to the active ingredients. Medicines containing plant material combined with chemically defined active substances, including chemically defined, isolated constituents of plants, are not considered to be herbal medicines. Exceptionally, in some countries herbal medicines may also contain, by tradition, natural organic or inorganic active ingredients which are not of plant origin."

Increasing popularity of medicinal plants

The high costs of western pharmaceuticals put modern health care services out of reach of most of the world's population, which relies on traditional medicine and medicinal plants to meet their primary health care needs. Even where modern medical care is available and affordable, many people prefer more traditional practices. This is particularly true for First Nations and immigrant populations, who have tended to retain ethnic medical practices. In the last decade, there has been considerable interest in ressurecting medicinal plants in western medicine, and integrating their use into modern medical systems. The reasons for this interest are varied, and include: a) low cost: herbals are relatively inexpensive and the cost of pharmaceuticals to governments and individuals is rising; b) drug resistance: the need for alternative treatments for drug-resistant pathogens; c) limitations of medicine: the existence of ailments without an effective pharmaceutical treatment; d) medicinal value: laboratory and clinical corroboration of safety and efficacy for a growing number of medicinal plants; e) cultural exchange: expanding contact and growing respect for foreign cultures, including alternative systems of medicine; and f) commercial value: growing appreciation of trade and other commercial economic opportunities represented by medicinal plants. However, the pace of re-adopting the use of traditional medicinal plants is by no means uniform in western medicine (Duke 1993, Cox and Balick 1994). In parts of Europe, especially in Germany, herbal medicine (or phytomedicine) is much more popular than is the case in North America. Some 67,000 different herbal products are available in Germany (Foster 1995). The already well-established medicinal plant trade of Europe is increasing at an annual rate of about

10%. In Canada, and the US, the regulatory climate has been much less receptive to herbal medicines (Tyler 1993b). This is because lack of proper scientific evaluation, limited regulation, absence of quality control, limited education of many herbal practitioners, and the presence of "snake-oil salesmen" have all combined to give herbal medicine a bad reputation. However, in response to public demand for "alternative" or "complementary" medicine, this situation is changing. At least 20% of Canadians have used some form of alternative therapy, such as herbalism, naturopathy, acupuncture, and homeopathy (Kozyrskyj 1997). Herbs are the fastest-growing part of the pharmacy industry of North America, with an annual growth variously estimated as 15 to 20%, and thousands of herbal products are now available to Canadians (Carmen-Kasparek 1993). Herbal remedies have been estimated to have a current value of between two and ten billion dollars in North America, depending on how comprehensively the category of medicinal herbs is interpreted (Marles 1997). Foster (1995) predicted that with appropriate research and regulation, "herbal medicine will regain its rightful status as an important and integral aspect of classical medicine."

Nutraceuticals and functional foods

Medicinal plants are finding a new, expanding market as herbal components of health foods and preventative medicines, especially under the marketing term "nutraceuticals" (about as frequently spelled nutriceuticals) (Insight Press, 1996a, 1996b; also see Childs 1997). An economic analysis of nutraceuticals in Canada is found in Culhane (1995) and a good general discussion is in Spak (1998). Essentially synonymous phrases include "medical foods," "pharma foods," "phytofoods" and "functional foods[1]." All of these terms are applied to substances that may or may not be considered foods or parts of food, but provide health benefits when eaten.

The most widely used of these terms, nutraceutical, was coined by Dr. Stephen DeFelice of the Foundation for Innovative Medicine, a New Jersey based industry group. His definition was "a food derived from naturally occurring substances which can and should be consumed as part of the daily diet, and which serves to regulate or otherwise affect a particular body process when ingested." The term is now commonly applied to an extremely wide variety of preparations with perceived medicinal value but not necessarily with apparent food value (such as amino acids, essential fats, dietary fibers and fiber-enriched foods, plant and animal pigments, antioxidants, vitamins, minerals, sugar and fat substitutes, fatless meat, skim milk, genetically engineered designer foods, herbal products and processed foods such as cereals, soups and beverages).

Some have contended that fruits and vegetables should be included in "functional foods" because they are so nutrient-packed, while others would reserve the term for foods fortified in some fashion for health (in this sense, the first functional food seems to have been calcium-fortified orange juice). The term "phytonutrient," which should be used for plant materials that by definition have nutritional value, has been applied to medicinal plant preparations without apparent food value. "Phytomedicines" have been defined as therapeutic agents derived from plants or parts of plants, or preparations made from them, but not isolated chemically pure substances, such as menthol from peppermint (Foster 1995).

Unlike pharmaceuticals, which are usually potentially toxic medications that can only be prescribed by a medical doctor, nutritional supplements for the most part can be purchased from a health food store, herbal practitioner or independent distributor. Because they are much less expensive than drugs, herbal preparations or extracts, as additions to diet, have been advanced as a new, cost-effective health care system. Plant-based vitamins and a wide variety of chemical constituents in fruits and vegetables provide many of the benefits of medicinal plants (fruits and vegetables *are* medicinal plants, although rarely thought of as such), and concentrated extracts from them are commonly marketed today as nutriceuticals. The expression "an apple a day keeps the doctor away" reflects the essential medical wisdom of a sensible plant-based diet. This wisdom is quite ancient, as reflected by the saying of Hippocrates (460?-?377 B.C., Greek

[1] In September 1996, Health Canada made available a discussion paper entitled "Recommendations for defining and dealing with functional foods." This contained the following working definitions, which are far more restrictive than found in common usage: "A functional food is similar in appearance to conventional foods, is consumed as part of a usual diet, and has demonstrated physiological benefits and (or) reduces the risk of chronic diseases beyond basic nutritional functions." "A nutraceutical is a product produced from food but sold in pills, powders (potions) and other medicinal forms not generally associated with food and demonstrated to have a physiological benefit or provide protection against chronic disease." For further information, see http://www.legalsuites.com/Buletins%20Jun%2097.htm.

physician, considered to be the father of medicine) "Let food be thy medicine."

Pharmaceutical compounds from plants

Above, we have discussed the use of medicinal plants in the form of raw herbs and crude extracts. Modern pharmacology, however, relies on refined chemicals - either obtained from plants, or synthesized. The first pure medicinal substance derived from plants was morphine, extracted from the opium poppy at the turn of the 19th century. Often, chemicals extracted from plants are altered to produce drugs. For example, diosgenin is obtained from various yam (*Dioscorea*) species of South America, and is converted to progesterone, the basis of the oral contraceptive pill. Aspirin-like chemicals were once obtained from willows (*Salix* species) and European meadowsweet (*Filipendula ulmaria*), but aspirin is now synthesized in the laboratory. Numerous medicines in use today are extracted from plants. About 50 to 60% of pharmaceutical drugs are either of natural origin or obtained through use of natural products as starting points in their synthesis (Verlet 1990, Balandrin et al. 1993). The commercial value of biologically active compounds from plant sources has been estimated to approach $30 billion annually worldwide (Deans and Svoboda 1990). Higher plants have given rise to about 120 commercial drugs and 10–25% of all prescription drugs contain at least one active compound from a higher plant (Duke 1993, Cox and Balick 1994).

The tradition for developing plant-based drugs in modern Western medicine is largely based on a paradigm (model) that there is a single active ingredient in medicinal plants, or at least a primary chemical, that is responsible for the medical effectiveness. However, it may be that many preparations used in traditional herbal medicine are effective because of synergistic (interactive) therapeutic effects of several ingredients. Certainly many traditional herbal drug preparations are compounded from several plants. Such drug mixtures are not of interest to pharmaceutical firms, because they generally cannot be patented (although under some conditions natural products can secure patent protection). On the other hand, as a visit to a pharmacy or "health-food" store quickly reveals, numerous companies are marketing plant mixtures as "dietary supplements," which are in fact being utilized as non-prescription drugs, although there is generally limited or no modern research proof of effectiveness. Since the private sector has limited interest in this issue, there is a clear need for public supported (government) research.

Seeking new drugs from plants

For several decades, the pharmaceutical industry has debated the relative merits of seeking new drugs by synthesis in the laboratory or by screening and testing chemical constituents of plants. The majority of commercial refined plant-based drugs come from only about a hundred plant species. On the whole, laboratory-based chemistry has been supplanting the search for natural drugs, because testing plants is comparatively labor-intensive and random plant testing has been shown to have a relatively low rate of return. For example, many thousands of plants have been tested for drugs effective for treating cancer, but the success rate of finding an effective chemical or chemical derivative, like taxol from the Pacific yew (discussed in this work), was found to be only one in several hundred. Worse, the chance that a pharmaceutical company's investment in plant-based drugs will produce a profitable drug has been estimated as perhaps no better than one in several thousand. In the United States, bringing a new drug to market costs $125 million (Mendelsohn and Balick 1995), so it is no wonder that drug research is undertaken with great caution. [See Feinsilver and Chapela (1996) for the viewpoint that prospecting for new pharmaceuticals among wild plants "has little probability of success." Also see http://www.economist.com/editorial/freeforall/current/st4723.html for an analysis concluding that recent pharmaceutical prospecting in the tropical rainforests has been unsuccessful when guided by traditional ethnobotanical knowledge; it is suggested that simply screening large numbers of plants is a more promising approach.]

Nevertheless, a number of companies have invested in the last several years in the search for plant-based drugs. There are several reasons for this. First, there is a need to study biodiversity, especially in third world countries, and the traditional (folklore) medicinal knowledge that native peoples have, before advancing civilization destroys both the plant species and knowledge of their use. Ethnobotany is the branch of biology specifically dedicated to researching the economic relationships between plants and so-called "primitive" human societies. Second, improvements in automation and robotics have facilitated laboratory evaluation of large samples in a short time. Third, synthetic chemists have proven to need examples of effective natural drugs from plants as structure-function models in order to rationally design analogous drugs on the basis of molecular structure; having a natural example of how a novel plant-derived enzyme functions on human receptors may enable the engineering of analogous synthesized molecules with predicted biological activity.

Economic opportunities

Most of the world's supply of medicinal herbs is obtained by wild collection (often called "wildcrafting"), not by cultivation. Harvesting renewable wild resources is perfectly legitimate so long as this is conducted in a sustainable fashion that does not eliminate populations or degrade the habitat where the plants grow. There are still many minor medicinal plant species in Canada that are abundant in nature and can be collected sustainably. However, because of shrinking wild resources and a strenthening sentiment that biodiversity should be preserved, cultivation is becoming increasingly important. When a plant is (or becomes) popular medicinally, its commercial value is likely to lead to overcollection. Many very important Canadian drug plants grow in the shade of trees (for examples, ginseng, goldenseal, Mayapple, and Pacific yew) and, because they grow very slowly, are especially susceptible to overcollecting. Such nontimber forest resources are of importance to the forest industry, which is looking for alternative crops. Ginseng has been overcollected to the point that the wild Canadian reserve has been designated as "threatened." Native supplies of Pacific yew (which furnishes the anti-cancer drug taxol) are decreasing, and can no longer meet market demand. Sometimes cultivation is preferable even when there is a wild supply, because of the advantages of growing certain cultivars (e.g., uniform maturation or consistency of chemical concentrations), proximity of supply, or quality considerations (e.g., being able to certify that a product has been grown organically). Cultivation offers the possibility of not only preserving economically important wild plants in their natural habitats, but also of providing farmers with new crops.

Domestic and foreign markets for medicinal plants are growing rapidly and provide important opportunities for the development and diversification of Canadian agriculture. Currently, ginseng dominates the medicinal crops of Canada. Ginseng (including both the American and Asian species) is the world's most widely used medicinal plant, and Canada's most important medicinal crop, contributing about $100 million annually to the Canadian economy. Canadian farmers, entrepreneurs and pharmaceutical companies have increasingly been searching to exploit additional medicinal plants that can be grown in Canada, but have been limited by the difficulty of acquiring information on the many promising possibilities that exist.

In the following chapters on selected native Canadian medicinal plants, we provide summary information intended to improve the utilization of economically important plants of Canada, by providing a guide to critical and relevant information. Except for ginseng, information resources are limited and often difficult to obtain. Although our primary focus is economic, we have also tried to include information of general interest, since the topic of medicinal plants is both crucial and fascinating.

Achillea millefolium (yarrow)

Achillea millefolium L. Yarrow

Pronounce the genus name a-KILL' ee-ah or a-KIL-lee-a, and the specific epithet mil-le-FOE-lee-um.

English Common Names

Yarrow, milfoil. The name "yarrow" is said to have originated from Scots Gaelic, where it means "rough stream," and is the name of Scotland's Yarrow River, as well as a place in the county of Selkirkshire. "Yarrow" refers chiefly to *A. millefolium*, but is also sometimes used for related species. The name milfoil, which refers to the finely dissected foliage, is a corruption of *millefolium*, meaning thousand-leaved, originally from Latin. Uncommon or archaic names include: band man's plaything, bloodwort, carpenter's weed, devil's nettle, devil's plaything, field hop, nose bleed, old man's pepper, sanguinary, soldier's woundwort, staunchweed, thousand-leaf, thousand weed, and white yarrow.

French Common Names

Achillée millefeuille, herbe à dinde, persil à dinde.

Morphology

Yarrow plants are strongly aromatic, long-lived perennial herbs, 10–100 (typically 30–60) cm tall, with highly dissected leaves to 15 cm long, and flowers in a flat-topped inflorescence. As with most members of the daisy family, what appears to be a single, small flower about 5 mm across, is actually a flower head, often including outer ray flowers and inner disc flowers. The blooms are usually white, with pink, magenta, and red occasionally found. The flowers, appearing in July and August, are self-incompatible and pollinated by insects. The "seeds," actually small, dry indehiscent fruits about 2 mm long, mature in August and September, and are probably (like many other plants apparently lacking special dispersal adaptations) disseminated by adhesion to wet animals, by blowing over snow surfaces, and in other ways. They are not equipped with the parachute-like pappus that enables effective wind dispersal in many members of the daisy family. Yarrow overwinters in cold climates as a dormant rosette. The plant spreads from an extensive, much-branched rhizome system.

Classification and Geography

Achillea millefolium occurs mostly in temperate and boreal zones of the Northern Hemisphere, and to a lesser extent in more southern regions. It is a very variable and widespread species, for which a satisfactory infraspecific classification is not available. Diploid, tetraploid, and hexaploid plants (respectively with 18, 36, and 54 chromosomes) are known. In Canada the hexaploids occur mainly in the East, and appear to represent weedy plants introduced from Europe. Indigenous Canadian *A. millefolium* is often segregated into several species. Two such segregates, *A. lanulosa* Nutt. (= *A. millefolium* var. *lanulosa* (Nutt.) Piper) and *A. borealis* Bong. (= *A. millefolium* var. *borealis* (Bong.) Farw.), occur mostly as tetraploid (also as hexaploid) plants across Canada. Unlike the former, the latter variant is mostly absent from the southern prairies and southern Ontario and Quebec, while more frequent in the north, in alpine regions, and along both seacoasts. Neither chromosome number nor morphological features are consistently helpful in distinguishing the major indigenous Canadian variants, which are best simply assigned to *A. millefolium sensu lato* (i.e., in the broad sense). As the venerable Harvard botanist M.L. Fernald put it a half century ago, yarrow is still "sadly in need of a well-balanced study, its intricacies not properly understood." One of the more interesting variants that recent studies have supported as distinct is ssp. *megacephala* (Raup) Argus. It is one of several taxa (including *Stellaria arenicola* Raup, *Deschampsia mackenzieana* Raup, and *Salix silicicola* Raup) that are endemic to the Athabasca sand dunes on the south shore of Lake Athabasca in northwestern Saskatchewan. The distinctive taxa occurring in this region are believed to have evolved recently, during the Holocene (10,000 B.P. to the present).

Ecology

Yarrow occurs in a wide variety of natural habitats, such as tundra meadows, saline flats, salt marshes, sand dunes, edges of woods, rocky outcrops, and cliffs. This species appears to be Canada's second most common weed (although not really noxious), surpassed only by the dandelion. It is often found as a weed of open areas, such as pastures, meadows, lawns, roadsides, and waste ground. It does not tolerate shade well, but grows very well on poor soils. Yarrow may occur beside

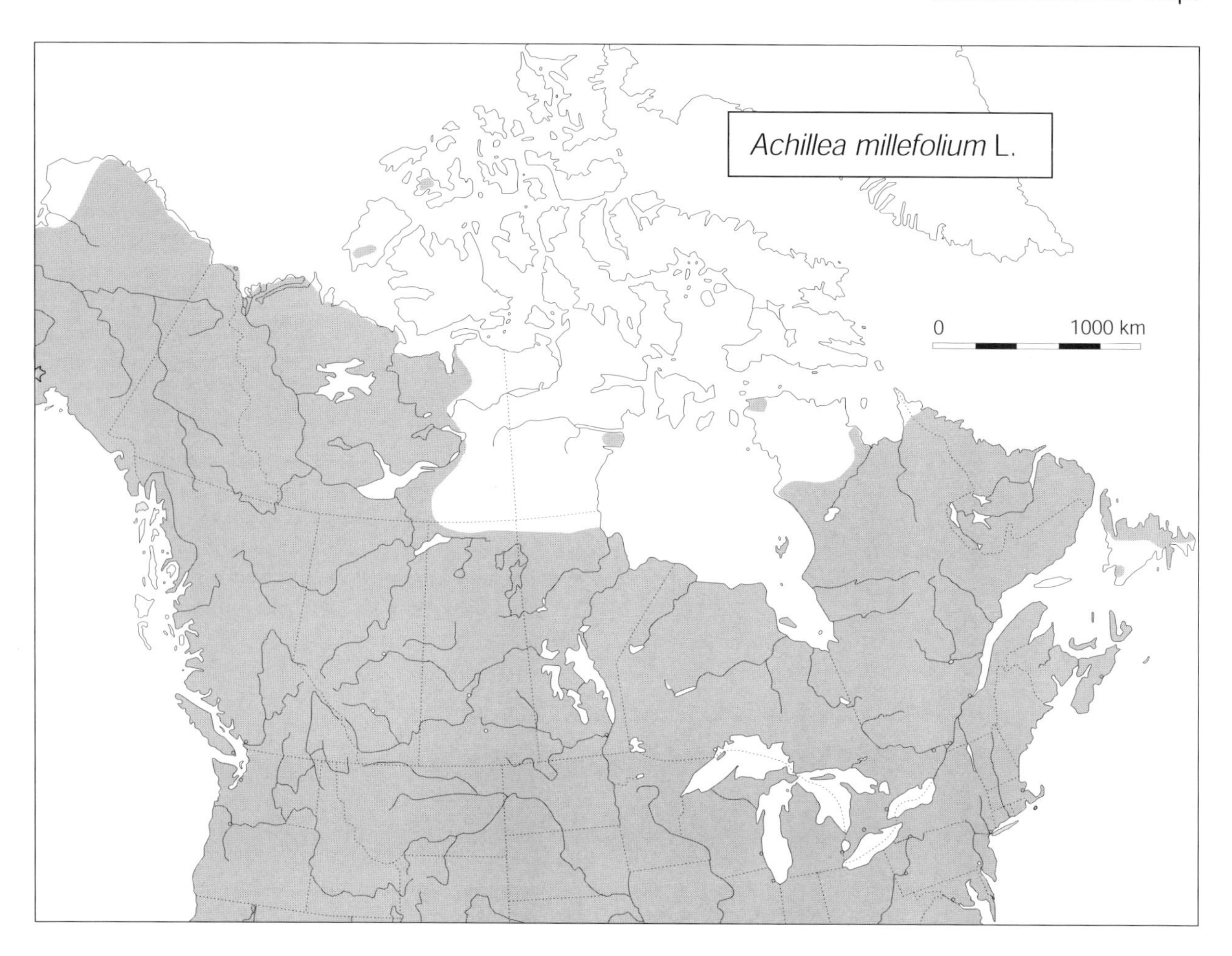

lake shores and stream edges, but is quite drought-tolerant. Its ability to withstand dry conditions is due in part to a deep, extensive root system. Yarrow has been a favorite subject of investigation of students of genecology, who have demonstrated numerous ecotypes, highly specialized to existence in particular habitats. For example genetically-short plants occur in some frequently cut lawns and classic studies of climatic races in plants were based on *Achillea*.

Medicinal Uses

Yarrow has been used medicinally for millennia in both North America and Europe. Its reputation for healing wounds led to such common names as stanchweed, bloodwort, soldier's woundwort, nosebleed, and carpenter's grass. Principal traditional uses besides stopping the flow of blood from wounds included treatment of fevers, the common cold, diarrhea, dysentery, and hypertension. Yarrow has also been used in folk medicine as a cure for toothache, earache, and diseases of the lungs, bladder and kidneys. Today, it is employed internally (as a tea, tincture or pill) to treat gastrointestinal complaints (inflammation, diarrhea, flatulence, cramps), and as a bitter aromatic (to stimulate appetite); and externally in poultices, lotions and bath preparations. Yarrow is present in more than 20 pharmaceutical products marketed in Canada, and is very popular in commercial European herbal remedies.

Toxicity

A reported case of the death of a calf following ingestion of a single plant is often cited, but this is possibly an incorrect interpretation, since the plant does appear to be consumed regularly by grazing animals. "Yarrow dermatitis" involves itching and inflammatory changes in the skin with the formation of vesicles. This is sometimes accompanied by development of photosensitivity (i.e., the condition is triggered by sunlight). Those with allergies to the Asteraceae (Compositae) family, such as susceptibility to ragweed hayfever, should be on guard for allergic reactions to any type of exposure to yarrow – even simply drinking a herbal tea. Except for occasional cases of hayfever and dermatitis, yarrow is generally considered non-toxic.

Nevertheless, as with most medicinal herbs, large doses and prolonged use are inadvisable and can be dangerous. Yarrow should be avoided during pregnancy, because it may stimulate the uterus. High doses may interfere with anticoagulant and hypo- or hypertensive therapies. Caution has also been recommended for epileptic patients.

Chemistry

Well over a hundred chemicals have been characterized in yarrow. Of greatest interest are the lactones, present in a volatile oil. A metabolic derivative of these, azulene, was once thought to be the constituent primarily responsible for the anti-inflammatory and antipruritic properties of yarrow; however, the medicinal value could be due to chamazulene, the sesquiterpene lactones, or other constituents such as tannins, menthol, camphor, sterols and triterpenes. The antispasmodic activity of yarrow could be due to its flavonoids. The alkaloid achilleine is an active hemostatic agent, and may explain the traditional uses of checking bleeding of wounds and sores. It has been hypothesized that the salicylic acid derivatives eugenol, menthol, or similar compounds may produce local analgesia and reduction of fever. The presence of thujone, a known abortifacient, might explain some traditional uses of yarrow associated with the female reproductive system (however, thujone is usually present only in limited amounts). There is evidence of taxonomic and geographical differences in content of these chemicals, but documentation with herbarium vouchers has been poor, and considerable additional analysis is needed. The important constituent chamazule appears to be present in tetraploid plants only. Some have claimed that American plants are more effective medicinally than European plants.

Non-medicinal Uses

Yarrow is a very popular ornamental plant, and there are numerous attractive cultivars (many of which are hybrids) with deeply colored flowers. In some areas, it is recommended as a groundcover to control soil erosion on slopes and hillsides. Its capacity to spread by rhizomes makes it valuable for this application. It is ironic that yarrow is also recommended as a low-maintenance, infrequently mowed lawn, as vigorous attempts are often made to eliminate it from lawns, where low-growing ecotypes are capable of blooming under the lawn mower blades. Yarrow is widely used for dried and fresh flower arrangements, valued for the feathery, fernlike foliage and pungency.

Achillea millefolium (yarrow)

Despite its bitter taste, some domestic livestock (notably sheep) and deer consume yarrow. The French name herbe à dinde reflects previous use of the plant as chicken feed. Cows grazing on yarrow may produce dairy products with an undesirable flavor, but cattle seem to avoid it. Young yarrow leaves are sometimes consumed (cooked or fresh) in salads (large amounts are said to turn urine brown). The leaves and flowers are used to flavor liqueurs, and were once substituted for hops to flavor beer.

In addition to food and ornamental usage, yarrow has been employed as a tobacco, snuff, and hair rinse reputed to brighten blonde hair. Yarrow also has insecticidal constituents, and this is consistent with its reputation as an insect-repelling garden plant that dissuades visits from some ants, flies, and beetles.

Agricultural and Commercial Aspects

Most commercial supplies of medicinal yarrow are obtained from Europe. Yarrow is considered to be a minor essential oil crop, but nevertheless the annual world production of oil is substantial –

about 800 tonnes, estimated to have a value of (US$) 88 million.

Features of yarrow make it easy to adapt as a crop. Germination percentages are generally high, and the seeds only need to be scattered on the soil surface. Seeds sown in the autumn germinate in spring, produce sturdy rosettes the first year and reach mature flowering size in the second year. Since they are perennial, the plants grow back after harvesting of above-ground parts. An additional potential benefit is that some populations yield sufficient nectar for honey production. With its capability for regional adaptation and ability to grow in poor soils of various moisture regimes, Yarrow is a relatively undemanding crop, that can be grown throughout much of Canada. Effective selection of cultivars will require careful attention to genotypic variation, especially with regard to chemical composition.

Myths, Legends, Tales, Folklore, and Interesting Facts

- The generic name *Achillea* is usually interpreted as a reference to Achilles, the legendary Greek hero of the Trojan War (about 1200 B.C.) as reported in Homer's epic poem, Iliad. He is said to have used the foliage of yarrow to stanch the flow of blood from wounded fellow soldiers. Achilles, the mortal son of Thetis, was dipped by his heel into a sacred fire to burn away his mortality. Unfortunately his heel, not bathed in the fire, remained vulnerable to injury. Subsequently he was struck in the heel by an arrow, which killed him, thereafter providing a metaphor for an area of weakness in something that is otherwise invulnerable. A less romantic interpretation of the genus name is that it commemorates a Greek doctor named *Achilles* who recorded the medicinal uses of the plant.
- Ancient Chinese sages are said to have selected yarrow stalks at random as a means of consulting the oracles of the I Ching (Book of Changes; a compendium alleged to contain the wisdom of thousands of years of human history).
- Yarrow was once used in Ireland for love divination: young girls would cultivate a yarrow plant and subsequently place it beneath their pillow so that they would dream of their sweetheart. It was brought by bridesmaids to weddings to ensure seven years of love. The closest that research has come to supporting such uses is the finding that the volatile oil of yarrow causes a sexual response in male cockroaches.
- In France and Ireland yarrow is one of the herbs of St. John (John the Baptist, Christian martyr born on June 24), and on St. John's Eve (June 23, time of traditional European midsummer celebration) the Irish hang it in their houses to avert illness.
- Unearthed fossil pollen from ancient burial caves has been interpreted as evidence of prehistoric use of the herb.

Selected References

Armitage, A.M. 1992. Field studies of *Achillea* as a cut flower: longevity, spacing, and cultivar response. J. Am. Soc. Hortic. Sci. **117**: 65–67.

Bélanger, A., and Dextraze, L. 1992. Variability of chamazulene within *Achillea millefolium*. Acta Hortic. **330**: 141–146.

Bélanger, A., and Dextraze, L. 1993. Variation intraspécifique de la composition chimique de l'huile essentielle d'Achillée millefeuille. Rivista Italiana, EPPOS **4**: 708–712.

Bélanger, A., Dextraze, L., Lachance, Y., and Savard, S. 1991. Extraction et caractérisation de l'achillée millefeuille (*Achillea millefolium* L.) cultivé au Québec. Rivista Italiana, EPPOS **2**: 574–580.

Bourdot, G.W., Saville, D.J., and Field, R.J. 1984. The response of *Achillea millefolium* L. (yarrow) to shading. New Phytol. **97**: 653–663.

Bugge, G. 1991. Investigation on the content of azulene and the chromosome number of the taxa of the *Achillea millefolium* complex. Angew. Bot. **65**: 331–339. [In German.]

Chandler, R.F., Hooper, S.N., and Harvey, M.J. 1982. Ethnobotany and phytochemistry of yarrow, *Achillea millefolium*, Compositae. Econ. Bot. **36**: 203–232.

Chandler, R.F., Hooper, S.N., Hooper, D.L., Jamieson, W.D., Flinn, C.G., and Safe, L.M. 1991. Herbal remedies of the Maritime Indians: sterols and triterpenes of *Achillea millefolium* L. (yarrow). J. Ethnopharmacol. **33**: 187–191.

Clausen, J., Hiesey, W.M., and Nobs, M. 1955. Diploid, tetraploid and hexaploid hybrids of *Achillea*. Carnegie Inst. Wash. Year Book **54**: 182–183.

Clausen, J., Keck, D.D., and Hiesey, W.M. 1940. Experimental studies on the nature of species. I. The effect of varied environments on western North American plants. Carnegie Inst. Wash. Publ. **520**: 296–324.

Clausen, J., Keck, D.D., and Hiesey, W.M. 1948. Experimental studies on the nature of species. III. Environmental responses of climatic races of *Achillea*. Carnegie Inst. Wash. Publ. **581**: 1–129.

Connelly, K. 1991. A yarrow lawn. Pac. Hortic. (San Francisco) **52**(3): 28–30.

Davies, M.G., and Kersey, P.J. 1986. Contact allergy to yarrow and dandelion. Contact Dermatitis **14**: 256–257.

Falk, A.J.C., Bauer, L., and Bell, C.L. 1974. The constituents of the essential oil from *Achillea millefolium* L. **37**: 598–602.

Field, R.J., and Jayaweera, C.S. 1985. Regeneration of yarrow (*Achillea millefolium*) rhizomes as influenced by rhizome age, fragmentation and depth of soil burial. Plant Protect. Q. **1**: 71–73.

Figueiredo, A.C., and Pais, M.S.S. 1994. Ultrastructural aspects of the glandular cells from the secretory trichomes and from the cell suspension cultures of *Achillea millefolium* L. ssp. *millefolium*. Ann. Bot. **74**: 179–190.

Figueiredo, A.C., Barroso, J.G., Pais, M.S.S., and Scheffer, J.J.C. 1992. Composition of the essential oils from leaves and flowers of *Achillea millefolium* ssp. *millefolium*. Flavour Fragrance J. **7**: 219–222.

Gervais, C. 1977. Cytological investigation of the *Achillea millefolium* complex (Compositae) in Quebec. Can. J. Bot. **55**: 796–808.

Goldberg, A.S., Mueller, E.C., Eigen,E., and Desalva, S.J. 1969. Isolation of the anti-inflammatory principles from *Achillea millefolium* (Compositae). J. Pharm. Sci. **58**: 938–941.

Guedon, D., Abbe, P., and Lamaison, J.L. 1993. Leaf and flower head flavonoids of *Achillea millefolium* L. subspecies. Biochem. Syst. Ecol. **21**: 607–611.

Gurevitch, J. 1988. Variation in leaf dissection and leaf energy budgets among populations of *Achillea* from an altitudinal gradient. Am. J. Bot. **75**: 1298–1306

Hausen, B.M., Breuer, J., Weglewski, J., and Rucker, G. 1991. Alpha-Peroxyachifolid and other new sensitizing sesquiterpene lactones from yarrow (*Achillea millefolium* L., Compositae). Contact Dermatitis **24**: 274–280.

Hiesey, W.M. 1953. Comparative growth between and within climatic races of *Achillea* under controlled conditions. Evolution **7**: 297–316.

Hiesey, W.M., and Nobs, M. 1970. Genetic and transplant studies on contrasting species and ecological races of the *Achillea millefolium* complex. Bot. Gaz. **131**: 245–259.

Higgins, S.S., and Mack, R.N. 1987. Comparative responses of *Achillea millefolium* ecotypes to competition and soil type. Oecologia (Berlin) **73**: 591–597.

Hofmann, L., and Fritz, D. 1993. Genetical, ontogenetical and environmental caused variability of the essential oil of different types of the *Achillea millefolium* 'complex.' Acta Hortic. **330**: 147–157.

Kokkalou, E., Kokkini, S., and Hanlidou, E. 1992. Volatile constituents of *Achillea millefolium* in relation to their infraspecific variation. Biochem. Syst. Ecol. **20**: 665–670.

Krupinska, A.A. 1986. Distribution of azulene-containing and azulene-free forms of yarrow (*Achillea millefolium*) sensu lato in northwestern Poland. Herba Pol. **31**: 39–46. [In Polish.]

Lamaison, J.L., and Carnat, A.P. 1988. Study of azulene in three subspecies of *Achillea millefolium* L. Ann. Pharm. Fr. **46**(2): 139–143. [In French.]

Lawrence, B.M. 1984. Progress in essential oils. Perfum. Flavorist **9**(4): 37–41, 43, 46–48.

Lawrence, W.E. 1947. Chromosome numbers in *Achillea* in relation to geographical distribution. Am. J. Bot. **34**: 538–545.

Michler, B., Preitschopf, A., Erhard, P., and Arnold, C.G. 1992. *Achillea millefolium*: relationships among habitat factors, ploidy, occurrence of proazulene and the content of chamazulene in the essential oil. Pz (Pharmazeutische Zeitung) Wissenschaft **137**: 23–29. [In German.]

Mittich, L.W. 1990. Yarrow - the herb of Achilles. Weed Technol. **4**: 451–453.

Mulligan, G.A., and Bassett, I.J. 1959. *Achillea millefolium* complex in Canada and portions of the United States. Can. J. Bot. **37**: 73–79.

Pireh, W., and Tyrl, R.J. 1980. Cytogeography of *Achillea millefolium* in Oklahoma and adjacent states. Rhodora **82**: 361-367.

Purdy, B.G., and Bayer, R.J. 1996. Genetic variation in populations of the endemic *Achillea millefolium* ssp. *megacephala* from the Athabasca sand dunes and the widespread ssp. *lanulosa* in western North America. Can. J. Bot. **74**: 1138–1146.

Robocker, W.C. 1977. Germination of the seeds of common yarrow, *Achilea millefolium* and its herbicidal control. Weed Sci. **25**: 456–459.

Saukel, J., and Laenger, R. 1992. The *Achillea millefolium* — group (Asteraceae) in central Europe. 1. Introduction, evaluation of characters and plant material. Phyton Ann. Rei. Bot. Horn. **31**: 185–207. [In German.]

Saukel, J., and Laenger, R. 1992. The *Achillea millefolium* — group (Asteraceae) in central Europe: 2. Comparison of populations, multivariate classification and biosystematic comments. Phyton Ann. Rei. Bot. Horn. **32**: 47–78. [In German.]

Scheffer, M.C., Ronzelli, P., and Koehler, H.S. 1993. Influence of organic fertilization on the biomass, yield and composition of the essential oil of *Achillea millefolium* L. Acta Hortic. **331**: 109–114.

Schneider, A. 1984. Our honey flora: *Achillea millefolium* L. [Taxonomic description]. Unsere Bienenflora: die gemeine Schafgarbe *Achillea millefolium* L.). Allg. Dtsch. Imkerztg. **18**: 323. [In German.]

Stahl, E., and Wollensah, A. 1986. Observations on the function of the glandular hairs of yarrow. Effects of selective herbicides on the glandular hairs and tissue of the florets. J. Plant Physiol. **122**: 93–96.

Terziiski, D., Yurukova-Grancharova, P., Daskalova, T., and Robeva, P. 1995. Apomixis in the morphological complex of *Achillea millefolium* (Asteraceae). Dokl. B"lgarskata Akad. Nauk. **48**(3): 53–56.

Tewari, J.P., Srivastava, M.C., and Bajpai, J.L.1974. Phytopharmacologic studies of *Achillea millefolium* Linn. Indian J. Med. Sci. **28**: 331–336.

Tunon, H., Thorsell, W., and Bohlin, L. 1994. Mosquito repelling activity of compounds occurring in *Achillea millefolium* L. (Asteraceae). Econ. Bot. **48**: 111–120.

Tozyo, T., Yoshimura, Y., Sakurai, K., Uchida, N., Takeda, Y., Nakai, H., and Ishii, H. 1994. Novel antitumor sesquiterpenoids in *Achillea millefolium*. Chem. Pharm. Bull. (Tokyo). **42**: 1096–1100.

Tyrl, R.J. 1975. Origin and distribution of polyploid *Achillea* [*millefolium*] (Compositae) in western North America. Brittonia **27**: 187–196.

Ustyuzhanin, A.A., Konovalov, D.A, Shreter, A.I., Konovalova, O.A., and Rybalko, K.S. 1987. Chamazulene content in *Achillea millefolium* L. sensu lato in the European part of the USSR. Rastit. Resur. **23**: 424–429 [In Russian.]

Valant-Vetschera, A.K.M., and Wollenweber, E. 1988. Leaf flavonoids of the *Achillea millefolium* group: Part II. Distribution patterns of free aglycones in leaf exudates. J. Biochem. Syst. Ecol. **16**: 605–614.

Wallner, E., Weising, K., Rompf, R., Kahl, G., and Kopp, B. 1996. Oligonucleotide fingerprinting and RAPD analysis of *Achillea* species: characterization and long-term monitoring of micropropagated clones. Plant Cell Rep. **15**: 647–652.

Warwick, S.I., and Black, L. 1982. The biology of Canadian weeds. 52. *Achillea millefolium* L. s.l. Can. J. Plant Sci. **62**: 163–182.

Warwick, S.I., and Briggs, D. 1979.The genecology of lawn weeds. III. Cultivation experiments with *Achillea millefolium* L., *Bellis perennis* L., *Plantago lanceolata* L., *Plantago major* L., and *Prunella vulgaris* L. collected from lawns and contrasting grassland habitats. New Phytol. **83**: 509–536.

Warwick, S.I., and Briggs, D. 1980.The genecology of lawn weeds. VI. The adaptive significance of variation in *Achillea millefolium* L. as investigated by transplant experiments. New Phytol. **85**: 451–460.

World Wide Web Links

(Warning. The quality of information on the internet varies from excellent to erroneous and highly misleading. The links below were chosen because they were the most informative sites located at the time of our internet search. Since medicinal plants are the subject, information on medicinal usage is often given. Such information may be flawed, and in any event should not be substituted for professional medical guidance.)

Achillea disease problems, Michigan State University Extension Ornamental Plants:
http://www.msue.msu.edu/msue/imp/modop/00000041.html

Wholesale nursery growers providing common yarrow, *Achillea millefolium*, wholesale nursery growers' plant listings. Sales to the trade only:
http://www.growit.com/plants/growers/CN/56.htm

AGIS Ethnobotany Database:
http://probe.nal.usda.gov:8300/cgi-bin/webace?db=ethnobotdb&class=Taxon&obect=Achillea +millefolium

Phytochemicals of *Achillea millefolium*:
http://probe.nal.usda.gov:8300/cgi-bin/table-maker?db=phytochemdb&definition+file=chems-in-taxon& arg1=Achillea+millefolium

A modern herbal by M. Grieve:
http://www.botanical.com/botanical/mgmh/y/yarrow02.html

Health Centre - Herb monographs:
http://www.healthcentre.org.uk/hc/alternatives/herbal_monographs/yarrow.htm

Achillea millefolium Yarrow:
http://www.fsl.orst.edu/coops/ntc/anreport/yarrow.htm

Yarrow- HealthWorld Online:
http://www.healthworld.com/LIBRARY/Books/Hoffman/MateriaMedica/yarrow.htm

Achillea millefolium - common yarrow, Michigan State University Extension:
http://www.msue.msu.edu/msue/imp/modop/00000038.html

Fire effects information system:
http://svinet2.fs.fed.us/database/feis/plants/forb/achmil/

Acorus calamus (sweet flag)

Acorus calamus L. Sweet Flag

English Common Names

Sweet flag, sweet sedge, calamus, ratroot (rat root), calamus root, flag root, sweet calomel, sweet myrtle, myrtle flag, sweet cane, sweet rush, beewort, muskrat root, pine root.

Material from sweet flag used as a medicinal agent or food additive is usually referred to as "calamus," which is also one of the common names of the plant. The "flag" in the name is a reference to the iris-like leaves (i.e., like those of yellow flag, *Iris pseudacorus* L., or blue flag, *I. versicolor* L.), while the "sweet" refers to the pleasantly aromatic odor and (bittersweet) taste of most parts of the plant, especially the rhizome. The native North American variety may be called "American sweet flag" whereas the variety introduced from Europe may be called "European sweet flag." The name rat root for sweet flag reflects the fact that the rodent consumes copious quantities of the root. Calomel is mercurous chloride, which was used medicinally in early times, and gave its name to plants used for the same purpose (also see May-apple, *Podophyllum peltatum*, which was referred to as "vegetable calomel").

French Common Names

Belle angélique (belle-angélique), acorus roseau, acore odorant.

Morphology

Sweet flag is a perennial herb. The erect, sword-shaped leaves up to 2 m long emerge from a tortuous, branched, underground rhizome with V-shaped leaf scales. The rhizome is whitish-pink internally, cylindrical, 1–2 cm thick and up to a metre long. The numerous yellow and green flowers are on a spike-like spadix, which is subtended by a leaf-like spathe.

Classification and Geography

Although traditionally placed in the Araceae (Arum family), recent studies have suggested that *Acorus* deserves to be separated into its own monotypic family, the Acoraceae. It has been contended that *A. calamus* may represent the oldest extant lineage of monocotyledons (one of the two great groups of flowering plants). Most authors consider *Acorus* to have just one species, but recent studies suggest that two taxa exist in North America and at least three worldwide.

Sweet flag is found in temperate to subtemperate regions of Eurasia and the Americas. The diploid (with 24 chromosomes) *A. calamus* var. *americanus* (Raf.) Wulff [sometimes treated as a species, *A. americanus* (Raf.) Raf.] occurs from North America to Siberia; the tetraploid (with 48 chromosomes) *A. c.* var. *angustatus* Bess. occupies eastern and tropical southern Asia; and a sterile triploid (with 36 chromosomes), *A. c.* var. *calamus*, is in Europe, temperate India, the Himalayan region, and eastern North America. Variety *calamus* is differentiated from var. *americanus* by its lack of fruit and aborted pollen that fails to stain in standard viability tests. This triploid is believed to have been introduced from Asia to Europe and North America. A hexaploid form (with 72 chromosomes) has been reported from the Kashmir area. American sweet flag, the fertile diploid, occurs in every province of Canada, with a possibly introduced collection recorded from the District of Mackenzie. European sweet flag, the sterile triploid, is relatively uncommon in Canada, but has been recorded in Ontario, Quebec, Nova Scotia, New Brunswick, and Prince Edward Island. In the US, sweet flag occurs as far south as Florida, Texas and Colorado. Sweet flag may have been widely dispersed around the United States by Native Americans who planted it along their migratory paths to be harvested as needed. The species can often be found growing close to the sites of Indian villages, camping areas or trails.

Ecology

Sweet flag is semi-aquatic, occurring in swamps and the edges of streams, marshes, ponds and lakes.

Medicinal Uses

The rhizome of sweet flag has been employed primarily as medicine, almost everywhere the species occurs. Such usage often evolved independently. Ancient Egyptians and classical Chinese, Indian, Greek, and Roman civilizations all appear to have used sweet flag, mostly medicinally. North American Indians also used it extensively for medicinal purposes for a wide variety of illnesses, and often as a panacea. Early Europeans, Chinese, Arabs, and Indians considered sweet flag to be a strong aphrodisiac, and incorporated it into love potions. In North America and New Guinea,

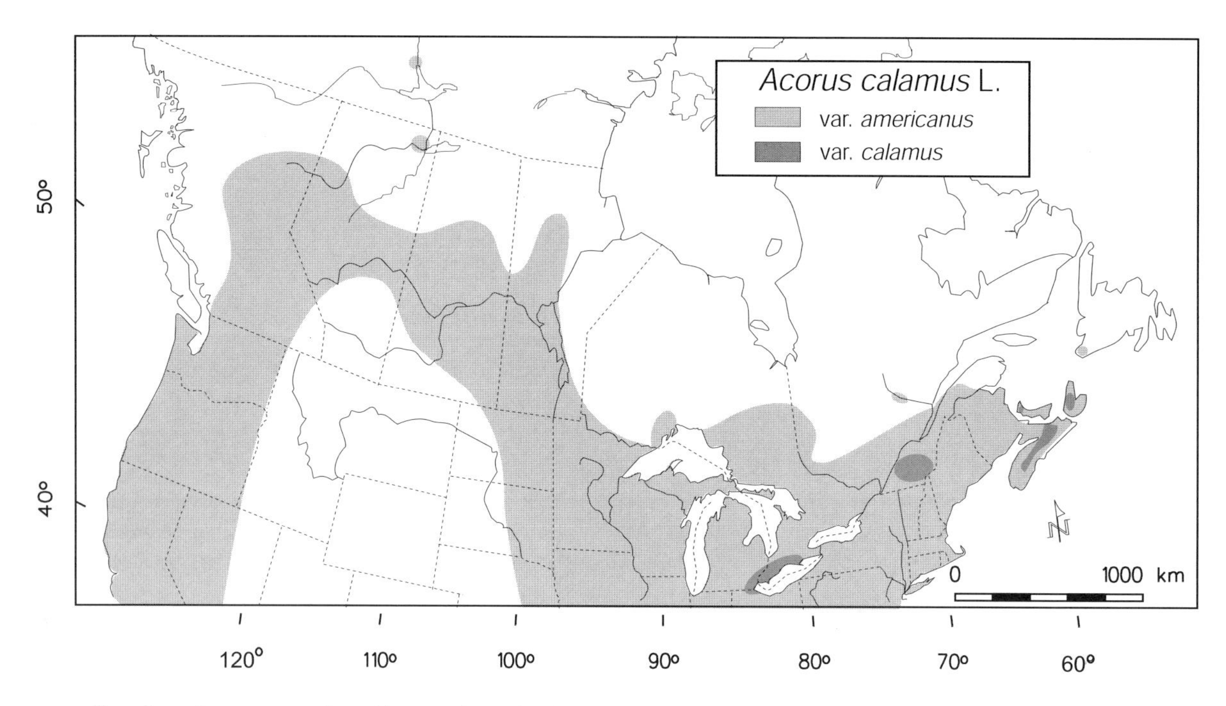

sweet flag has been occasionally used to induce abortion.

The oil of sweet flag has been established to have antibacterial, antifungal, and antiamebic properties. Not surprisingly then, sweet flag has been used frequently for antibiotic purposes: as a vermifuge, antiseptic, antiprotozoal agent, and to treat diverse diseases caused by microorganisms.

At least until the middle part of this century, calamus was accepted as a legitimate pharmacological agent in Western medicine, employed primarily to treat digestive upsets and fevers. Calamus is still used to a minor extent by modern doctors, mostly in Eurasia, and is encountered in several multi-ingredient commercial drug preparations marketed in Canada. The antispasm property of calamus may be the basis of relieving digestive disorders and coughs, as traditionally prescribed. Many experimental studies have established that one of the medical virtues of calamus is its ability to reduce spasms. The North American variety appears to have a greater antispasmodic effect than the other varieties.

Toxicity

There has been concern over the safety of sweet flag, which is currently banned in food products in North America. This prohibition was based on cancerous tumors developed in laboratory animals treated with sweet flag high in content of carcinogenic β-asarone. Carcinogenic β-asarone is present in large amounts in Asian plants, and in limited amounts in European plants. The apparent absence of this and other toxic phenylpropane derivatives in the plants of North America may represent a means of using sweet flag relatively safely. However, it is important to realize that even if North American sweet flag lacks the carcinogenic substances found in the plants of the Old World, under certain conditions it is still reputed to be poisonous, producing disturbed digestion, gastroenteritis, constipation, and bloody diarrhea. The volatile oil causes dermatitis on contact with the skin in some individuals. It should be emphasized in view of the toxic potential of the plant, that any kind of personal use without the supervision of an informed physician is hazardous.

Anyone contemplating using sweet flag as food should also be cautioned that aquatic plants growing in contaminated water may harbor harmful chemicals and organisms acquired from the water. The possibility also exists for those collecting rhizomes in aquatic habitats that the seriously poisonous water hemlock (*Cicuta maculata* L.) could be mistakenly collected.

This herb was listed in a 1995 Health Canada document as unacceptable as a nonprescription drug product for oral use (see "Herbs used as nonmedicinal ingredients in nonprescription drugs for human use," world wide web site http://www.hc-sc.gc.ca/hpb-dgps/therapeut/drhtmeng/policy.html).

Chemistry

Sweet flag oil has been found to have hundreds of compounds, particularly phenylpropanes, monoterpenes, and sesquiterpenoids. Oil of the tetraploid is very high in the carcinogenic β-asarone (often over 90%), while the triploids have less than 5% and the diploids have none.

Non-medicinal Uses

The fragrant oil of sweet flag has been used for many centuries in perfumes. Indeed, the value of calamus used by the North American fragrance industry has exceeded $30,000,000.00 in some recent years. Occasionally, sweet flag has been used as an edible plant. Some North American Indians roasted the rhizome as a vegetable. The rhizome was candied as a confection by Europeans and early American colonists. Wild food collectors sometimes use the young leaves in salads. Up until the Second World War, sweet flag was employed in North America to flavor food products, tonics, and tooth powders. Calamus oil is still used in Europe as a flavoring in alcoholic beverages. The fragrant leaves were once employed to remove disagreeable odors and deter insects. The oil of sweet flag has insecticidal properties, and has been used as a flea repellent, moth repellent, and ant repellent, and has some potential for protecting stored food products against insect pests. Sweet flag leaves were also used to weave mats and reinforce the rims of bark containers.

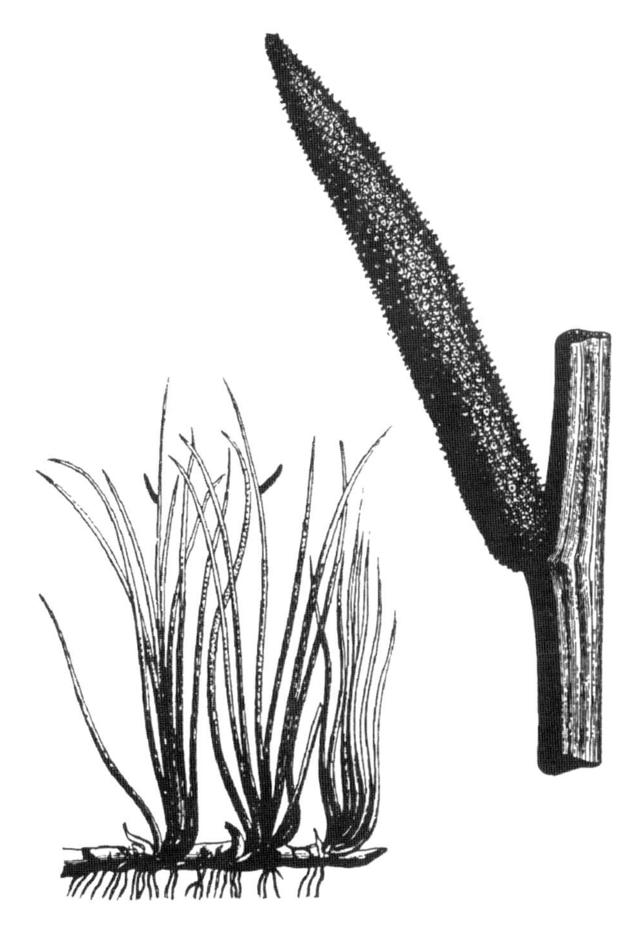

Acorus calamus L.

Agricultural and Commercial Aspects

Sweet flag has commercial promise as a natural pesticide, antifungal and antibacterial agent, flavoring ingredient, perfume component, and medicine. It has been commercially cultivated for its products in various parts of Europe and Asia, and is currently cultivated as an ornamental. As a cultivated crop, the plant has the advantage of rapid propagation by rhizomes, which can be harvested within 2 years of planting. The recent finding that native Canadian plants appear free of carcinogenic β-asarone suggests that the food and medicinal uses that have been thought unwise require reconsideration. However, additional phytochemical and pharmacological study is needed. From an agricultural viewpoint, a semi-aquatic crop would not be easy to manage, but offers the possibility of creating multi-use wetlands.

Myths, Legends, Tales, Folklore, and Interesting Facts

- Moses related how God instructed him to prepare a sacred oil with "calamus" and other sweet-smelling herbs to anoint important ritual items:

  > "Take thou also unto thee the chief spice, of flowing myrrh five hundred shekels, and of sweet calamus two hundred and fifty, and of cassia five hundred, after the shekel of the sanctuary, and of olive oil a kin. And thou shall make it a holy anointing oil, an essence compounded after the art of the perfumer; it shall be a holy anointing oil." (*Exodus 30:22–25*)

  Whether or not the calamus of the bible is sweet flag or some other herb has been debated.
- Cardinal Wolsey of London, England was notorious for extravagant expenditures to obtain pleasant-smelling calamus from distant locations to strew on the floors of cathedrals during festivals. Up until the 20th century, "strewing herbs" were widely used in households and public buildings for sanitary and deodorant purposes.
- The omnivorous Brer Rabbit of Joel Chandler Harris' Uncle Remus fables exclaimed that "I done got so now dat I can't eat no chicken 'ceppin she's seasoned up wid calamus root."

- In India sweet flag was employed to narcotize cobras.
- North American colonists covered their floors with lemony-smelling sweet flag leaves in order to mask the poor sanitation of the times.
- A powder made from sweet flag rhizomes used to be smoked or chewed as a cure (because of its mild sedative effect) for tobacco addiction.
- In medieval Europe, it came to be appreciated that sweet flag is psychotomimetic (mood-altering), and indeed it was believed to be one of the ingredients in the hallucinogenic "flying ointments" used by witches. The use of sweet flag in North America is analogous to the use of coca leaves (*Erythroxylon coca* Lam.) in South America, to combat fatigue, ward off hunger, and increase stamina. The Cree Indians of Alberta used to say that they could consume sweet flag and "travel great distances without touching the ground." Canadian trappers working for the Hudson Bay Company, also used sweet flag as a stimulant, chewing a small piece when tired. It is the asarones in the oil that are psychoactive, whereas other components in the oil relax smooth muscle tissue. The narcotic capacity is much too subtle to have attracted use as a recreational inebriant.
- Walt Whitman's "Leaves of Grass" contains 45 ballads under the title "Calamus." He referred repeatedly to sweet flag and is said to have hidden descriptions of the mental effects in the poetry.
- "Orders for very large quantities of calamus root or extract might arouse suspicion as it is fairly easily converted by amination to TMA-2, which is scheduled" (http://www.Lycaeum.org/~iamklaus/acorus.htm). (TMA-2, a controlled drug in the US, is a hallucinogen with at least 10 times the potency of mescaline. Asarone is naturally converted to TMA-2 in the body by amination shortly after ingestion.)
- "Calamus is also an aphrodisiac, especially when used as an additive in your bathing-water" (http://nepenthes.lycaeum.org/Plants/Acorus/calamus.html).
- "AAAGHHHH! The taste is horrible!" (http://www.hyperreal.org/drugs/natural/calamus.info).

Selected References

Bucher, M., and Kuhlemeier, C. 1993. Long-term anoxia tolerance. Multi-level regulation of gene expression in the amphibious plant *Acorus calamus* L. Plant Physiol. **103**: 441–448.

Bown, D. 1987. *Acorus calamus* L.: a species with a history. Aroideana [International Aroid Society, South Miami, Fla.] **10**(3): 11–14.

Carlquist, S., and Schneider, E.L. 1997. Origins and nature of vessels in monocotyledons. I. *Acorus*. Int. J. Plant Sci. **158**: 51–56.

Duvall, M.R., Learn, G.H., Jr., Eguiarte, L.E., and Clegg, M.T. 1993. Phylogenetic analysis of rbcL sequences identifies *Acorus calamus* as the primal extant monocotyledon. Proc. Nat. Acad. Sci. USA **90**: 4641–4644.

El-Nahal, A.K.M., Schmidt, G.H., and Risha, E.M. 1989. Vapours of *Acorus calamus* oil - a space treatment for stored-product insects. J. Stored Prod. Res. **25**: 211–216.

Evstatieva, L.N., Todorova, M.N., Ognyanov, I.V., and Kuleva, L.V. 1996. Chemical composition of the essential oil in *Acorus calamus* L. (Araceae). Fitologija **48**: 19–23.

Harikrishnan, K.N., Martin, K.P., Anand, P.H.M., and Hariharan, M. 1997. Micropropagation of sweet flag (*Acorus calamus*): A medicinal plant. J. Med. Aromatic Plant Sci. **19**: 427–429.

Lander, V., and Schreier, P. 1990. Acorenone and gamma-asarone: indicators of the origin of calamus oils (*Acorus calamus* L.). Flavour Fragrance J. **5**(2): 75–80.

Lawrence, B.M. 1997. Progress in essential oils. Perfum. Flavor. **22**(2): 59–67.

Mazza, G. 1984. Identification of oxidation products of beta-asarone in *Acorus calamus* L. by gas chromatography and mass spectrometry. Sci. Aliment. (Paris) **4**: 437–482.

Mitchell, R. 1968. *Acorus calamus*. The Beaver (Hudson's Bay Company) **1968**(Spring): 24-26. [An account of Indian use of sweet flag in Manitoba.]

Packer, J.G., and Ringius, G.S. 1984. The distribution and status of *Acorus* (Araceae) in Canada. Can. J. Bot. **62**: 2248–2252.

Nawamaki, K., and Kuroyanagi, M. 1996. Sesquiterpenoids from *Acorus calamus* as germination inhibitors. Phytochemistry **43**: 1175–1182.

Panchal, G.M., Venkatakrishna-Bhatt, H., Doctor, R.B., and Vajpayee, S. 1989. Pharmacology of *Acorus calamus* L. Indian J. Exp. Biol. **27**: 561–567.

Mathur, A.C., and Saxena B.P. 1975. Induction of sterility in male houseflies by vapors of *Acorus calamus* L. oil. Naturwissenschaften **62**: 576–577.

Mazza, G. 1985. Gas chromatographic and mass spectrometric studies of the constituents of the rhizome of calamus (*Acorus calamus*): 1.Volatile constituents of the essential oil. J. Chromatogr. **328**: 179–194.

Mazza, G. 1985. Gas chromatographic and mass spectrometric studies of the constituents of the rhizome of calamus (*Acorus calamus*): 2. Volatile constituents of alcoholic extracts. J. Chromatogr. **328**: 195–206.

Motley, T.J. 1994. The ethnobotany of sweet flag, *Acorus calamus* (Araceae). Econ. Bot. **48**: 397–412.

Menon, M.K., and Dandiya P.C. 1967. The mechanism

of the tranquillizing action of asarone from *Acorus calamus* Linn. J. Pharm. Pharmacol. **19**: 170–175.

Schmidt, G.H., Risha, E.M., and El-Nahal, A.K.M. 1991. Reduction of progeny of some stored-product coleoptera by vapours of *Acorus calamus* oil. J. Stored Prod. Res. **27**: 121–127.

Risha, E.M., El-Nahal, A.K.M., and Schmidt, G.H. 1990. Toxicity of vapours of *Acorus calamus* L. oil to the immature stages of some stored-product Coleoptera. J. Stored Prod. Res.**26**: 133–137.

Su, H.C.F. 1995. Laboratory evaluation of toxicity of calamus oil against four species of stored-product insects. J. Entomol. Sci. **26**: 76–80.

Todorova, M.N., Ognyanov, I.V., and Shatar, S. 1995. Chemical composition of essential oil from Mongolian *Acorus calamus* L. rhizomes. J. Essent. Oil Res. **7**: 191–193.

Vohora, S.B., Shah, S.A., and Dandiya, P.C. 1990. Central nervous system studies on an ethanol extract of *Acorus calamus* rhizomes. J. Ethnopharmacol. **28**: 53–62.

Weber, M., and Braendle, R. 1996. Some aspects of the extreme anoxia tolerance of the sweet flag, *Acorus calamus* L. Folia Geobot. Phytotaxon. **31**(1): 37–46.

World Wide Web Links

(Warning. The quality of information on the internet varies from excellent to erroneous and highly misleading. The links below were chosen because they were the most informative sites located at the time of our internet search. Since medicinal plants are the subject, information on medicinal usage is often given. Such information may be flawed, and in any event should not be substituted for professional medical guidance.)

Acorus calamus, sacred plant of the native Cree:
http://www.lycaeum.org/~iamklaus/acorus.htm

Wholesale nursery growers providing: *Acorus* spp. Sales to the trade only:
http://www.growit.com/plants/growers/SN/59.htm

Acorus calamus [has some useful links]:
http://nepenthes.lycaeum.org/Plants/Acorus/calamus.html

A modern herbal by M. Grieve:
http://www.botanical.com/botanicaL/mgmh/s/sedges39.html

Edible Wild Plants of Southeastern Ohio [has some links]:
http://www.plantbio.ohiou.edu/epb/facility/edibleplants/wild.html

Sweet flag:
http://home.luna.nL/~lachen/sweetflag.html

Sweet flag:
http://www.rook.org/earL/bwca/nature/aquatic/acorus.html

Arctostaphylos uva-ursi (bearberry)

Arctostaphylos uva-ursi (L.) Spreng. Bearberry

Arctostaphylos is Greek for bear's bunch of grapes, while *uva-ursi* is Latin for bear's grape, as is the French raisin d'ours.

English Common Names

Bearberry, common bearberry, kinnikinnick, tinnick, mealberry, chipmunk's apples, uva-ursi, crowberry, foxberry, hog cranberry, mountain cranberry, sandberry, bear's grape, red bear's grape, arberry, mountain box, mountain tobacco, red bearberry, upland cranberry.

Cascara sagrada (*Rhamnus purshianus*), also treated in this work, is sometimes called bearberry. "kinnikinnick" is Algonquin for "mixture," a reference to use in a smoking mixture with tobacco.

French Common Names

Raisin-d'ours, raisin d'ours commun.

Morphology

The bearberry is a prostrate evergreen shrub which creeps over the ground and can reach several metres in length, although the upright branches rarely attain heights of more than 20 cm. Papery bark characteristically peels off the older reddish-brown or gray branches. The oval leaves (1–3 cm long) are firm and leathery. Small (6 mm long) white or pink, bell-shaped flowers in small clusters (3–15 flowers) are produced in the spring, and pollinated by bumblebees. Red berry-like drupes (fleshy fruits with a stony seed) 6–10 mm in diameter, with dry insipid pulp and usually five seeds, ripen in the autumn (the time depending on location and altitude) and persist on the plant through the winter.

Classification and Geography

In both Europe and North America bearberry has long been considered difficult to classify. The most recent authoritative taxonomic analysis of bearberry in Canada, by J.G. Packer and K.E. Denford, recognized four taxa (ssp. *uva-ursi* var. *uva-ursi*, ssp. *uva-ursi* var. *coactilis*, ssp. *longipilosa*, and ssp. *stipitata*). Most of these occur throughout the Canadian range, the exception being ssp. *stipitata* which is exclusively western. The species does appear to be quite plastic, developing differently depending on habitat, but additionally there is ecotypic variation. Variation in the Rocky mountains is especially extensive.

Bearberry is a widely distributed circumboreal species, especially common in Canada and the northern US, but also found across Eurasia. In North America it is encountered from the northern half of California north to Alaska, across Canada and the northern US to new England and Newfoundland. It ranges south in the Rocky Mountains to New Mexico in the west, and in the east extends south along the Atlantic Coast to New Jersey, and in the Appalachian Mountains to Virginia. Rare, disjunct populations are known in Georgia. This hardy plant is found in most of Canada but is localized in many areas, such as southwestern Ontario, where natural open habitats are limited.

Ecology

Bearberry occurs in open, dry habitats including shorelines, dunes, rocky barrens and slopes, sandy barrens and prairies. It is frequently dominant on beaches, dunes and mountains, and may play a major role in preventing erosion. Bearberry occupies a wide range of soil textures and soil pH, but is especially common on dry, nutrient-poor soils with limited clay and silt. It is a shade-intolerant species, growing best in open situations where it forms a compact mat. Fires may help to maintain optimal habitats. The plants regenerate rapidly from dormant buds after fires, if the crown has not been killed. The seeds, which are dispersed by animals and gravity, may be fire resistant and require cold stratification for germination. The plants are extremely cold hardy. This species regenerates primarily vegetatively.

Medicinal Uses

In the Old World, medical use of bearberries was appreciated as far back as the early Romans. Bearberry was also widely used by Indigenous Peoples of North America for various medicinal purposes. Early settlers used it to treat kidney stones and other diseases of the urinary system.

Medicinal preparations of bearberry are made from the leaves, which are collected in the fall. The aqueous extract of the dried leaves has been designated by the pharmacological binomial "*Uvae-ursi Folium*," and more commonly in the commercial pharmacological trade as "*uva ursi*." The drug preparation is used mostly as an anti-inflammatory disinfectant to counteract bacterial infections of the urinary

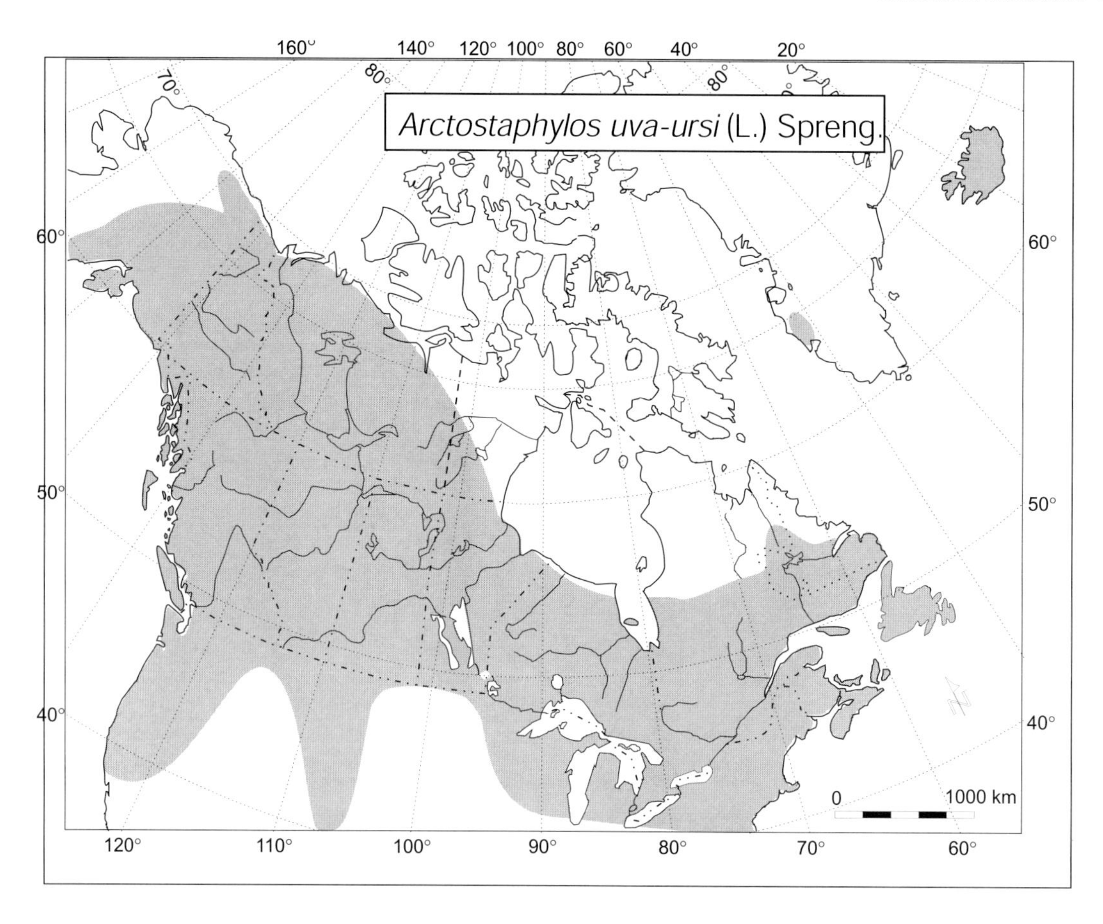

tract. Bearberry is the best known antibacterial herb for the treatment of such urinary tract infections as urethritis and cystitis (inflammation of the bladder). Use has decreased with the development of sulfa drugs and antibiotics. The herb is found in almost all of the teas marketed in Europe for treatment of kidney and bladder conditions.

Knowledge of the mechanism of action of bearberry is indispensable if medicinal benefits are to be realized. It is commonly claimed that bearberry is a diuretic, and indeed the triterpene derivative ursolic acid and the flavonoid pigment isoquercitrin promote urination, although not nearly to the extent of many other herbs. Astringent chemicals, such as the tannins, are well known to exert healing action on the body, but this too docs not seem to be an important remedial mode for bearberry. Moreover, the very high concentrations of tannin (15–20%) tends to result in digestive disturbances. Since tannins are extracted with hot water, the normal method of tea preparation should not be followed. To minimize tannin content of the beverage to be consumed, it is best combined with cold water and allowed to stand 10–24 hours before drinking. Bearberry is effective only if the urine is alkaline. Arbutin from bearberry is hydrolysed in the intestinal tract to produce hydroquinone, resulting in an antibacterial effect in the urine if the pH is above 8. To develop alkaline urine, one can consume 6 – 8 g of sodium bicarbonate per day; or eat a diet rich in milk and non-acidic vegetables (such as potatoes), while avoiding acid-rich foods such as many fruits and their juices, sauerkraut and vitamin C. These requirements are an obvious hurdle to the use of bearberry.

Toxicity

Unlike several of the other medicinal herbs discussed in this publication, there is enough of a gap between the effective dose and the toxic or fatal dose that consumers who have joined the fashion of hcrbal sclf-mcdication are more likely to benefit than to suffer. The recommended dose varies from 1 g, 3–6 times daily to about 10 g (1/3 oz) daily. The latter, approximating 400–700 mg of arbutin, has not been associated significantly with negative effects, but bearberry should not be utilized for more than a few days.

The tannins of bearberry can cause nausea and other effects. Excessive use of bearberry can lead to stomach distress, vomiting, tinnitus (ringing in the ears) and eventually delirium, convulsions, collapse, and even death. These symptoms have been attributed specifically to arbutin. Bearberry has been used as a vasoconstrictor for the endometrium of the uterus. However, constricting the blood supply to the uterus may damage a foetus, and so bearberry should not be used during pregnancy.

This herb was listed in a 1995 Health Canada document (see "Herbs used as non-medicinal ingredients in nonprescription drugs for human use," http://www.hc-sc.gc.ca/hpb-dgps/therapeut/drhtmeng/policy.html) as a herb that is unacceptable as a nonprescription drug product for oral use.

Chemistry

The main active chemical of interest is the phenolic glycoside arbutin, a hydroquinone, usually making up 5–12%, sometimes more than 15%, of the dry weight of the foliage. The leaves also contain the pharmacologically active monotropein (an iridoid), and numerous other constituents, including trace amounts of aspirin. The leaves contain large amounts of tannins, so much so that the plants were once commonly employed to tan leather. A use that persists to this day in Scandinavia.

Arctostaphylos uva-ursi (bearberry)

Non-medicinal Uses

In North America bearberry is primarily cultivated to stabilize slopes prone to erosion and as an ornamental ground cover. The species is very useful as an attractive erosion control plant along highway embankments, especially on coarse-textured soils low in nutrients. It grows well on steep as well as gentle slopes. It also makes an excellent cover for sunny rock walls, rockeries, parking strips, and other urban niches. Bearberry tolerates low summer moisture, and some garden selections withstand salt spray or grow well in semishade. Fruiting branches are used as decorations, especially for Christmas.

The berries are usable as an emergency food, and were once stewed with venison by the Ojibwa, while other Native North Americans fried them or dried them for use in pemican. Today, the fruit still finds use, in jelly, jam, and sauces.

True to their name, the berries of bearberry are relished by both black and grizzly bears, in the autumn, but especially in the early spring when they may be critical to survival. Although the fruits are of low quality, persistence through the winter makes them an invaluable food. The berries are also eaten by a variety of other wildlife, including small mammals, bighorn sheep, mountain goat, deer, elk, songbirds, and gamebirds (including grouse and wild turkey). Hummingbirds have been observed taking nectar from the flowers. However, cattle and other domestic stock avoid bearberry. Bearberry appears to be resistant to livestock trampling and, in some heavily grazed locations, becomes dominant.

Agricultural and Commercial Aspects

Two kinds of cultivars have been, and continue to be selected: ornamental cultivars which may have attractive habit, showy flowers, abundant fruit and glossy leaves, such as Vancouver Jade and Tom's Point; and medicinal cultivars that produce high yields of arbutin, such as the non-flowering Arbuta, released in Czechoslovakia in 1981. Since it is the leaves not the fruits that are medicinally effective, the non-flowering Arbuta is desirable as a medicinal crop as the plant is not diverting its energy into berry production. Medicine is obtained from both wild and cultivated plants. Generally, ornamental cultivars can benefit from more fertilization than can medicinal cultivars. Because bearberry is useful for many purposes, its growth and propagation have been studied in detail and trials have been conducted on mechanized harvesting. In the wild, good seed crops may be periodic, occurring only once in 5 years. Establishing seedlings is relatively difficult, although commercial seed is available. Taking stem cuttings in the fall is considered the best method of propagation. The roots usually form ectendomycorrhizae[1], although cuttings may be inoculated with endomycorrhizal fungi before rooting.

With rapid commercial growth of medicinal herbal products in North America, it is likely that bearberry will regain some of its importance. Over 50 commercial pharmacological products sold in Canada contain bearberry. The well known Richters herb catalogue (1999 issue) offers dried leaves of bearberry at 25 g for $3.00, 1 kg for $40.00. There appears to be increasing interest in Europe in cultivating bearberry for medicinal harvest, while in North America medicinal bearberry has been gathered from the wild. In some regions of Europe the supply of wild plants has been exhausted, suggesting potential profit in medicinal bearberry cultivation in North America.

Myths, Legends, Tales, Folklore, and Interesting Facts

- Consumption of bearberry at medicinally effective dosages results in the strange side effect of turning the urine green.
- One of the obsolete names of bearberry, "mountain tobacco," reflects its very common past use in smoking mixtures. For Native Americans, smoking bearberry in a sacred pipe was a way of carrying prayers to the Great Spirit.

Selected References

Dittberner, P.L., and Olson, M.R. 1983. The plant information network (PIN) data base: Colorado, Montana, North Dakota, Utah, and Wyoming. Washington, DC. Department of the Interior, Fish and Wildlife Service. 786 pp. [Rating of degree of use of bearberry by livestock.]

Fromard, F. 1987. Systématique du taxon *Arctostaphylos uva-ursi*, Ericaceae, en Europe: données nouvelles concernant les populations pyrénéennes et circumpyrénéenes. Can. J. Bot. **65**: 687–695.

Gastler, G.F., McKean, A.L., and William, T. 1951. Composition of some plants eaten by deer in the Black Hills of South Dakota. J. Wildl. Manage. **15**: 352–357.

Gawlowska, J. 1969. Seminatural cultivation of economically important plant species growing in the wild state. Biol. Conserv. **1**: 151–155. [Bearberry leaves used medicinally in Europe.]

Hart, J. 1976. Montana-native plants and early peoples. Montana Historical Society, Helena, MT. 75 pp.

Komissarenko, A.N., and Tochkova, T.V. 1995. Biologically active substances in the leaves of *Arctostaphylos uva-ursi* (L.) Spreng., and their quantitative analysis. Rastit. Resur. **31**(1): 37–44. [In Russian.]

Kruckeberg, A.R. 1982. Gardening with native plants of the Pacific Northwest. University of Washington Press, Seattle, WA. 252 pp. [Use of bearberry as a ground cover.]

Lutz, H.J. 1956. Ecological effects of forest fires in the interior of Alaska. US Dep. of Agricult., For. Serv., Tech. Bull. No. 1133. Washington, DC. 121 pp.

Matsuda, H., Nakamura, S., Shiomoto, H., Tanaka, T., and Kubo, M. 1992. Pharmacological studies on leaf of *Arctostaphylos uva-ursi* (L.) Spreng.: IV. Effect of 50 percent methanolic extract from *Arctostaphylos uva-ursi* (L.) Spreng. (bearberry leaf) on melanin synthesis. Yakugaku Zasshi **112**: 276–282.

M'kada, J., Dorion, N., and Bigot, C. 1991. In vitro micropropagation of *Arctostaphylos uva-ursi* (L.) Sprengel: comparison between two methodologies. Plant Cell Tissue Organ Cult. **24**: 217–222.

Molina, R., and Trappe, J.M. 1982. Lack of mycorrhizal specificity by the ericaceous hosts *Arbutus menziesii* and *Arctostaphylos uva-ursi*. New Phytol. **90**: 495–509.

Mukhina, V.F. 1994. Grinding of medicinal raw material of *Arctostaphylos uva-ursi* (L.) Spreng., and *Vaccinium vitis-idaea* L. as a method of acceleration of its drying. Rastit. Resur. **30**(4): 47–53. [In Russian.]

Mukhina, V.F. 1995. Methods of forecasting the yield of *Arctostaphylos uva-ursi* (L.) Spreng. in Central Yakutia. Rastit. Resur. **31**(1): 94–100. [In Russian.]

Mukhina, V.F. 1996. Seed regeneration of *Arctostaphylos uva-ursi* (L.) Spreng. in Central Yakutia. Rastit. Resur. **32**(1–2): 17–40. [In Russian.]

Nikolaev, S.M., Shantanova, L.N., Mondodoev, A.G., Rakshaina, M.Ts., Lonshakova, K.S., and Glyzin, V.I. 1996. Pharmacological activity of the dry extract from the leaves of *Arctostaphylos uva-ursi* L. in experimental nephropyelitis. Rastit. Resur. **32**(3): 118–123. [In Russian.]

Packer, J.G., and Denford, K.E. 1974. A contribution to the taxonomy of *Arctostaphylos uva-ursi*. Can. J. Bot. **52**: 743–753.

Remphrey, W.R., and Steeves, T.A. 1984. Shoot ontogeny in *Arctostaphylos uva-ursi* (bearberry): origin and early development of lateral vegetative and floral buds. Can. J. Bot. **62**: 1933–1939.

[1] The roots of almost all higher plants are known to form mutualistic symbioses with fungi, termed mycorrhizae ("fungus roots," from the Greek *mykes* = mushroom or fungus and *rhiza* = root). The mycelium of the fungus is more extensive than the roots of the host plant, and the fungus is able to enhance nutrient uptake from the soil for the plant. Mycorrhizae can be broadly classified into three groups, ecto-, ectendo- and endo-mycorrhizae. *Ectomycorrhizae* are characterized by forming an external sheath of mycelium around the root tips, and hyphal cells do not penetrate the cell walls (intercellular) although they may go between cells in the cortex (forming a Harting Net). *Endomycorrhizae* are characterized by the lack of an external sheath around the root tip and the penetration of cortical cells (intracellular) by the fungus mycelium. *Ectendomycorrhizae* seem to be intermediate between ecto- and endomycorrhizae; the mycelium sheath around the root is reduced, or may even be absent, but Hartig Net is usually well developed as in ectomycorrhizae, and hyphal cells may penetrate the cortical cells as in endomycorrhizae.

Remphrey, W.R., Steeves,T.A., and Neal, B.R. 1983. The morphology and growth of *Arctostaphylos uva-ursi* (bearberry): an architectural analysis. Can. J. Bot. **61**: 2430–2450.

Rosatti, T.J. 1981. A new chromosome number in *Arctostaphylos uva-ursi*. Can. J. Bot. **59**: 272–273.

Rosatti, T.J. 1982. Trichome variation and the ecology of *Arctostaphylos* in Michigan. Mich. Bot. **21**: 171–180.

Rosatti, T.J. 1987. Field and garden studies of *Arctostaphylos uva-ursi* (Ericaceae) in North America. Syst. Bot. **12**: 61–77.

Rosatti, T.J. 1988. Pollen morphology of *Arctostaphylos uva-ursi* (Ericaceae) in North America. Grana **27**: 115–122.

Rowe, J.S. 1983. Concepts of fire effects on plant individuals and species. *In* SCOPE 18: The role of fire in northern circumpolar ecosystems. *Edited by* R.W. Wein and D.A. MacLean. John Wiley & Sons., NY. pp. 135–154.

Schmidt, W.C., and Lotan, J.E. 1980. Phenology of common forest flora of the northern Rockies, 1928 to 1937. Res. Pap. Int. 259. US Department of Agriculture, Forest Service, Intermountain Forest and Range Experiment Station. Ogden, UT. 20 pp.

Severson, K.E., and Uresk, D.W. 1988. Influence of ponderosa pine overstory on forage quality in the Black Hills, SD. Great Basin Nat. **48**(1): 78–82.

Sperka, M. 1973. Growing wildflowers: a gardener's guide. Harper & Row, NY. 277 pp.

Tiffney, Jr., W.N., Benson, D.R., and Eveleigh, D.E. 1978. Does *Arctostaphylos uva-ursi* (bearberry) have nitrogen-fixing root nodules? Am. J. Bot. **65**: 625–628.

Watson, L.E., Parker, R.W., and Polster, D.F. 1980. Manual of plant species suitability for reclamation in Alberta. Vol. 2. Forbs, shrubs and trees. Land Conservation and Reclamation Council. Edmonton, AB.

Wells, P.V. 1988. New combinations in *Arctostaphylos* (Ericaceae). Annotated list of changes in status. Madrono **35**: 330–341.

Williams, C.K., and Lillybridge, T.R. 1983. Forested plant associations of the Okanogan National Forest. R6-Ecol-132b. Department of Agriculture, Forest Service, Pacific Northwest Region. Portland, OR. 116 pp. [Bearberry response to distribution.]

Zak, B. 1976. Pure culture synthesis of bearberry mycorrhizae. Can. J. Bot. **54**: 1297–1305.

Zaitseva, N.L., and Litinskaya, N.L. 1993. *Arctostaphylos uva-ursi* (L.) Spreng. in Karelia and ways of improvement state of its coenoformations. Rastit. Resur. **29**(4): 31–39. [In Russian.]

World Wide Web Links

(Warning. The quality of information on the varies from excellent to erroneous and highly misleading. The links below were chosen because they were the most informative sites located at the time of our internet search. Since medicinal plants are the subject, information on medicinal usage is often given. Such information may be flawed, and in any event should not be substituted for professional medical guidance.)

Arctostaphylos uva-ursi 'Vancouver Jade':
http://www.nats-nursery.com/gc/vanjade.htm

Arctostaphylos uva-ursi 'Massachusetts':
http://www.nats-nursery.com/gc/massa.htm

Bearberry: multi-tribal uses:
http://indy4.fdl.cc.mn.us/~isk/food/bearuses.html

Bearberry Phytochemicals:
http://indy4.fdl.cc.mn.us/~isk/food/bearfood.html

Bearberry:
http://www.eyeonmedicine.com/atoz/bearberry.htm

A modern herbal by M. Grieve:
http://www.botanical.com/botanical/mgmh/b/bearbe22.html

Bearberry [excellent!]:
http://www.rook.org/earl/bwca/nature/shrubs/arctouvaursi.html

Arctostaphylos (L.) Spreng. bearberry:
http://bbg.org/NYMF/encyclopedia/eri/arc0010b.htm

Bearberry — HealthWorld Online:
http://www.healthy.net/library/books/hoffman/materiamedica/bearberry.htm

Bearberry [good photograph]:
http://www.mikebaker.com/plants/Arctostaphylos_uva-ursi.html

Fire effects information system [excellent!]:
http://svinet2.fs.fed.us/database/feis/plants/shrub/arcuva/

Arnica montana (European arnica)

Arnica species

Arnica

A. cordifolia Hook
A. fulgens Pursh
A. sororia Greene

The European *Arnica montana* L. is the most widely used source of arnica preparations, but the three North American species named above also supply material used in drug products and are officially listed as sources of arnica in the American Pharmaceutical Association's "National Formulary." All of these species are reported to have similar properties. Medicinally, the three North American species are generally not distinguished from each other or from the European species. *Arnica angustifolia* Vahl both from North America and Eurasia has also been reported to be medicinal. The extent to which the North American *A. chamissonis* Less. is a drug source is unclear. In the following, all of the drug species are referred to collectively as "arnica."

English Common Names

Arnica, wolf's-bane.

The European plant is called arnica, European arnica, mountain daisy, mountain tobacco, mountain snuff, leopard's bane (leopard's-bane), sneezewort, and fall-kraut. The Mexican antimicrobial herb *Heterotheca inuloides* Cass. is sometimes also referred to as "Arnica." "Arnica" from Brazil might be *Solidago microglossa* DC., which has been recommended as a substitute for *A. montana*.

Arnica sororia *Arnica fulgens*

French Common Names

Arnica.

Arnica cordifolia

Morphology

Arnicas are erect, perennial herbs 10–70 cm tall. The leaves are opposite, with smooth or toothed margins. Both simple and glandular (i.e., gland-tipped) hairs are often present. The basal leaves, 4–20 cm long, are the longest, while the uppermost leaves are short and stalkless. The leaves are particularly useful in distinguishing the three North American drug species. *Arnica cordifolia* has long-stalked basal leaves that are broadly ovate and more or less heart-shaped, whereas *A. fulgens* and *A. sororia* have lanceolate leaves that are tapered at the base and have relatively short stalks. *Arnica fulgens* has dense tufts of brown hair in axils of old leaves whereas *A. sororia* has at most a few whitish hairs in the axils. The one to seven yellow flower heads of the drug species are 1.5–3 cm across and are produced in July and August. The seeds, actually achenes (one-seeded dry fruits), are

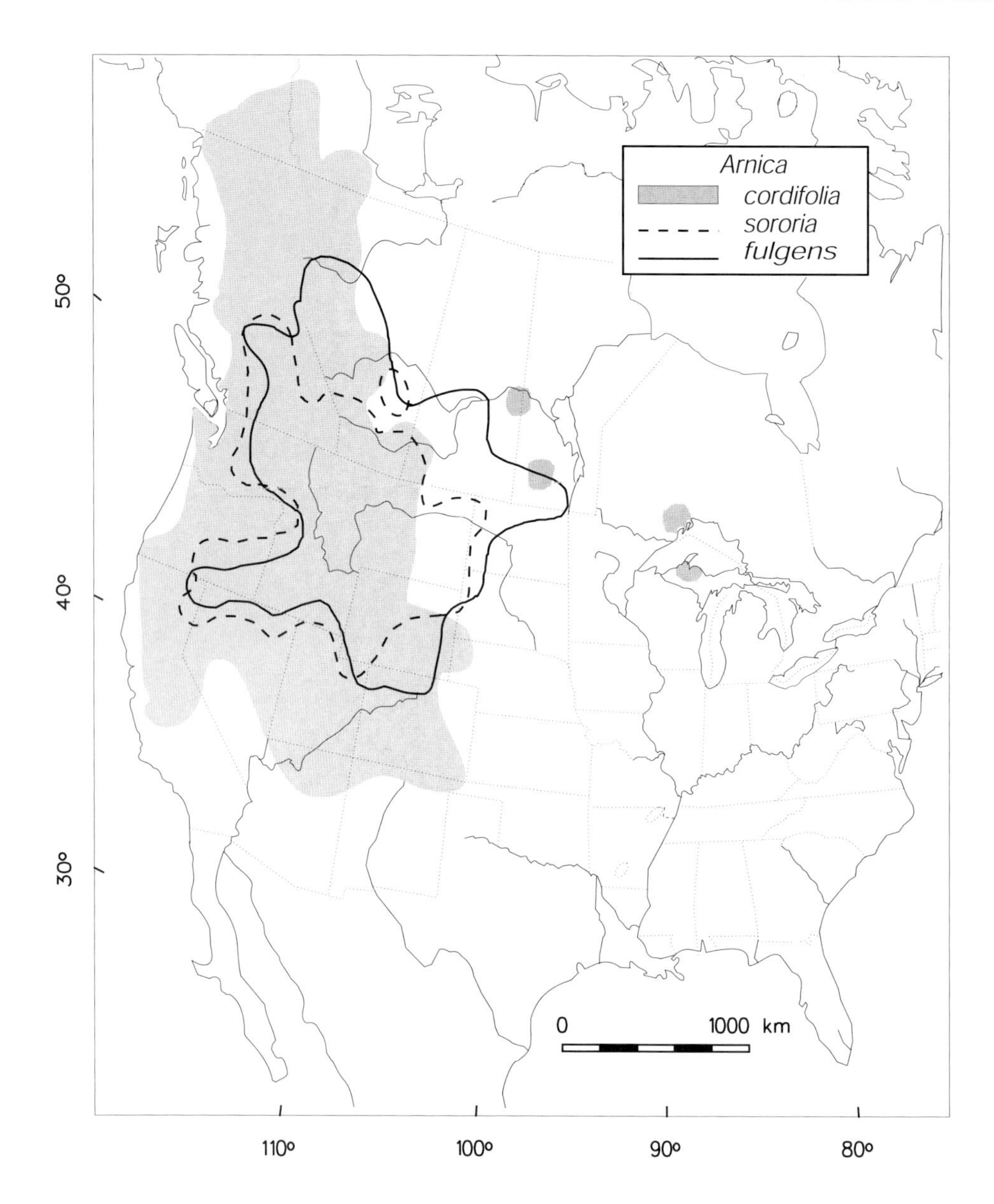

3–10 mm long with a whitish or brownish pappus. (The pappus refers to the hairs or bristles at the top of the achene, a feature possessed by many members of the daisy family, which serves for dispersal, for example by wind or attachment to animals.) The rhizomes are short and thick in the grassland species (*A. fulgens* and *A. sororia*), but long and creeping in *A. cordifolia*.

Classification and Geography

The genus *Arnica* includes about 28 species of the north temperate region, most occurring in the montane region of western North America. All three North American species that supply drugs occur in the western cordillera. *Arnica fulgens* and *A. sororia* are found largely on the high plains and interior valleys. They were once thought to be best treated as varieties of a single species, but recent studies have shown that they are quite distinct, although more closely related to each other than to other species of the genus.

The wide ranging *A. cordifolia* is found in the western American mountains, with isolated populations in Manitoba and the Lake Superior region. The northern populations differ from southern populations in flavonoid composition. Studies suggest that *A. cordifolia* gave rise to a number of narrowly distributed species of Oregon and California. Within *A. cordifolia* the alpine or subalpine var. *pumila* (Rydb.) Maguire is distinguished by its relatively small size, narrower leaves and more glandular achenes. *Arnica cordifolia* has five races differing in chromosome number, with the ancestral types (the diploids, with the lowest chromosome number) restricted to the unglaciated territory in northeastern Oregon and southern Yukon. Isolated eastern occurrences of *A. cordifolia* in the Lake Superior region were first discovered on the Kewenaw Peninsula and the plants were described as the new species, *A. whitneyi* Fernald, but subsequent study suggested that they were better combined with the western *A. cordifolia*. The species was recently discovered in Sibley Park on the shore of Lake Superior. There are about 50 other arctic-alpine vascular plant species that were left as isolated disjuncts on or near to the cold Lake Superior shore following the glacial retreat 10 000 years ago.

Ecology

Arnica cordifolia occurs in woodlands and along woodland edges. Cutting and thinning of forests have resulted in population increase. Prairies are the primary habitats of *A. fulgens* and *A. sororia.* The latter two species reproduce sexually and are self-incompatible. Many different insects including bees, flies and butterflies visit the flowers and serve as pollinators. In contrast, *A. cordifolia* produces seeds without fertilization, but like many other such apomicts, it can presumably also take advantage of occasional fertilization following pollination by insects. A study in The Netherlands suggested that the European *A. montana* is largely self-incompatible and has an outcrossing mating system.

Low levels of grazing have been found to be necessary to maintain *Arnica montana* in some European grasslands where it is otherwise outcompeted by grasses. Acidification of nutrient-poor soils in The Netherlands due to air pollution has been related to a decline in populations of *A. montana*.

Medicinal Uses

Arnica montana, the European species, was used in folk medicine in Europe for a variety of purposes, just as North American Indians employed indigenous American species medicinally. This old remedy was used particularly as a counterirritant, an agent applied locally to produce superficial inflammation in order to reduce pain in deeper adjacent areas. Bactericidal and fungicidal properties also led to use in treatment of abrasions and gunshot wounds in Europe. Arnica was also used to treat tumors. Most of these treatments are outmoded.

Arnica Montana (European arnica)

Arnica is still primarily used as a counterirritant, usually as a hydroalcoholic extract, to reduce the inflammation and pain of bruises, sprains, and aches. Its use declined with the development of effective synthetic painkillers. It is more popular in Europe, where it is available in creams, than in North America. Application to damaged skin seems unwise considering the high irritation capability and, as noted below, it is sufficiently toxic that internal use is unwarranted.

Toxicity

Arnica should not be taken internally as it is sufficiently toxic that it can cause fatal poisoning. Unidentified substances that it contains can produce stress on the nervous system, the digestive system and the circulatory system, resulting in muscular weakness, collapse and even death. Experiments with small animals have confirmed that arnica causes cardiac toxicity and large increases in blood pressure. This herb was listed in a 1995 Health Canada document as unacceptable as a nonprescription drug product for oral use (see "Herbs used as non-medicinal ingredients in nonprescription drugs for human use," http://www.hc-sc.gc.ca/hpb-dgps/therapeut/drhtmeng/policy.html). Arnica is a popular homeopathic remedy, i.e., it is orally consumed in very diluted doses that are said to be too weak to cause harm (see *Homeopathic* in Glossary for reference to an information source with the viewpoint that homeopathy is without scientific merit). External use may cause contact dermatitis due to the allergenic helanin, a sesquiterpene lactone (which has medicinal properties), and other constituents. Arnica should not be applied to broken or damaged skin. Given the potential for allergenic reactions, it seems unwise to use this herb except under qualified medical supervision.

Chemistry

In some cases entire plants or the roots are utilized, but most often drugs are obtained from the dried flower heads which yield a yellowish-brown powder containing arnicin, volatile oil, resin and tannin. Numerous chemical constituents have been reported, but it was only recently discovered that certain sesquiterpenoid lactones were responsible for some of the beneficial effects, particularly the anti-inflammatory action. Helenalin, dihydrohelenalin, and their esters are among the most active principles. Isomeric alcohols such as arnidiol and foradiol have been reported to contribute to counterirritant action. Much of what is known of the medicinal properties and their chemical basis has been published in Europe.

Arnica montana and *A. chamissonis* ssp. *foliosa* have been adulterated by blending with *Heterotheca inuloides*, but a chromatographic method has been developed to rapidly detect this in Arnica drugs. A number of other species of the aster family have also been used to adulterate arnica preparations.

Non-medicinal Uses

Arnica montana yields an oil which has been used in perfumery. *Arnica montana*, *A. cordifolia*, and *A. fulgens* are all grown as garden ornamentals.

Agricultural and Commercial Aspects

Germany markets as many as 300 drug preparations containing arnica extract, while Canada has about 20 such products. Arnica is obtained from both wild and cultivated sources. The US is apparently a more important supplier than Canada, with material collected in the wild from Montana, Wyoming and the Dakotas. Decline of the wild sources and the likely capability of both the North American and European plants to grow throughout much of Canada, as well as demonstrated ease of cultivation, make arnica a potentially interesting diversification crop for Canada.

Arnica montana has become particularly scarce in some parts of Europe, making the cultivated crop more significant. This species is now cultivated as a drug source in parts of Europe and in northern India. Highly productive clones have been produced in Bavaria, with over a hundred flower heads and more than 1% of both flavonoids and sesquiterpene lactones in the harvested drug preparations. Optimal cultivation conditions, including substrates, fertilization and climate, for productivity of *A. montana* in Germany and Russia have been determined. Studies in the Venetian Alps have suggested that *Tephritis arnicae*, a troublesome, widespread fly pest that feeds on the ovaries in the flower heads of *A. montana*, might be controlled effectively with natural parasites. Related species of *Arnica*, such as *A. foliosa* and *A. chamissonis*, have been found to be useful as alternative drug sources.

Some characteristics of the North American species of *Arnica* adapt them to cultivation. These include rapid maturation (i.e., flowering) in the second year after growth from seed; rapid multiplication by rhizomes; upper parts harvest readily, leaving the rhizomes for subsequent growth; self-seeding by some species, so replanting is not necessary.

Myths, Legends, Tales, Folklore, and Interesting Facts

- One of the stranger old-time medicinal preparations is "toad ointment," used to treat sprains, strains, lame back, rheumatism, caked breast, caked udders, etc. A recipe is: put four good-sized toads into boiling water and cook until very soft; remove them and boil the water down to 1/2 pint; add 1 pound of fresh churned, unsalted butter and simmer; add 2 ounces of tincture of arnica.
- Although not safe for internal consumption by humans, *Arnica cordifolia* has been identified as an important constituent of the diet of elk and mule deer.
- Applied to the bald scalp, arnica is said to make hair grow (although skin irritation is the more likely result).
- Large quantities of arnica were used to treat soldiers during World War II.
- Arnica, from the European *Arnica montana*, is said to be one of the most frequently used homeopathic remedies for sports injuries, including "Tennis Elbow." Swiss mountain climbers have sought out the herb and chewed it to relieve sore, tired muscles (a dangerous practice in view of its poisonous and allergenic properties).

- *Arnica montana* was attractively pictured (along with crowberry) on a 1995 Swedish postal stamp (see http://www.posten.postnet.se/stamps/frim_utg/1995/fjallblo/arnika.htm).

Selected References

Baillargeon, L., Drouin, J., Desjardins, L., Leroux, D., and Audet, D. 1993. The effects of *Arnica montana* on blood coagulation. Randomized controlled trial. Can. Fam. Physician **39**: 2362–2367. [In French.]

Bomme, U., and Daniel, G. 1994. First results on selection breeding of *Arnica montana* L. Gartenbauwissenschaft **59**(2): 67–71

Bomme, U., Rinder, R., and Voit, K. 1991. Influence of substrates and fertilization on raising transplants of *Arnica montana* L. Gartenbauwissenschaft **56**(3): 106–113.

Cayouette, J. 1999. Kohlmeister et Kmoch, deux moraves en Ungava, en 1811. Flora Quebeca **3**(3): 7-8. [Mentions medicinal use of *Arnica angustifolia*.]

Conchou, O., Nichterlein, K., and Voemel, A. 1992. Shoot tip culture of *Arnica montana* for micropropagation. Planta Med. **58**: 73–76.

Downie, S.R. 1988. Morphological, cytological, and flavonoid variability of the *Arnica angustifolia* aggregate (Asteraceae). Can. J. Bot. **66**: 24–39.

Downie, S.R., and Denford, K.E. 1986. The taxonomy of *Arnica frigida* and *Arnica louiseana* (Asteraceae). Can. J. Bot. **64**: 1355–1372.

Downie, S.R., and Denford, K.E. 1986. The flavonoids of *Arnica frigida* and *Arnica louiseana* (Asteraceae). Can. J. Bot. **64**: 2748–2752.

Downie, S.R., and Denford, K.E. 1987. The biosystematics of *Arnica fulgens* and *Arnica sororia* (Asteraceae). Can. J. Bot. **65**: 559–570.

Downie, S.R., and Denford, K.E. 1988. Taxonomy of *Arnica* (Asteraceae) subgenus *Arctica*. Rhodora **90**: 245–276.

Downie, S.R., and Denford, K.E. 1988. Flavonoid variation in *Arnica* subgenus *Arctica*. Biochem. Syst. Ecol. **16**: 133–138.

Ediger, R.I., and Barkley, T.M. 1978. *Arnica*. *In* North American Flora. Series II, Part 10. *Edited by* C.T. Rogerson. New York Botanical Garden, New York, NY.

Fennema, F. 1992. Sulphur dioxide and ammonia deposition as possible causes for the extinction of *Arnica montana*. Water Air Soil Pollut. **62**: 325–336.

Fernald, M.L. 1935. Critical plants of the upper Great Lakes region of Ontario and Michigan. Rhodora **37**: 324–341.

Gervais, C., Grandtner, M.M., Doyon, D., and Guay, L. 1990. Nouvelles stations d'*Arnica lanceolata* Nutt. et d'*A. chamissonis* Less. au Québec: notes cytologiques et écologiques. Naturaliste can. **117**: 127-131.

Given, D.R., and Soper, J.H. 1981. The arctic-alpine element of the vascular flora of Lake Superior. Natl. Mus. Nat. Sci. Publ. Bot. **10**: 70 pp.

Gruezo, W.S., and Denford, K.E. 1994. Taxonomy of *Arnica* L. subgenus *Chamissonis* Maguire (Asteraceae). Asia Life Sci. **3**: 89–212.

Gruezo, W.S., and Denford, K.E. 1995. A cytogeographic investigation of *Arnica* L. subgenus *Chamissonis* Maguire (Asteraceae) in western North America. Asia Life Sci. **4**: 95–124.

Gruezo, W.S., and Denford, K.E. 1995. Foliar flavonoid variation in *Arnica* L. subgenus *Chamissonis* Maguire (Asteraceae) in western North America. Asia Life Sci. **4**: 151–170.

Hausen, B.M. 1978. Identification of the allergens of *Arnica montana* L. Contact Dermatitis **4**: 308.

Hausen, B.M. 1980. Arnica allergy. Hautarzt **31**(1): 10–17. [In German.]

Herrmann, H.D., Willuhn, G., and Hausen, B.M. 1978. Helenalinmethacrylate, a new pseudoguaianolide from the flowers of *Arnica montana* L., and the sensitizing capacity of their sesquiterpene lactones. Planta Med. **34**: 299–304.

Hocking, G.M. 1945. American arnica in medicine. Chem. Dig. **4**: 10–12.

Jenelten, U., and Feller, U. 1992. Mineral nutrition of *Arnica montana* L., and *Arnica chamissionis* ssp. *foliosa* Maguire: differences in the cation acquisition. J. Plant Nutr. *15*: 2351–2361.

Kalemba, D., Gora, J., Kurowska, A., and Zadernowski, R. 1986. Comparisons of the chemical composition of inflorescences of *Arnica* spp. Herba Pol. **32**(1): 9–18.

Kating, H., and Seidel, F. 1967. Cultivation experiments with *Arnica* species. II. Vegetative propagation of *Arnica montana* L. Planta Med. **15**: 420–429. [In German.]

Kating, H., Rinn W., and Willuhn, G. 1970. Studies on the substance of species of *Arnica*. 3. Fatty acids in etheric oils of the flowers of various species of *Arnica*. Planta Med. **18**: 130–146. [In German.]

Kaziro, G.S. 1990. Metronidazole (Flagyl) and *Arnica montana* in the prevention of post-surgical complications, a comparative placebo controlled clinical trial. Br. J. Clin. Pract. **44**: 619–621.

Labadie, R.P. 1968. *Arnica montana* L. Pharm. Weekbl. **103**: 769–781. [In Dutch.]

Levin, W., and Willuhn, G. 1987. Sesquiterpene lactones from *Arnica chamissonis* Less. VI. Identification and quantitative determination by high performance liquid and gas chromatography. J. Chromatogr. **41**: 329–342.

Luijten, S.H., Gerard, J., Oostereijer, B., van Leeuwen, N.C., and den Nijs, H.C.M. 1996. Reproductive success and clonal genetic structure of the rare *Arnica montana* (Compositae) in the Netherlands. Plant Syst. Evol. **201**: 15–30.

Maguire, B. 1943. A monograph of the genus *Arnica*. Brittonia **4**: 386–510.

Marquis, R.J., and Voss, E.G. 1981. Distributions of some western North American plants in the Great Lakes region. Mich. Bot. **20**: 53–82.

Merfort, I. 1988. Acetylated and other flavonoid glycosides from *Arnica chamissonis*. Phytochemistry **27**: 3281–3284.

Merfort, I. 1992. Caffeoylquinic acids from flowers of *Arnica montana* and *Arnica chamissonis*. Phytochemistry **31**: 2111–2113.

Merfort, I., and Wendisch, D. 1992. New flavonoid glycosides from Arnicae flos DAB 9. Planta Med. **58**: 355–357.

Merfort, I., and Wendisch, D. 1993. Sesquiterpene lactones of *Arnica cordifolia*, subgenus *Austromontana*. Phytochemistry **34**: 1436–1437.

Merfort, I., Marcinek, C., and Eggert, A. 1986. Flavonoid distribution in *Arnica* subgenus *Chamissonis*. Phytochemistry **25**: 2901–2903.

Passreiter, C.M., Willuhn, G., and Roeder, E. 1992. Tussilagine and isotussilagine: two pyrrolizidine alkaloids in the genus *Arnica*. Planta Med. **58**: 556–557.

Pietta, P.G., Mauri, P.L., Bruno, A., and Merfort, I. 1994. MEKC as an improved method to detect falsifications in the flowers of *Arnica montana* and *A. chamissonis*. Planta Med. **60**: 369–372.

Rinn, W. 1970. Isobutyric acid thymylester — main constituent of etheric oil of rhizomes and roots of *Arnica chamissonis*. Planta Med. **18**: 147–149. [In German.]

Rudzki, E., and Grzywa, Z. 1977. Dermatitis from *Arnica montana*. Contact Dermatitis **3**: 281–282.

Scaltriti, G.P. 1985. The insects of medicinal plants: *Arnica montana* L., and two of its phytophagous insects. Redia **68**: 355–364.

Schroeder, H., Loesche, W., Strobach, H., Leven, W., Willuhn, G., T ill, U., and Schroer, K. 1990. Helenalin and 11-alpha, 13-dihydrohelrnalin, two constituents from *Arnica montana* L., inhibit human platelet function via thiol-independent pathways. Thromb. Res. **57**: 839–846.

Schulte, K.E., Rucker, G., and Reithmayr, K. 1969. Certain constituents of *Arnica chamissonis* and other *Arnica* species. Lloydia **32**: 360–368. [In German.]

Schwabe, A. 1990. Syndynamic processes in Nardo callunetea communities: changes in fallow land after renewed cattle grazing and life history of *Arnica montana* L. Carolinea **48**: 45–68.

Speight, P. 1980. *Arnica*, the wonder herb; the remedy that should be in every home. C. W. Daniel Company Limited, Saffron Walden, Essex, England. 45 pp.

Torres, L.M.B., Akisue, M.K., and Roque, N.F. 1987. Quercitrin from *Solidago microglossa* DC., the Arnica of Brazil. Rev. Farm. Bioquim. univ. Sao Paulo **23**(1): 33–40.

Vanhaelen, M. 1973. Identification of carotenoids in *Arnica montana*. Planta Med. **23**: 308–311. [In German.]

Willuhn, G. 1972. Studies on components of *Arnica* species. V. Content and content differences of volatile oil in various organs of *Arnica* species. Planta Med. **21**: 221–245. [In German.]

Willuhn, G. 1972. Studies on the components of *Arnica* species. VI. Characterization and preparative separation of volatile oils from roots, rhizoma, leaves and flowerheads of various *Arnica* species. Planta Med. **21**: 329–342. [In German.]

Willuhn, G. 1972. Studies on the components of *Arnica* species. VII. Composition of the volatile oil from subterranian organs and flower heads of various *Arnica* species. Planta Med. **22**: 1–3. [In German.]

Willuhn, G 1972. Fatty acids of the essential oil from leaves of *Arnica montana* and *Arnica longifolia*. Z. Naturforsch. B. **27**: 728. [In German.]

Willuhn, G., Kresken, J., and Leven, W. 1990. Further helenanolides from the flowers of *Arnica chamissonis* ssp. *foliosa*. Planta Med. **56**: 111–114. [In German.]

Woerdenbag, H.J., Merfort, I., Passreiter, C.M., Schmidt, T.J., Willuhn, G., Van-Uden, W., Pras, N., Kampinga, H.H., and Konings, A.W.T. 1994. Cytotoxicity of flavonoids and sesquiterpene lactones from *Arnica* species against the GLC-4 and the COLO 320 cell lines. Planta Med. **60**: 434–437.

Wolf, S.J. 1980. Cytogeographical studies in the genus *Arnica* (Compositae: Senecioneae). I. Am. J. Bot. **67**: 300–308.

Wolf, S.J. 1987. Cytotaxonomic studies in the genus *Arnica* (Compositae: Senecioneae). Rhodora **89**: 391–400.

Wolf, S.J., and Denford, K.E. 1983. Flavonoid variation in *Arnica cordifolia*: an apomictic polyploid complex. Biochem. Syst. Ecol. **11**: 111–114.

Wolf, S.J., and Denford, K.E. 1984. Taxonomy of *Arnica* (Compositae) subgenus *Austromontana*. Rhodora **86**: 239–309.

Wolf, S.J., and Denford, K.E. 1984. Flavonoid diversity and endemism in *Arnica* subgenus *Austromontana*. Biochem. Syst. Ecol. **12**: 183–188.

Wolf, S.J., and Whitkus, R. 1987. A numerical analysis of flavonoid variation in *Arnica* subgenus *Austromontana*. Am. J. Bot. **74**: 1577–1584.

Ziegler, B., Michler, B., and Arnold, C.G. 1992. *Arnica montana* L.: a protected plant as a source of a pharmacopoeia drug. Pharm. Zeit. Wissen. **137**: 198–201.

World Wide Web Links

(Warning. The quality of information on the internet varies from excellent to erroneous and highly misleading. The links below were chosen because they were the most informative sites located at the time of our internet search. Since medicinal plants are the subject, information on medicinal usage is often given. Such information may be flawed, and in any event should not be substituted for professional medical guidance.)

A modern herbal by M. Grieve:
http://www.botanical.com/botanicaL/mgmh/a/arnic058.html

Arnica the wonder herb [an advertisement for an enthusiastic book by a practitioner of homeopathy; see *Homeopathy* in Glossary for a critical view]:
http://www.minimum.com/doc00076.htm

Arnica — HealthWorld Online:
http://www.healthy.net/library/books/hoffman/materiamedica/arnica.htm

A list of *Arnica* names for inclusion in Flora North America:
http://arnica.csustan.edu/herbarium/taxa.htm

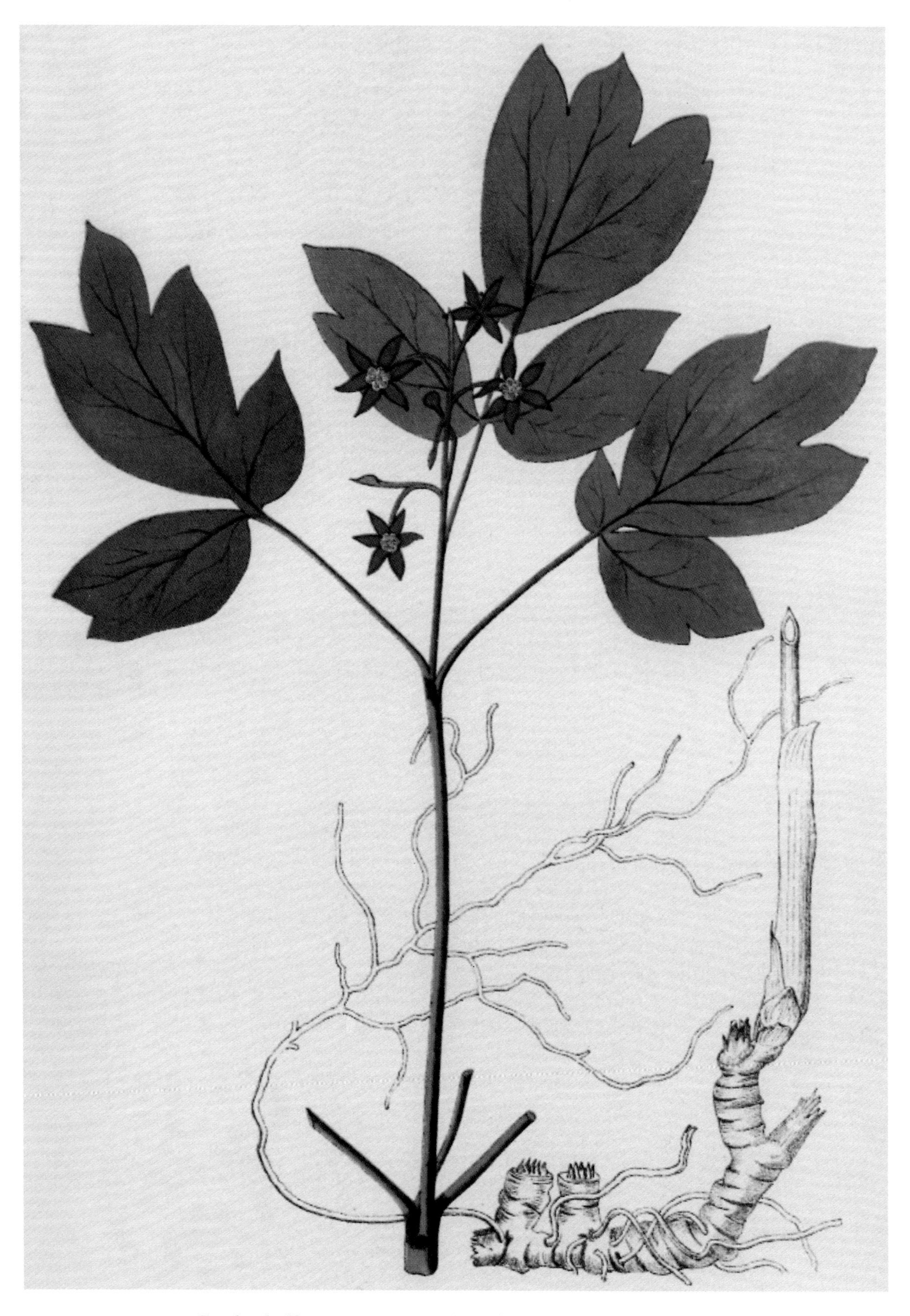

Caulophyllum giganteum (purple flowered blue cohosh)

Caulophyllum species

Blue Cohosh

C. thalictroides (L.) Michx. = yellow-flowered blue cohosh
C. giganteum (Farwell) Loconte & W.H. Blackwell = purple-flowered blue cohosh

These species have only recently been recognized as separate, and much of what is known and written about them is applicable to both. The name "blue cohosh" in the following discussion refers to both species.

English Common Names

Blue cohosh, papoose root (papoose-root), squaw root (squaw-root), blue ginseng, yellow ginseng, blue berry, blueberry root, beechdrops.

Blue cohosh should not to be confused with black cohosh, *Cimicifuga racemosa*, also discussed in this work. The latter is apparently a more widely utilized and possibly safer medicinal plant. The names papoose root and squaw root are believed to be derived from use by indigenous people to ease pain associated with childbirth. "Squaw root" has also been applied to *Cimicifuga racemosa*. Blue cohosh is the only name commonly encountered.

French Common Names

Caulophylle faux-pigamon, caulophylle, léontice faux-pigamon, faux-pigamon, cohoche bleu, graines à chapelet.

Morphology

This erect perennial produces bluish-purple clumps of young shoots in April. The flowers, 1–2 cm across, begin to open in April and May while the leaves are still folded. Authors have differed in their interpretation of floral parts of blue cohosh. What some have called sepals, others have called petals, and some have thought that the nectaries originated from either petals or anthers. The flowers have six prominent sepals (5–6 mm long), six nectaries probably derived from the stamens, and six stamens. The purple-flowered blue cohosh flowers up to a week or two earlier than the yellow-flowered species in many, but not all localities. In addition to its distinctive flower color, the purple-flowered plant has styles 1–1.5 mm long whereas those of the yellow-flowered species are 0.1–0.7 mm long. In other respects the two species are very similar. The several to many flowers are borne in branching clusters. By the time the forest canopy has fully developed in late spring the stems have reached their maximum height of between 30 and 75 cm. Each stem bears two compound leaves, one large, centrally located, and three times divided into leaflets, and a smaller leaf just below the inflorescence. Although definitely green, the mature leaves retain a bluish-purple cast and to some extent a whitish bloom, and are smooth, with leaflets that are 2- or 3-lobed (not serrated as in some similar species). These characters help to distinguish blue cohosh leaves from those of baneberries (*Actaea* spp.), meadow-rues (*Thalictrum* spp.) and black cohosh. By late summer the leaves deteriorate, leaving stems with what appear to be dark blue berries 1–1.5 cm in diameter. These are naked seeds with a fleshy blue covering. The horizontal rootstock is matted and knotty, yellow-brown externally and whitish to yellow internally, with many stem scars and numerous cylindrical branching roots The rootstock tastes bittersweet and acrid, and has a slightly pungent fragrant odor.

Classification and Geography

Prior to 1964 it was believed that only one kind of blue cohosh existed in North America, but that year well known Canadian plant taxonomist William Dore wrote a paper entitled "Two kinds of blue cohosh." In his article Dore related the earlier observations of Harold Minshall, an expert on flowering phenology, that some plants of blue cohosh flower almost 2 weeks earlier in the spring than others. Dore demonstrated that the early flowering plants had purple flowers with long styles, while the later blooming plants had yellowish-green or creamy flowers with short styles. He also noted differences in geographic distribution within southern Ontario. Based on these differences Dore distinguished the two kinds using the available varietal names: var. *thalictroides* for the short-styled plant; and var. *giganteum* Farwell for the long-styled plant. However Dore was of the opinion that they should be treated as different species. Later experts agreed and the var. *giganteum* was elevated to the rank of species in 1981 with additional supporting data published in 1985. The two species can be distinguished even late in the year because styles persist on aborted flowers.

Purple-flowered blue cohosh occurs in the northern Appalachian and eastern Great Lakes region and is the more common and widespread of the two species in southern Ontario. Interestingly however, it has a rather restricted total range. In

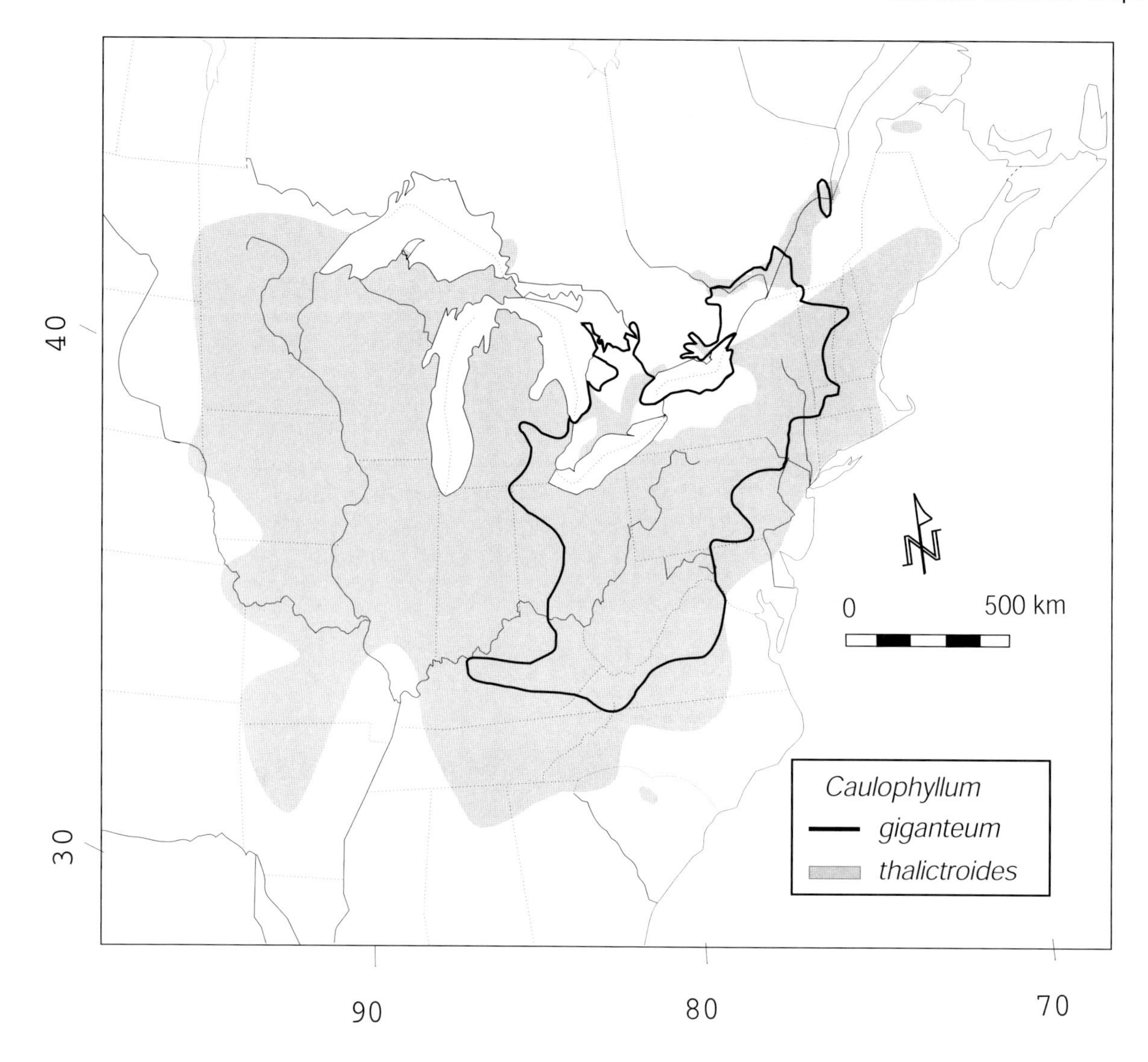

contrast yellow-flowered blue cohosh has a broad range extending further to the north, south, east, and west, and includes a large portion of eastern and midwestern North America. The genus *Caulophyllum* provides another example of the floristic relationship between eastern Asia and eastern North America (like May-apple, ginseng and goldenseal, all discussed in this work). The eastern Asian representative is *C. robustum* Maxim. (most closely related to *C. thalictroides* and once treated as a variety of the latter), which differs from the North American plants in having longer inflorescences on longer stems with more flowers.

Ecology

The species of blue cohosh grow in rich, wet to mesic, shady woods. In Canada they are most frequent in maple woods on limestone, in rocky, calcareous and organic substrates. A Michigan study suggested that seed production requires cross-pollination by insects, but visitation by insects, mostly flies and small bees, was sporadic. However, successful self-pollination was found to be relatively low. The two species are reproductively isolated due to differences in flowering time and other factors.

Medicinal Uses

Blue cohosh is a traditional woman's herb. The best known use is as a parturifacient, i.e., a substance that induces uterine contractions to speed delivery of a baby. Teas and root extracts of blue cohosh were used in the past by Indians and settlers to ease delivery at birth, reduce labor pains, and regulate menstruation. Blue cohosh was used by indigenous North Americans and early settlers to induce abortion, often in conjunction with black cohosh (*Cimicifuga racemosa*). Side effects included sleepiness, headaches, frequent urination and vomiting as well as arm and leg pains. Herbal abortion may be based on stimulating blood flow to the pelvic area and uterus, or stimulating a hormone responsible for uterine contractions. Blue cohosh has been associated with the latter mechanism. Although it promotes delivery, the extent to which blue cohosh causes abortion is unclear. Herbal abortion may be limited in effectiveness and is generally considered dangerous. Less commonly, the herb was used to reduce spasms, relieve stomach cramps, expel parasitic worms, and treat lung ailments, asthma, bronchitis, nervous disorders, urinary tract ailments, rheumatism, arthritis, breast pain, nervous cough, epilepsy, gout, gonorrhea, hysteria, and bee stings. The leaves of blue cohosh have been applied externally to treat the dermatitis induced by poison ivy and related species. In recent times, blue cohosh tea was even recommended for runners to ease the symptoms of muscle spasms and leg cramps. Although not widely used today in drug products, blue cohosh is available in natural herbal supplements, especially for women. Clinical studies are needed to establish the safety and efficacy of blue cohosh.

The Asian species of *Caulophyllum*, *C. robustum*, is used in China to treat rheumatism and menstrual disorders, the roots either being soaked in rice wine or decocted for tea.

Caulophyllum thalictroides
(yellow-flowered blue cohosh)

Toxicity

Although blue cohosh preparations are often recommended for female discomforts, these remedies should not be used during pregnancy, and not without a physician's guidance. Menstrual irregularities, for example, could have many undiagnosed causes, such as pregnancy, cancers, ovarian cysts and thyroid disorders, and herbal remedies may be quite inappropriate. The strong uterine contractions provoked by caulosaponin could endanger a pregnancy. Caulosaponin also constricts coronary blood vessels, and thereby has a toxic effect on cardiac muscle. Since blue cohosh can increase blood pressure, those already at risk from this condition probably should avoid the herb. Caulosaponin is also a suspected teratogen (i.e., causes birth defects). Nausea, vomiting and gastroenteritis are reported as a consequence of large doses or prolonged consumption. The dust of the powdered root is strongly irritating to mucous membranes, and therefore commercial powdered preparations must be handled carefully. The blue "berries" are insipid and various texts recommend that they not be eaten. There are reports of children being poisoned by them, although other reports indicate "low toxicity."

Blue cohosh was listed in a 1995 Health Canada document as a herb that is unacceptable as a nonprescription drug product for oral use (see "Herbs used as non-medicinal ingredients in nonprescription drugs for human use," http://www.hc-sc.gc.ca/hpb-dgps/therapeut/drhtmeng/policy.html).

Chemistry

The roots and rhizomes are collected in the autumn at which time they are reported to be richest in active chemicals. The medicinally significant substances are evidently the glycosides (saponins, particularly caulosaponin) and alkaloids, particularly methylcytisine (caulophylline), but also anagyrine, baptifoline, and magniflorine. Methylcytisine, which increases respiration, blood pressure, and intestinal motility (sometimes causing intestinal spasms), is reported to have effects similar to nicotine, although less pronounced. The glycosides have been associated with uterine stimulation, vasoconstriction of coronary blood vessels, and antifungal properties. Plant extracts have been shown to be antimicrobial and, in rats, anti-inflammatory and ovule-inhibitory action has been reported, the latter suggesting contraceptive potential. The Russian literature on the chemical composition of the Asian *C. robustum* is much more extensive than the information available on the North American species, and could prove useful because of the close relationship of the species.

Non-medicinal Uses

Blue cohosh is occasionally cultivated as a garden ornamental. Some texts indicate that the pea-sized seeds can be roasted to make a coffee-like beverage.

Agricultural and Commercial Aspects

Blue cohosh is harvested from the wild in some parts of North America and is considered at risk from overcollecting in some areas. Although little information is available on its cultivation, it might be grown and harvested in much the same way as ginseng. Plants could be propagated by either root division after flowering or by seeds. Blue cohosh could become a medicinal crop, and the climate and soil in parts of southern Canada including its natural range are well suited to its growth.

Myths, Legends, Tales, Folklore, and Interesting Facts

- In 1915 wild Canadian dried rootstock of blue cohosh was worth 3 – 5½¢ a pound. Other prices for comparison: pair of socks: 4¢; steel frying pan 8¢; skirt: $1.00; pair of shoes: $2.00; shotgun $5.00; man's suit: $10.00; bicycle: $15.00; piano: $100.00.
- Modern flowering plants are divided into two great groups, dicots (dicotyledons, with two seed leaves), and monocots (monocotyledons, with one seed leaf). Most of the plants treated in this work are dicots (sweet grass and sweet flag are monocots, and the kelps are not flowering plants). Blue cohosh is a dicot, but is very unusual in having floral parts in multiples of three, like most monocots (for example, lilies, grasses, sedges and orchids).
- The gymnosperms (mostly evergreen conifers like pines and spruces) are a more ancient lineage of plants, lacking true flowers. Another characteristic is naked seeds (gymnosperm is Greek for naked seed), and the naked seeds of the unusual *Caulophyllum* are curiously reminiscent of the quite unrelated gymnosperms.
- Using herbs to regulate birth was considered to be incontrovertible evidence of witchcraft during the witch hunts (1450–1700), and so at the time reliable information on blue cohosh and some other herbs used to treat gynecological conditions was difficult to obtain.

Caulophyllum giganteum
(purple-flowered blue cohosh)

- In 1856, Charles Darwin questioned Harvard botanist Asa Gray about how the plants of eastern Asia, widely separated from the plants of eastern North America, came to be very similar. Gray subsequently examined a rich collection from Japan and wrote that: "perhaps the most interesting and unexpected discovery of the expedition is that of *Caulophyllum thalictroides* separated by 140 degrees of longitude, are we to suppose independent origin?" Gray later developed the explanation of a previously more continuous temperate flora that was separated in ancient times by geological and climate change. This influenced Darwin's theory of evolution.

Selected References

Baillie, N., and Rasmussen, P. 1997. Black and blue cohosh in labour. N. Z. Med. J. **110**(1036): 20–21.

Boufford, D.E., and Spongberg, S.A. 1983. Eastern Asian – eastern North American phytogeographical relationships — a history from the time of Linnaeus to the twentieth century. Ann. MO Bot. Gard. **70**: 423–439.

Brett, J.F. 1981. The morphology and taxonomy of *Caulophyllum thalictroides* (L.) Michx. (Berberidaceae) in North America. M.Sc. thesis, University

of Guelph, ON.
Brett, J.F., and Posluszny, U. 1982. Floral development in *Caulophyllum thalictroides* (Berberidaceae). Can. J. Bot. **60**: 2133–2141.
Chandrasekhar, K., and Sarma, G.H. 1974. Observations of the effect of low and high doses of *Caulophyllum* on the ovaries and the consequential changes in the uterus and thyroid in rats. J. Reprod. Fertil. **38**: 236-237.
Dore, W.G. 1998. Two kinds of blue cohosh. Ont. Nat. **2**: 5–9.
Ernst, W.R. 1964. The genera of Berberidaceae, Lardizabalaceae, and Menispermaceae in the southeastern United States. J. Arnold Arbor. **45**: 1–35.
Flom, M.S., Doskotch, R.W., and Beal, J.L. 1967. Isolation and characterization of alkaloids from *Caulophyllum thalictroides*. J. Pharm. Sci. **56**: 1515–1517.
Gunn, T.R., and Wright, I.M. 1996. The use of black and blue cohosh in labour. N. Z. Med. J. **109**: 410–411.
Hannan, G.L., and Prucher, H.A. 1989. Reproductive biology and comparative reproductive success of *Caulophyllum thalictroides* (Berberidaceae) varieties in Michigan. Am. J. Bot. **76**(6 Suppl.): 104.
Hannan, G.L., and Prucher, H.A. 1996. Reproductive biology of *Caulophyllum thalictroides* (Berberidaceae), an early flowering perennial of Eastern North America. Am. Midl. Nat. **136**: 267–277.
Johnson, K.L. 1983. Rare plants of the eastern deciduous forest, IV: blue cohosh (*Caulophyllum thalictroides*). Bull. Manit. Nat. Soc. **6**(9): 13.
Jones, T.K., and Lawson, B.M. 1998. Profound neonatal congestive heart failure caused by maternal consumption of blue cohosh herbal mediation. J. Pediatrics **132**: 550-552.
Lee, N.S., Sang, T., Crawford, D.J., Yeau, S.H., and Kim, S.C. 1996. Molecular divergence between disjunct taxa in eastern Asia and eastern North America. Am. J. Bot. **83**: 1373–1378.
Loconte, H., and Estes, J.R. 1989. Generic relationships within Leonticeae (Berberidaceae). Can. J. Bot. **67**: 2310-2316.
Loconte, H., and Blackwell, W.H. 1981. A new species of blue cohosh (*Caulophyllum*, Berberidaceae) in eastern North America. Phytologia **49**: 483.
Loconte, H., and Blackwell, W.H. 1985. Intrageneric taxonomy of *Caulophyllum* (Berberidaceae). Rhodora **87**: 463–470.
Loconte, H. 1997. *Caulophyllum*. *In* Flora of North America north of Mexico, Vol. 3. *Edited by* Flora of North America Editorial Committee. Oxford University Press, New York, NY. pp. 274-275.
Meacham, C.A. 1980. Phylogeny of the Berberidaceae with an evaluation of classifications. Syst. Bot. **5**: 149–172.
Moore, R.J. 1963. Karyotype evolution in *Caulophyllum*. Can. J. Genet. Cytol. **5**: 384–388.
Olin, B.R. (*Editors*). 1992. Blue cohosh. The Lawrence Review of Natural Products Oct: 1-2.
Pringle, J.S. 1993. This native plant: blue cohosh causes classification dilemmas. Blue cohosh (*Caulophyllum thalictroides*). Pappus **12** (4): 8–9.
Strigina, L.I., Chetyrina, N.S., Isakov, V.V., Dzizenko, A.K., and Eliakov, G.B. 1974. Caulophyllogenin: a novel triterpenoid from roots of *Caulophyllum robustum*. Phytochemistry **13**: 479–480.
Telekalo, N.D., Gorovoi, P.G., Basargin, D.D., and Starchenko, V.M. 1981. Distribution of *Caulophyllum robustum* (Berberidaceae) in the Far East USSR. Bot. Zh. (Leningrad) **66**: 1311–1315. [In Russian.]
Terabayashi, S. 1987. Seedling morphology of the Berberidaceae. Acta Phytotaxon. Geobot. **38**: 63–74. [In Japanese.]
Woldemariam, T.Z., Betz, J.M., and Houghton, P.J. 1997. Analysis of aporphine and quinolizidine alkaloids from *Caulophyllum thalictroides* by densitometry and HPLC. J. Pharm. Biomed. Anal. **15**: 839–843.

World Wide Web Links

(Warning. The quality of information on the internet varies from excellent to erroneous and highly misleading. The links below were chosen because they were the most informative sites located at the time of our internet search. Since medicinal plants are the subject, information on medicinal usage is often given. Such information may be flawed, and in any event should not be substituted for professional medical guidance.)

A modern herbal by M. Grieve:
http://www.botanical.com/botanicaL/mgmh/c/caulop39.html

Blue cohosh - HealthWorld Online:
http://www.healthy.net/library/books/hoffman/materiamedica/bluecohosh.htm

Blue cohosh:
http://www.alternative-medicines.com/herbdesc/1bluecoh.htm

Cimicifuga racemosa (black cohosh)

Cimicifuga racemosa (L.) Nutt. Black Cohosh

The genus name (from the Latin *cimex*, bug, and *fugere*, to drive away) is derived from the European *C. europaea* Schipcz. (so-called "*C. foetida*"), whose strong odor proved useful to repel vermin. The name can be pronounced si-**mi**-si-**fue**-ga.

Cimicifuga racemosa (black cohosh)

English Common Names

Black cohosh, black snakeroot (less often: fairy candles, rattleweed, rattleroot, bugbane, bugwort, squaw root).

"Cohosh" is Algonquin for "rough," an allusion to the exterior of the rhizome. Blue cohosh is *Caulophyllum thalictroides*; it is also used medicinally, and is discussed in this work. The name "black snakeroot" for *Cimicifuga racemosa* originates from the use of the black, knotted rootstocks during pioneer times to treat snakebite. Many other plants are called "snakeroot." Members of the genus *Sanicula* in the Apiaceae are more often called "black snakeroot" than *Cimicifuga racemosa*. "Seneca snakeroot," also discussed in this work, is another of the many plants called snakeroot. "Virginia-snakeroot" or "common snakeroot" is *Aristolochia serpentaria* L. of the Aristolochiaceae, which occurs in the eastern United States south of Canada. Sampson's snakeroot, *Psoralea psoralioides* (Walt.) Cory, a member of the Fabaceae, also occurs south of Canada.

French Common Names

Actée à grappes.

Morphology

Black cohosh is an attractive herbaceous perennial, 1–2.6 m tall, with compound sharply toothed leaves. In mid- to late summer, branching feathery tapering racemes (occasionally as long as a meter) appear, with small, white, scented flowers blooming successively from the base upwards. The whiteness of the flowers is produced mainly by the white stamens, as the petals are minute and the sepals fall away as the bud opens. The stout, blackish rhizome is cylindrical, hard, and knotty, and older specimens have attached remains of numerous branches. The roots arise from the lower surface of the rhizome. (Note that the term "root" is often interpreted in non-botanical literature as the underground portion of a plant. Rhizomes are underground stems, to which roots are attached; that is, rhizomes are not true roots. Nevertheless, in most literature the rhizome of black cohosh, the most important medicinal part, is termed root.) The leaves of black cohosh resemble those of the more common baneberries (*Actaea* spp.), but are without elongate terminal teeth. The elongate nodding inflorescence (over 10 cm long) and the fruit being a follicle, rather than a berry, also distinguish black cohosh.

Classification and Geography

The genus *Cimicifuga* includes about 15 species of the north temperate zone. Compton et al. (1998) submerge the genus *Cimicifuga* under the genus *Actaea* (so that *C. racemosa* becomes *A. racemosa* L.). *Cimicifuga racemosa* is a native of eastern North America, occurring both in Canada and the US. In Canada it is native to a small portion of the Carolinian zone of Ontario, but is

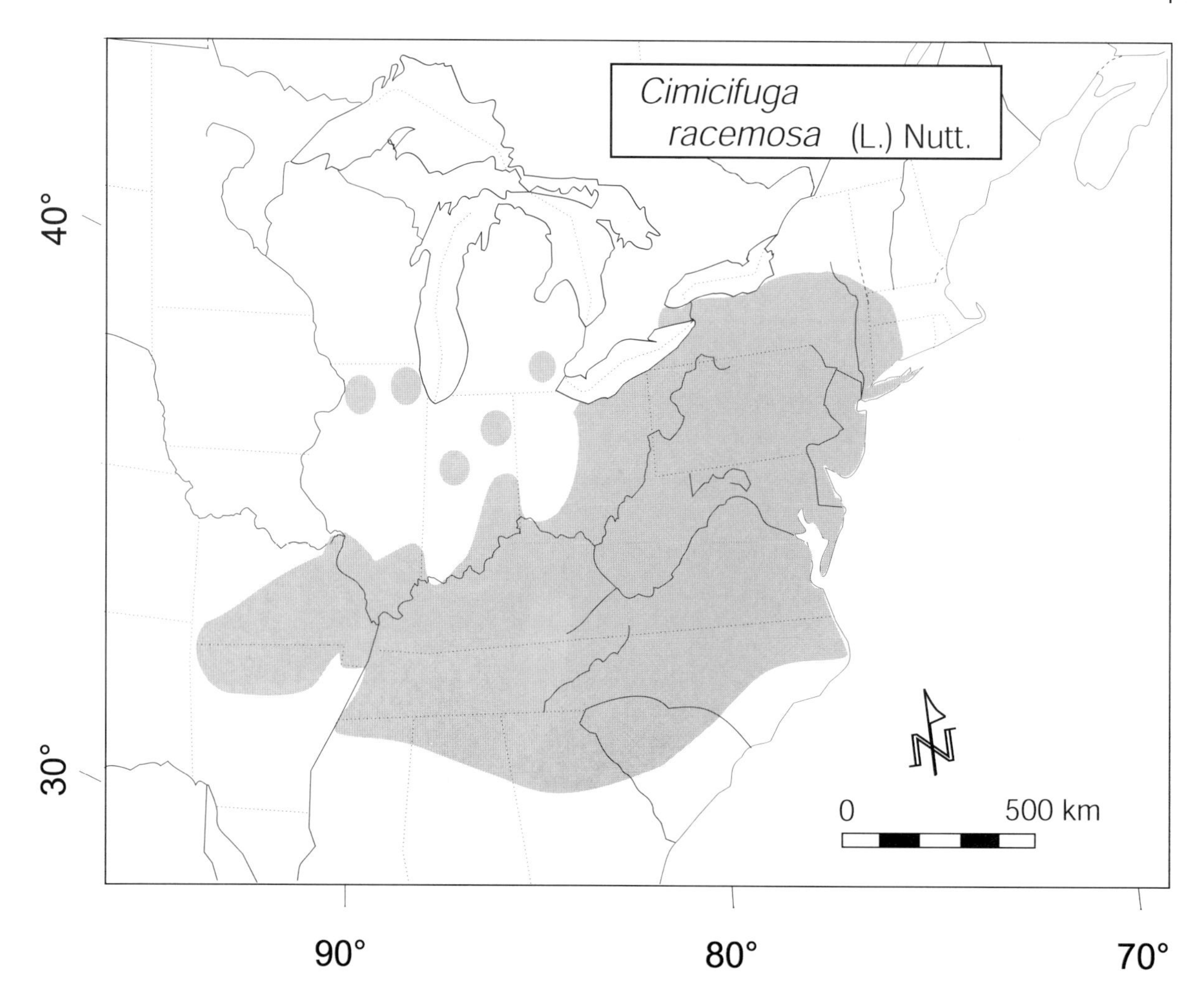

cultivated, and occasionally escapes outside of this region. The plant is considered rare in Canada. Five other species of *Cimicifuga* occur in North America. *Cimicifuga americana* Michx., American bugbane, is also used medicinally, but to a lesser extent. It occurs in eastern North America south of Canada, mostly in the Appalachians, and differs in having 3–8 stipitate pistils instead of 1–3 sessile pistils. *Cimicifuga elata* Nutt., tall bugbane, differing from *C. racemosa* in usually having nine leaflets instead of more than nine, occurs in SW British Columbia south through Washington to northwestern Oregon. It is the only other species of *Cimicifuga* found in Canada.

The leaves are deeply incised in *C. racemosa* forma *dissecta* (Gray) Fern., which is confined to the state of Delaware (this is recognized by Compton et al. 1998 as *Actaea racemosa* var. *dissecta* (Gray) J. Compton). Variety *cordifolia* (Pursh) Gray, with large and often cordate leaflets 1–2.5 dm long, occurs in the mountains of Virginia, North Carolina, and Tennessee. However, the taxa discussed in this paragraph are very rare, and their status requires study.

Ecology

A plant of moist or dry deciduous forests, black cohosh is found particularly on rich, wooded slopes. It seems adapted to partially shaded woodland openings and rocky thickets. It grows best in partial shade, in a moist soil with considerable organic matter and a pH of 5.0 to 6.0. Plants grown from seed may not flower until the third or fourth year.

Medicinal Uses

Cimicifuga racemosa is one of the more important medicinal plants. The drug-containing rhizomes and roots are collected in the fall after the fruits have matured and the leaves have died. The rhizome has a faintly disagreeable odor and a bitter, acrid taste.

Black cohosh has been very widely utilized by American Indians, who generally boiled the rhizome in water and drank the resulting infusion. In addition to considerable usage to treat female complaints (hence the name "squawroot"), native Americans used black cohosh for rheumatism, debility, sore throat, and other problems. The whole rhizomes were extracted with whiskey by the early settlers as a cure for rheumatism. Subsequently, Europeans used black cohosh to treat numerous ailments including diarrhea, bronchitis, measles, whooping cough, tuberculosis, high blood pressure, migraine headaches, neuralgia, arthritis and rheumatism.

Black cohosh is currently recommended for such uses as relieving depression and tinnitus (ringing of the ears), but is best known as a "women's plant" because of its usefulness in relieving menstrual cramps. This herb is a traditional remedy to treat menstrual problems, and facilitate labor and delivery. It is claimed to be effective as an alternative to estrogen replacement therapy for some symptoms of menstrual cessation, especially when estrogen replacement is not possible due to a history of uterine fibroids, fibrocystic breast cancer, etc. Extracts have been shown to suppress hot flashes in menopausal women by reducing the secretion of luteinizing hormone. Research has suggested sedative and anti-inflammatory effects, supporting use in treatment of arthritis and neuralgia. Medical evaluation of other uses of black cohosh is somewhat controversial. It has been considered useful for nerve and muscle pain because it lowers blood pressure and dilates blood vessels.

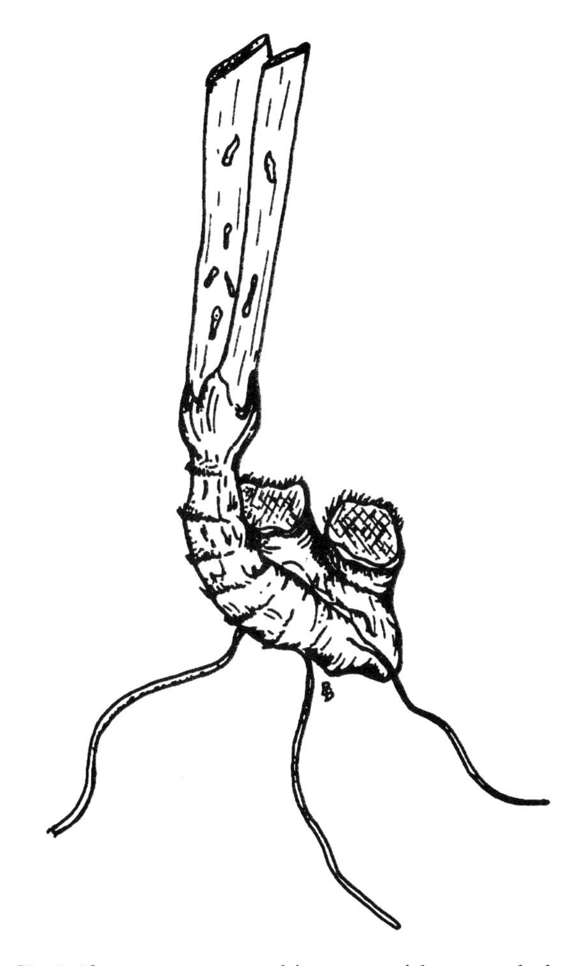

Cimicifuga racemosa rhizome, with roots below and portion of shoot above

Toxicity

Overdoses of black cohosh can result in intense headaches, nausea, vomiting, slow pulse rate, dizziness, and visual disturbances. Other possible complications that have been mentioned include abnormal blood clotting, liver problems, and the promotion of breast tumors. It is very strongly advised that consumption of black cohosh be avoided during pregnancy because it can precipitate a miscarriage (indeed, it has been administered to increase the intensity of uterine contractions during childbirth). It has also been recommended that anyone advised not to take contraceptive pills, or with heart disease should not use black cohosh. Treatment with a medicinal plant as potent as black cohosh should be of limited duration and monitored by an experienced physician.

This herb was listed in a 1995 Health Canada document (see "Herbs used as non-medicinal ingredients in nonprescription drugs for human use," http://www.hc-sc.gc.ca/hpb-dgps/therapeut/drhtmeng/policy.html) as a herb that is unacceptable as a nonprescription drug product for oral use.

Chemistry

Cimicifuga racemosa contains triterpene glycosides, resin, salicylates, isoferulic acid, sterols, and alkaloids. Salicylates are the forerunner of aspirin, and their presence in black cohosh provides a rationale for its early use to treat headache.

Non-medicinal Uses

Black cohosh is a striking garden plant for deep shade landscapes, and is often stocked by plant nurseries for this purpose. With tall stature and coarse foliage, it makes an excellent background plant. A variety of ornamental cultivars have been bred.

Agricultural and Commercial Aspects

Black cohosh is currently used in at least 29 Canadian drug products, with the supply coming exclusively from the wild, mostly from the Blue Ridge Mountains in the Appalachian chain of the US. It is used extensively in parts of Europe and Australia, where several million doses of Remifemin, a formulation of black cohosh, have been employed in recent years.

As a medicinal plant, black cohosh can be cultivated like other shade-loving, slow-growing woodland medicinal plants, such as ginseng and goldenseal. Consequently, it represents an interesting crop diversification opportunity. The Canadian populations are at the northern limit of the range of the species, and therefore are deserving of protection as germplasm for future development of a Canadian cultivated crop.

Myths, Legends, Tales, Folklore, and Interesting Facts

- Black cohosh was one of the components of "Lydia Pinkham's Vegetable Compound." This mixture was an old-time panacea for what was termed "female weakness" (menstrual discomforts) or "female hysteria" (menstrual complaints). One hesitates to define or question the treatment of "male hysteria." Lydia's compound is celebrated in verse:

  Widow Brown she had no children,
  Though she loved them very dear;
  So she took some Vegetable Compound,
  Now she has them twice a year.

Selected References

Baillie, N., and Rasmussen, P. 1997. Black and blue cohosh in labour. N. Z. Med. J. **110**: 20–21.

Baskin, J.M., and Baskin, C.C. 1985. Epicotyl dormancy in seeds of *Cimicifuga racemosa* and *Hepatica acutiloba*. Bull. Torrey Bot. Club **112**: 253–257.

Beuscher, N. 1995. *Cimicifuga racemosa* L. - black cohosh. Zeitschrift für Phytotherapie **16**: 301-310.

Compton, J.A., Culham, A., and Jury, S.L. 1998. Reclassification of *Actaea* to include *Cimicifuga* and *Souliea* (Ranunculaceae): phylogeny inferred from morphology, nrDNA, ITS, and cpDNA *trn*L-F sequence variation. Taxon **47**: 593–634.

Duker, E.M., Kopanski, L., Jarry, H., and Wuttke,W. 1991. Effects of extracts from *Cimicifuga recemosa* on gonadotropin release in menopausal women and ovariectomized rats. Planta Med. **57**: 420–424.

Einer-Jensen, N., Zhao, J., Andersen, K.P., and Kristoffersen, K. 1996. *Cimicifuga* and *Melbrosia* lack oestrogenic effects in mice and rats. Maturitas **25**: 149–153.

Foster, S. 1998. Black cohosh - *Cimicifuga racemosa*. Botanical Series No. 314. American Botanical Council, Austin, TX.

Gunn, T.R., and Wright, I.M. 1996. The use of black and blue cohosh in labour. N. Z. Med. J. **109**: 410–411.

Jarry, H., and Harnischfeger, G. 1985. Endocrine effects of the contents of *Cimicifuga racemosa*. 1. Influence on the serum concentration of pituitary hormones in ovariectomized rats. Plant Med. J. Med. Plant Res. **1985**: 46–49.

Jarry, H., Harnischfeger, G., and Duker, E. 1985. Endocrine effects of the contents of *Cimicifuga racemosa*. 2. In vitro binding of compounds to estrogen receptors. Plant Med. **51**: 316–319.

Koeda, M., Aoki, T., Sakurai, N., Kawai, K., and Magai, M. 1994. Three novel cyclolanostol xylosides from *Cimicifuga racemosa*. Chem. Pharm. Bull. **42**: 2205-2207.

Lehmann-Willenbrock, E., and Riedel, H.H. 1988. Clinical and endocrinological examinations concerning therapy of climacteric symptoms following hysterectomy with remaining ovaries. Zentralblatt für Gynakologie **110**: 611-618.

Linde, H. 1967. Contents of *Cimicifuga racemosa*. 2. On the structure of actein. Arch. Pharm. Ber. Dtsch. Pharm. Ges. **300**: 885–892. [In German.]

Linde, H. 1967. Contents of *Cimicifuga racemosa*. 3. On the constitution of the rings A, B and C of actein. Arch. Pharm. Ber. Dtsch. Pharm. Ges. **300**: 982–992. [In German.]

Linde, H. 1968. Contents of *Cimicifuga racemosa*. 5. 27-desoxyacetylacteol. Arch. Pharm. Ber. Dtsch. Pharm. Ges. **301**: 335–341. [In German.]

Liske, E. 1998. Therapeutic efficacy and safety of *Cimicifuga racemosa* for gynecological disorders. Advances Therapy **15**: 45-53.

Planer, F.R. 1972. Above ground stem infection caused by *Ditylenchus destructor*. Nematologica **18**: 417.

Ramsey, G.W. 1986. A biometrical analysis of terminal leaflet characteristics of the North American *Cimicifuga* (Ranunculaceae). VA. J. Sci. **37**(1): 1–8.

Ramsey, G.W. 1997. *Cimicifuga*. *In* Flora of North America north of Mexico, Vol. 3. *Edited by* Flora of North America Editorial Committee. Oxford University Press, New York, NY. pp. 177-181.

Ramsey, G.W. 1988. A comparison of vegetative characteristics of several genera with those of the genus *Cimicifuga* (Ranunculaceae). SIDA Contrib. Bot. **13**: 57–63.

Sakurai, N., and Nagai, M. 1996. Chemical constituents of original plants of *Cimicifuga* rhizoma in Chinese medicine. Yakugaku Zasshi **116**: 850-865.

Struck, D., Tegtmeier, M., and Harnischfeger, G. 1997. Flavones in extract of *Cimicifuga racemosa*. Planta Med. **63**: 289.

World Wide Web Links

(Warning. The quality of information on the internet varies from excellent to erroneous and highly misleading. The links below were chosen because they were the most informative sites located at the time of our internet search. Since medicinal plants are the subject, information on medicinal usage is often given. Such information may be flawed, and in any event should not be substituted for professional medical guidance.)

Black Cohosh - *Cimicifuga racemosa*, NCNatural's wildflower page:
http://ncnatural.com/wildflwr/cohosh.html

Black cohosh: A woman's herb comes of age:
http://www.qualitycounts.com/blackcohosh.html

The bright side of black cohosh by Varro E. Tyler:
http://www.qualitycounts.com/cohosh.html

Black Cohosh, A discourse from the Honest Herbal, reviewed from the publications of Varro E. Tyler:
http://www.sageways.com/sageline/0996/cohosh.html

A modern herbal by M. Grieve:
http://www.botanical.com/botanicaL/mgmh/c/cohblu84.html

Nature's field: black cohosh, repeller of darkness:
http://www.itsnet.com/~treelite/NF/BKCohosh.html

Cimicifuga racemosa - snakeroot, Michigan State University Extension Ornamental Plants:
http://www.msue.msu.edu/msue/imp/modop/00000367.html

Harbingers of fall part VI: *Cimicifuga*, gardening in shade [has several links]:
http://www.suite101.com/articles/article.cfm/3696

Cimicifuga by L. Perry:
http://pss.uvm.edu/pss123/percimic.html

Black cohosh - HealthWorld Online:
http://www.healthy.net/library/books/hoffman/materiamedica/blackcohosh.htm

Natural Products Monographs - black cohosh [an extensive amount of information, with references]:
http://www.anmp.org/monographs/bcohosh_7.html

Echinacea pallida var. *angustifolia* (purple coneflower)

Echinacea pallida (Nutt.) Nutt. var. *angustifolia* (DC.) Cronq.

Purple Coneflower

English Common Names

(Narrow-leaved) (purple) coneflower, echinacea, prairie purple coneflower.

Other names applied to *Echinacea* species: sampson root, black sampson.

The name *Echinacea* comes from the Greek *echinos*, referring to a spiny hedgehog or a sea-urchin, a reference to the spiny floral bracts of the species. The word "echinacea" may be used as a scientific plant name (the genus *Echinacea*), a common plant name, and the word for drug preparations from the plants. Species of *Echinacea* are known as "coneflowers" because of the raised, cone-like flower-heads (as in many other members of the Compositae or Asteraceae, what appear to be flowers are actually aggregates of tiny flowers). Species of two other North American compositae genera, *Rudbeckia* and *Ratibida*, are also known as coneflower.

French Common Names

Apparently unavailable (the English word is employed).

Morphology

The native Canadian plants of *E. pallida* are usually 15 – 50 cm in height. They have stout, more or less bristly-hairy stems and lance-shaped or linear-lanceolate leaves. Attractive flowering heads appear in late summer and autumn, with "petal" (ray flower) colors varying from whitish rose to pale purple. The phyllaries (subtending bracts) exceed the flowers in length and are spiny. In the fall, brown fruiting heads generally produce abundant seeds. The tap root is thick and blackish.

Classification and Geography

Echinacea is a genus of about nine species of perennial herbs, native to open woods and prairies of central and southeastern US, with one species extending into Canada. Native Canadian plants of *Echinacea* occur in southeastern Saskatchewan and southern Manitoba. We follow the taxonomy of A. Cronquist in recognizing the Canadian plants as an element of a polymorphic, widespread species, *E. pallida* (Nutt.) Nutt. (note map). The Canadian plants belong to var. *angustifolia* (DC.) Cronq., (known as the species *E. angustifolia* in almost all biological and pharmacological literature). This has yellow pollen, whereas var. *pallida* (called *E. pallida* in almost all biological and pharmacological literature), has white pollen, and is not native to Canada. The species is native throughout the prairie region as far south as Texas. Introduced populations have been established in eastern North America including parts of Ontario. Both varieties are used medicinally but var. *pallida*, known as pale-flowered echinacea, is considered much less desirable commercially. *Echinacea tenesseensis* (Beadle) Small is very closely related to *E. pallida*, and has sometimes been considered to be a component of the latter species. This endangered endemic is found only in a few populations on limestone glades near Nashville, Tennessee (see map). *Echinacea purpurea* (L.) Moench, a species of east-central US, is second only to *E. pallida* var. *angustifolia* as a medicinal source of echinacea.

Echinacea products have commonly been adulterated, and recently several manufacturers have made efforts to ensure that their product is genuine. Much research that has been conducted on echinacea is open to question because the classification of the group was not understood by the researchers and their identifications of material utilized were incorrect. There is a need to carry out rigorous chemical, clinical and pharmacological studies with the help of taxonomists.

Ecology

In Canada, *E. pallida* typically occurs on prairie slopes. Like other coneflowers, it grows best in a sunny location, in fertile, well-drained soil.

Medicinal Uses

The root is the predominant part of the plant used medicinally, but flowers and sometimes leaves are also employed. The chewed root causes an unusual, acrid, tingling sensation on the tongue.

Long before echinacea was considered useful for reducing the 2.4 colds per person per year typical of North America, Indians seem to have used echinacea as a remedy for more ailments than any other plant. Although archeological records show that echinacea is known to have been employed by indigenous North Americans at least since the 1600s, European settlers appear to have taken up such use only 2 centuries ago, with the first patent medicine produced about 1870. This was named

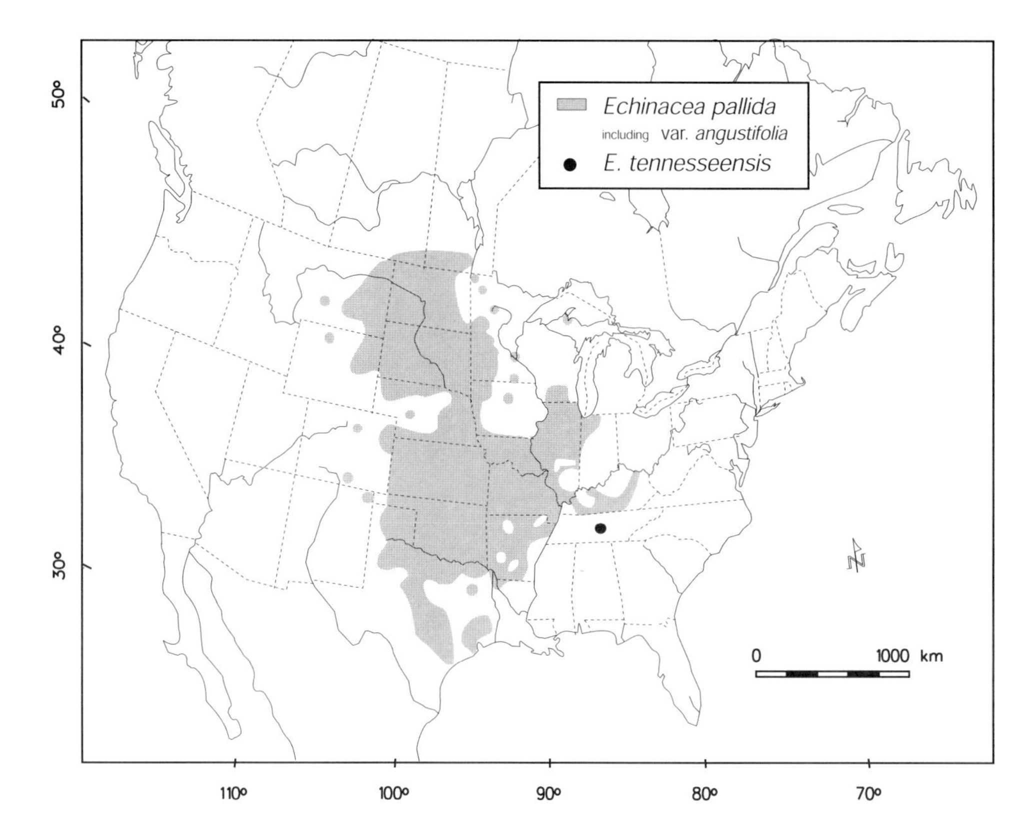

"Meyer's blood purifier" by the (German) Nebraskan lay physician H.C.F. Meyer, who created an echinacea mystique by widely publicizing his offer to allow a rattlesnake to bite him (which he never permitted) so that he could demonstrate the curative power of his miraculous medicine.

Echinacea today has become a star of the medicinal plant industry. Hundreds of scientific articles have been published about it, and many more non-scientific articles have extolled its virtues. As with several other herbal drugs that have become popular, there is some exaggeration concerning its benefits. Curiously, in the late 19th and early 20th centuries there was also a widespread conviction that echinacea was a wonder drug that would cure many illnesses, followed by an interval up to a decade ago during which the drug was thought to be ineffective. The present resurgence of interest is very large, so much so that some consider echinacea to be the most consumed herbal product in the US.

Root extracts of the medicinal species of *Echinacea* have been said to have cortisone-like antibiotic effects, antiviral properties, insecticidal capability, and a potential for stimulating the immune system. There is no doubt that echinacea affects white blood cells, apparently beneficially. It has been speculated that echinacea increases the ability of the body to produce white blood cells that destroy bacteria and viruses. Like ginseng, echinacea is often consumed not so much to cure as to prevent illness and promote well-being.

Most scientific studies of echinacea have been carried out in Germany, the Western country which leads the world in phytomedicinal research. More than 200 echinacea pharmaceutical preparations are marketed there, as extracts, salves, and tinctures for use on wounds, herpes sores, canker sores, throat infections, and as a preventative for influenza. In Canada, five products have been registered for *E. pallida* var. *angustifolia*, and one for *E. purpurea*.

There is considerable agreement that echinacea is useful internally for preventing and treating the common cold and associated conditions such as sore throat, as well as externally for treating superficial wounds.

Toxicity

Generally, there seem to be few undesirable metabolic effects of consuming echinacea. Since it is a member of the daisy family, those who are allergic to other members of the family (like ragweed) need to be aware of the possibility of cross sensitivity. Some have advised that echinacea should not be consumed when pregnant or suffering from diabetes. It has been recommended that echinacea not be used with such progressive systemic disorders as multiple sclerosis and HIV/AIDS-related illnesses.

Reports of hepatitis associated with echinacea have been noted in The Australian Adverse Drug Reactions Advisory Committee August 1993 bulletin.

Chemistry

A wide array of constituents of echinacea may contribute to its medicinal properties. The polysaccharides and the alkylamides are thought to be the most active ingredients. "Echinacosides" have been identified as active ingredients, but there is disagreement about their significance.

Non-medicinal Uses

Coneflowers are commonly grown in wildflower and perennial gardens, and harvested as cut flowers. *Echinacea purpurea* is an especially popular border plant, and there are numerous ornamental cultivars available. By contrast, there seem to be no cultivars of *E. pallida*; the so-called horticultural cultivar Strigosa appears to be based on the taxonomic variety *strigosa* of the southern US.

Agricultural and Commercial Aspects

In recent years in the US, over 50 tonnes of wild echinacea has been harvested for overseas shipment, at a time when the domestic market has been expanding. Wild *E. pallida* var. *angustifolia* has become more difficult to find, and cultivation is becomingly increasingly necessary to supply the demand so that various species or local races will not be exterminated. In Europe, *Echinacea* is extensively cultivated. There is scattered cultivation in Canada.

Substrate conditions are particularly important to perennial root crops like echinacea. It thrives in well drained loams and sandy loam soils with a pH of 6 to 7, and although wild plants seem adapted to dry soils, in cultivation adequate water should be supplied. Roots are harvested at 3 or 4 years of age. Flowers may also be harvested for sale to pharmacological firms, and seeds are still another saleable commodity. The cultivation of *Echinacea* appears to offer potential as a diversification crop in Canada, but as with other relatively undeveloped medicinal crops, caution, planning, and self-education are necessary to make it successful. In some recent years, the prices paid to farmers for growing *echinacea* have been low.

The wild Canadian populations are scattered, and doubtfully justify harvesting. Furthermore they could easily become endangered. These northernmost populations of the genus are of particular significance, since they are the ecotypes suited to the Canadian environment, and therefore constitute germplasm useful for improving future Canadian crops.

Echinacea pallida (purple coneflower)

Myths, Legends, Tales, Folklore, and Interesting Facts

- The Nebraska patent medicine purveyor Dr. H.C.F. Meyer, who offered to let himself be bitten by a rattlesnake to demonstrate the curative power of his echinacea formulation, has been credited as the source of the phrase "snake-oil salesman."
- The Meskwaki tribe of native Americans called *Echinacea* "the hairs of Grandmother Earth's head."
- There are various ways that have been recommended to pronounce echinacea ("ek-a-NAY-sha," "ek-in-EH-sha," "EHH-key-NAY-see-ya," "e-kin-na-sha"). The famous student of botanical Latin, W.T. Stearn, wrote on the subject of how to pronounce names of Latin origin "how they are pronounced really matters little provided they sound pleasant and are understood by all concerned."
- Magnus is a cultivar of *Echinacea purpurea* that was chosen as plant of the year by the Perennial Plant Association. It was selected by Magnus Nilsson near Paarp, Sweden, which happens to be just across the straits from Hamlet's legendary Danish castle at Helsingor.

Echineacea tennesseensis
(Tennessee coneflower)

Selected References

Awang, D.V.C., and Kindack, D.G. 1991. Herbal medicine: *Echinacea*. Can. Pharm. J. **124**: 512–516.

Baskauf, C.J., and Eickmeier, W.G. 1994. Comparative ecophysiology of a rare and widespread species of *Echinacea* (Asteraceae). Am. J. Bot. **81**: 958–964.

Baskauf, C.J., McCauley, D.E., and Eickmeier, W.G. 1994. Genetic analysis of a rare and a widespread species of *Echinacea* (Asteraceae). Evolution **48**: 180–188.

Baskin, J.M., and Baskin, C.C. 1982. Effects of vernalization and photoperiod on flowering in *Echinacea tennesseensis*, an endangered species. J. Tenn. Acad. Sci. **5**(2): 53–56.

Baskin, C.C., Baskin, J.M., and Hoffman, G.R. 1992. Seed dormancy in the prairie forb *Echinacea angustifolia* var. *angustifolia* (Asteraceae): after-ripening pattern during cold stratification. Int. J. Plant Sci. **153**: 239–243.

Baskin, J.M., Snyder, K.M., and Baskin, C.C. 1993. Nomenclatural history and taxonomic status of *Echinacea angustifolia*, *E. pallida*, and *E. tennesseensis* (Asteraceae). Sida Contrib. Bot. **15**: 597–604.

Bauer, R. 1996. Echinacea drugs — effects and active ingredients. Z. Arztl. Fortbild (Jena) **90**: 111–115. [In German.]

Bauer, R., and Foster, S. 1989. HPLC analysis of *Echinacea simulata* and *E. paradoxa* roots. Planta Med. **55**: 637.

Bauer, R., and Foster, S. 1991. Analysis of alkamides and caffeic acid derivatives from *Echinacea simulata* and *Echinacea paradoxa* roots. Planta Med. **57**: 447–449.

Bauer, R., and Wagner, H. 1987. Comments on the *Echinacea* problem. Am. Herb. Assoc. Q. **5**(3): 4.

Bauer, R., and Wagner, H. 1991. *Echinacea* species as potential immunostimulatory drugs. Econ. Med. Plant Res. **5**: 253-321

Bauer, R., Khan, I.A., and Wagner, H. 1988. TLC and HPLC analysis of *E. angustifolia* roots. Planta Med. **54**: 426–430.

Blumenthal, M. 1993. Echinacea highlighted as a cold and flu remedy. Herbalgram **29**: 8–9.

Bomme, U., Hoelzl, J., Hessler, C., and Stahn, T. 1992. What effect does the cultivar have on active ingredient content and yield of *Echinacea purpurea* (L.) Moench with regard to its pharmaceutical use? Bayer. Landwirtsch. Jahrb. **69**: 149–164. [In German.]

Bomme, U., Hoelzl, J., Hessler, C., and Stahn, T. 1992. How does variety influence active substance contents and crop yield of *Echinacea purpurea* (L.) Moench with regard to its pharmaceutical use? Bayer. Landwirtsch. Jahrb. **69**: 323–342. [In German.]

Cody, W.J., and Boivin, B. 1973. Purple coneflower, *Echinacea purpurea*, in Ontario. Can. Field-Nat. **87**: 70.

de Vries, B. 1975. Range extension of purple coneflower in southeastern Saskatchewan. Blue Jay **33**: 220–223.

Dorsch, W. 1996. Clinical application of extracts of *Echinacea purpurea* or *Echinacea pallida*. Critical evaluation of controlled clinical studies. Z. Arztl. Fortbild (Jena). **90**: 117–122. [In German.]

Drew, M.B., and Clebsch, E.E.C. 1995. Studies on the endangered *Echinacea tennesseensis* (Asteraceae): plant community and demographic analysis. Castanea **60**: 60–69.

Federal Register. 1979. Determination that *Echinacea tennesseensis* is an endangered species. **44**(110): 32604–5 (June 6).

Feghahat, S.M.J., and Reese, R.N. 1994. Ethylene-, light-, and prechill-enhanced germination of *Echinacea angustifolia* seeds. J. Am. Soc. Hort. Sci. **119**: 853–858.

Foster, S. 1991. Echinacea: nature's immune enhancer. Healing Arts Press, Rochester, VT. 150 pp.

Heinzer, F., Chavanne, M., Meusy, J.P., Maitre, H.P., Giger, E., and Baumann, T.W. 1988. The classification of therapeutically used species of the genus *Echinacea*. Pharm. Acta Helv. **63**: 132–136. [In German.]

Hemmerly, T.E. 1986. Life cycle strategy of the highly endemic cedar glade species: *Echinacea tennesseensis*. ASB Bull. **33**(4): 193–199.

Hobbs, C. 1994. Echinacea: a literature review. Herbalgram **30**: 33–49.

Hobbs, C. 1994. Echinacea: the immune herb. Botanica Press, Capitola, CA. 83 pp.

Houghton, P. 1994. Herbal products. 3. Echinacea. Pharm. J. **253**: 342–343.

Kindscher, K. 1989. Ethnobotany of purple coneflower (*Echinacea angustifolia*, Asteraceae). Econ. Bot. **43**: 498–507.

Leuszler, H.K., Tepedino, V.J., and Alston, D.G. 1996. Reproductive biology of purple coneflower in southwestern North Dakota. Prairie Nat. **28**(2): 91–102.

McGregor, R.L. 1968. The taxonomy of the genus *Echinacea* (Compositae). Univ. Kansas Sci. Bull. **48**: 113-142.

Mengs, U., Clare, C.B., and Poiley, J.A. 1991. Toxicity of *Echinacea purpurea*. Acute, subacute and genotoxicity studies. Arzneimittelforschung **41**: 1076–1081.

Oliver, A., Price, J., Li, T.S.C., and Gunner, A. 1995. *Echinacea*, purple coneflower. Specialty Crops Infosheet. Ministry of Agriculture, Fisheries and Food, BC. 8 pp.

Parmenter, G.A., and Littlejohn, R.P. 1997. Planting density effects on root yield of purple coneflower (*Echinacea purpurea* (L.) Moench). N.Z. J. Crop Hort. Sci. **25**: 169–175.

Parmenter, G.A., Burton, L.C., and Littlejohn, R.P. 1996. Chilling requirement of commercial *Echinacea* seed. N.Z. J. Crop Hort. Sci. **24**: 109–114.

Perry, N.B., Van Klink, J.W., Burgess, E.J., and Parmenter, G.A. 1997. Alkamide levels in *Echinacea purpurea*: A rapid analytical method revealing differences among roots, rhizomes, stems, leaves and flowers. Planta Med. **63**: 58–62.

Quarterman, E, and Hemmerly, T.E. 1971. Rediscovery of *Echinacea tennesseensis* (Beadle) Small. Rhodora **73**: 304–305.

Samfield, D.M., Zajicek, J.M., and Cobb, B.G. 1990. Germination of *Coreopsis lanceolata* and *Echinacea purpurea* seeds following priming and storage. Hortscience **25**: 1605–1606.

Scaglione, F., and Lund, B. 1995. Efficacy in the treatment of the common cold of a preparation containing an *Echinacea* extract. Int. J. Immunother. **11**(4): 163–166.

Schulthess, B.H., Giger, E., and Baumann, T.W. 1991. *Echinacea*: anatomy, phytochemical pattern, and germination of the achene. Planta Med. **57**: 384–388.

Smith-Jochum, C., and Albrecht, M.L. 1988. Transplanting or seeding in raised beds aids field establishment of some *Echinacea* species. Hortscience **23**(6 Part 1): 1004–1005.

Smith-Jochum, C.C., and Davis, L.C. 1991. Variation in the hexane extracted oils of three *Echinacea* spp. Trans. Kans. Acad. Sci. **94**: 12–21.

Snyder, K.M., Baskin, J.M., and Baskin, C.C. 1994. Comparative ecology of the narrow endemic *Echinacea tennesseensis* and two geographically widespread congeners: relative competitive ability and growth characteristics. Internat. J. Plant Sci. **155**: 57–65.

Somers, P. 1983. Recovery plan for a cedar glade endemic, the Tennessee coneflower, *Echinacea tennesseensis* (Asteraceae). Nat. Areas J. **3**(4): 56–58.

Viles, A.L., and Reese, R.N. 1996. Allelopathic potential of *Echinacea angustifolia* DC. Environ. Exp. Bot. **36**: 39–43.

Wagner, H., Stuppner, H., Schäfer, W., and Zenk, M. 1988. Immunologically active polysaccharides of *Echinacea purpurea* cell cultures. Phytochemistry **27**: 119–126.

Wartidiningsih, N., and Geneve, R.L. 1994. Seed source and quality influence germination in purple coneflower [*Echinacea purpurea* (L.) Moench.]. Hortscience **29**: 1443–1444.

Wartidiningsih, N., Geneve, R.L., and Kester, S.T. 1994. Osmotic priming or chilling stratification improves seed germination of purple coneflower. Hortscience **29**: 1445–1448.

World Wide Web Links

(Warning. The quality of information on the internet varies from excellent to erroneous and highly misleading. The links below were chosen because they were the most informative sites located at the time of our internet search. Since medicinal plants are the subject, information on medicinal usage is often given. Such information may be flawed, and in any event should not be substituted for professional medical guidance.)

Crop & Food Research - *Echinacea*: the purple coneflowers. Description of research into the commercial production of echinacea and echinacea products in New Zealand, by Crop & Food Research scientists:
http://www.crop.cri.nz/broadshe/echinace.htm

Echinacea pallida var. *angustifolia*, Agriculture & Agri-Food Canada, Southern Crop Protection & Food Research Centre:
http://res.agr.ca/lond/pmrc/study/newcrops/echinacea.html

Echinacea - A new crop with potential [for Oklahoma]?:
http://www.kerrcenter.com/nwsltr/news23–2.htm

A modern herbal by M. Grieve:
http://www.botanical.com/botanicaL/mgmh/e/echina01.html

Medical attributes of *Echinacea* spp. coneflowers:
http://wilkes1.wilkes.edu/~kklemow/Echinacea.html

The development of *Echinacea* as a new crop for Alberta:
http://itsd-s3.agric.gov.ab.ca/research/ari/matching/97–98/97–0750.html

HealthWorld - *Echinacea* by Hoffman:
http://www.healthy.net/library/books/hoffman/MateriaMedica/echinacea.htm

Medicinal Herbs Online - *Echinacea*:
http://www.egregore.com/herb/echinacea.html

Top herbal products encountered in drug information requests (Part 1) by J.L. Muller and K.A. Clauson [requires registration (free) with Medscape; one of the herbals discussed is *Echinacea*]:
http://www.medscape.com/SCP/DBT/1998/v10.n05/d3287.mulL/d3287.mull-01.html

Dr. James Downey's herbal research & healing - *Echinacea* [has considerable information]:
http://www.herbsinfo.com/pages/echin.htm

Epilobium angustifolium (fireweed).

Epilobium angustifolium L. Fireweed

The name of the genus derives from the Greek words *epi* (upon) and *lobos* (a pod), a reference to the relatively advanced (evolutionarily) position of the floral parts on top of the young pod-like ovaries. Fireweed is frequently called *Chamaenerion angustifolium* (L.) Scop. in the European literature.

English Common Names

Fireweed, common fireweed, perennial fireweed, narrow-leaved fireweed, great willow-herb, spiked willow-herb, rosebay willow-herb, blooming Sally, wild asparagus, purple rocket, wickup, wicopy.

The name "fireweed" refers to *E. angustifolium* in North America, but in some parts of the world other species quickly colonize fire-ravaged areas, and are also known as fireweed. The name wild asparagus reflects consumption of the young shoots which are sometimes eaten like asparagus. The "willow" in some variants of the name refers to the willow-like shape of the leaves; in Europe the plant was known as flowering willow, French willow, Persian willow, and rosebay willow. The name Sally is a corruption of *Salix* (the willow genus), still another reference to willow-like leaves. The name rosebay is a reference to the rosy flowers and the bay-like leaves (bay is the culinary plant *Laurus nobilis* L.).

French Common Names

Épilobe à feuilles étroites, bouquet rouge.

Morphology

Fireweed is a robust perennial herb, 1–3 m tall, topped by a long inflorescence of very attractive purple or pink (rarely white) flowers with petals 1–2 cm long. The alternate, lance-shaped leaves are 3–20 cm long, and characteristically reticulate-veiny on the lower surfaces. Unlike most other plants, the flowers have four petals, like evening primrose, which is also in the Onagraceae family. The slender fruit is 5 to 8 cm long, and contains many seeds. The seeds are 1–1.3 mm long, and have a tuft of long hairs at one end, which serves as a sail for wind distribution.

Classification and Geography

Epilobium is a large genus (ca. 200 species) of mostly perennial herbs of temperate climates and tropical mountains. *Epilobium angustifolium* and *E. latifolium* L. (a smaller low-growing plant of river shores and sand bars) form a distinctive subgroup in the genus, with large flowers having petals of unequal size; this subgroup is sometimes assigned to the genus *Chamaenerion*.

Fireweed occurs in all Canadian provinces and territories. It is also widely distributed in the US, except the southeastern states and Texas. The species is circumboreal, occurring widely in Eurasia. The plants vary geographically in size and leaf shape, leading to past recognition of poorly defined varieties. In North America, *E. angustifolium* ssp. *angustifolium* is characterized by small to medium-sized leaves lacking pubescence on the abaxial ribs, triporate (3-pored) pollen grains, and a chromosome number of $2n=36$. By contrast, ssp. *circumvagum* Mosquin has small to very large leaves with glabrous to densely pubescent abaxial leaf ribs, quadriporate (4-pored) pollen as well as triporate, and a chromosome number of $2n=72$. Forms with white petals are occasionally seen.

The European medicinal species *E. parviflorum* Schreber has been introduced to North America and is well established in southern Ontario.

Ecology

Fireweed dominates many plant communities undergoing succession. It is common in streamside and upland habitats, and in logged and burned areas. This species is adapted to rapidly colonize newly disturbed habitats, especially where moist mineral soil is laid bare and considerable light is available. Although fireweed can tolerate considerable shade, it grows well only in open locations. It characteristically grows on acidic soils, pH varying from as low as 3.5 to circumneutral. Favorite habitats include coniferous and mixed forests, aspen parklands, grasslands, and muskegs, disturbed regions such as cut-over or burned forests and swamps, recently deglaciated territory, avalanche zones, riverbars, embankments of highways and railways, waste places and old fields.

Despite the "weed" in fireweed, this species is not usually a significant weed. Competition with conifer seedlings in revegetating burned land, and serving as the alternate host of conifer rusts have been noted as potential problems for forest managers. Fireweed has also been observed as a weed of some vegetable crops in northern regions.

Inbreeding depression (reduced fitness following self-fertilization) is extreme in fireweed, making cross-pollination very important. Individual

flowers mature their pollen before the stigma is receptive, to prevent selfing. Bees, bumblebees, moths and butterflies are important pollinators of fireweed. They characteristically move from the bottom towards the top of the flowering stem; the insects first deposit pollen on the lower flowers, which have receptive stigmas but anthers without pollen which was released earlier. Then, they acquire pollen from the upper flowers, which have not yet opened their stigmas.

The extremely large number of seeds produced can be carried by wind for hundreds of kilometers. A Swedish study revealed that up to half of the seed produced disperses over the landscape more than 100 m above the ground. Humidity expands the diameter of the seed hairs, decreasing loft, an adaptation which tends to deposit the seeds in humid areas and during wet periods, thus ensuring adequate moisture for germination. The seeds are nondormant and short-lived, rarely remaining viable for more than 3 years, and generally germinating as soon as a suitable site is found. Although seeds account for the remarkable ability of fireweed to colonize new areas, once a seed germinates, vegetative reproduction by rhizome spread becomes more important than sexual reproduction for propagating the plant at a given site. Fragmenting the rhizomes stimulates production of shoots, as with dandelions, contributing to the reputation of fireweed as a hard-to-eradicate weed. Most of the rhizomes and roots occur in the top 5 cm of mineral soils, and this underground portion can survive relatively intense fires. Indeed, the rhizomes sprout new shoots vigorously a few weeks following destruction of the old shoots. Young plants overwinter as rosettes. The aboveground shoots of older plants are killed by frost, but the plants overwinter as rhizomes.

Medicinal Uses

Native Americans employed fireweed juice to soothe skin irritation and burns, a practice also in European herbal medicine. In both European and North American folk medicine, fireweed has been used as tea to relieve stomach upset, respiratory complaints, and constipation. These uses could be explained by the high tannin content (and therefore astringency) of fireweed (but see below for possible presence of antiseptic, healing compounds).

Wichtl (1994; see General References) provides a fairly comprehensive discussion of the current medicinal usage of "willow-herb" in Europe, where several species of *Epilobium* are used (*E. parviflorum*, *E. montanum* L., *E. roseum* Schreber, *E. collinum* C.C. Gmelin, and others), and *E. angustifolium* is considered an adulterant (but one with similar properties). The European drug is wild-collected, consists mostly of pieces of stem, leaves, and some flower and fruit fragments, and is consumed as an infusion and used to treat benign prostate conditions and associated problems of difficult urination. *Epilobium* extracts have shown some capacity to counteract inflammation and fever. In experimental studies, two macrocyclic ellagitannins from *Epilobium* have been found to reduce benign prostatic hyperplasia (a noncancerous enlargement of the prostate that can interfere with urination) by suppressing the enzymes (5-α-reductase and aromatase) that contribute to this condition. A recent TV program in Australia alleging the value of "epilobium tea" (from *E. parviflorum*) for prostate cancer resulted in a sell-out of local stock. Extracts specifically from *E. angustifolium* have been shown to be capable of reducing the swelling from edema.

In 1994 a Canadian company put extracts with medicinal properties from fireweed on the market. According to the manufacturer, Fytokem Products Inc. of Saskatoon, "Canadian willowherb extract has been shown to be an effective anti-irritant and a mild sunscreen, as well as inhibiting microbial growth. It has application in creams, lotions, after sun products, after shave products and baby care products."

Toxicity

Fireweed is not considered toxic.

Chemistry

There is evidence of healing antiseptic compounds in fireweed. A novel flavonoid with strong anti-inflammatory effects, myricetin 3-O-β-D-glucuronide, has been found in the foliage. This active principle reaches its maximum concentration during and shortly after the plants flower. According to Fytokem Products Inc., their willowherb extract "is known to contain an array of sitosterol and flavonoid derivatives, along with a large polysaccharide content (about 50%)."

Epilobium angustifolium (fireweed)

Non-medicinal Uses

Possibly the most important agricultural significance of fireweed is as a honey plant, providing considerable nectar to bees. Indigenous people in both the Old and New Worlds have used fireweed as food, for example the young shoots consumed as greens, the leaves used to make tea, the petals made into jelly, and the roots eaten as a vegetable. Numerous wild mammals (including moose, caribou, elk, deer, mountain goats, muskrats, and hares), and some wild birds forage on fireweed, which is moderately nutritious and palatable. Fireweed will also be consumed by domestic livestock, although it is considered only fair as forage. The species is widely used for revegetation of northern and alpine disturbed sites, including roadways and logged areas. It is also useful for providing plant cover where oil spills have occurred, as well as for strip-mined areas and mine spoil deposits. Fireweed is occasionally grown as an ornamental, admired for its easy cultivation, spectacular floral display, and deep red fall foliage, although its weedy tendencies sometimes make it undesirable. An antiquated use of the plant was the incorporation of the down of the seeds into cotton and fur to make clothing.

Agricultural and Commercial Aspects

At present, the only major cultivation of fireweed is for soil stabilization. Its economic value as a medicinal plant is still limited. The success of Fytokem Products Inc. in marketing fireweed extracts for skin care products provides an instructive example of how, with appropriate research, Canada can develop industries based on native medicinal plants long thought to be economically obsolete.

Myths, Legends, Tales, Folklore, and Interesting Facts

- Fireweed is the floral emblem of the Yukon Territory.
- The Russian name "Ivan's tea" reflects the use of fireweed for tea in Russia. In Kamchatka, ale made with fireweed is said to have been rendered more intoxicating by adding the mushroom *Agaricus muscarius* (fly agaric, death cap; widely used historically as a hallucinogen, although dangerous).
- Fireweed sprouted quickly from surviving rhizomes, following the volcanic eruption on Mount St. Helens, Washington in 1980. One year after the explosion, 81% of all seedlings present were from fireweed.
- A fruit of fireweed can contain as many as 500 seeds. A single plant can produce as many as 80 000 seeds per year.
- It has been suggested that mutations from the radiation near uranium deposits results in a relatively high frequency of white-flowered variants of fireweed, and that this phenomenon can be used to find uranium.

Selected References

Baum, D.A., Sytsma, K.J., and Hoch, P.C. 1994. A phylogenetic analysis of *Epilobium* (Onagraceae) based on nuclear ribosomal DNA sequences. Syst. Bot. **19**: 363–388.

Brenchley, W.E., and Heintze, S.G. 1933. Colonization by *Epilobium angustifolium*. J. Ecol. **21**: 101–120.

Broderick, D.H. 1990. The biology of Canadian weeds: 93. *Epilobium angustifolium* L. (Onagraceae). Can. J. Plant Sci. **70**: 247–260.

Bult, C.J., and Zimmer, E.A. 1993. Nuclear ribosomal RNA sequences for inferring tribal relationships within Onagraceae. Syst. Bot. **18**: 48–63.

Ducrey, B., Wolfender, J.L., Marston, A., and Hostettmann, K. 1995. Analysis of flavonol glycosides of thirteen *Epilobium* species (Onagraceae) by LC-UV and thermospray LC-MS. Phytochemistry **38**: 129–137.

Ducrey, B., Marston, A., Goehring, S., Hartmann, R.W., and Hostettmann, K. 1997. Inhibition of 5-alpha-reductase and aromatase by the ellagitannins oenothein A and oenothein B from *Epilobium*. Planta Med. **63**(2): 111–114.

Galen, C., and Plowright, R.C. 1985. The effects of nectar level and flower development on pollen carry-over in inflorescences of fireweed *(Epilobium angustifolium)* (Onagraceae). Can. J. Bot. **63**: 488–491.

Gensac, P. 1986. The communities of *Epilobium angustifolium* and the colonizing role of this species in the mountain (District of Aime, Savoie, France). Bull. Soc. Bot. Fr. Lett. Bot. **133**: 179–188. [In French.]

Henderson, G., Holland, P.G., and Werren, G.L. 1979. The natural history of a subarctic adventive: *Epilobium angustifolium* L. (Onagraceae) at Schefferville, Quebec. Nat. Can. **106**: 425–437.

Hetherington, M., and Steck, W. 1997. Natural chemicals from northern prairie plants. Fytokem Publications, Saskatoon, SK. 279 pp. (Advertisement: http://www.fytokem.com/book.html)

Hiermann, A, and Bucar, F. 1997. Studies of *Epilobium angustifolium* extracts on growth of accessory sexual organs in rats. J. Ethnopharmacol. **55**: 179–183.

Hiermann, A., Reidlinger, M., Juan, H., and Sametz, W. 1991. Isolation of the antiphlogistic principle from *Epilobium angustifolium*. Planta Med. **57**: 357–360. [In German.]

Hoch, P.C., Crisci, J.V., Tobe, H., and Berry, P.E. 1993. A cladistic analysis of the plant family Onagraceae. Syst. Bot. **18**: 31–47.

Husband, B.C., and Schemske, D.W. 1995. Timing and magnitude of inbreeding depression in diploid *Epilobium angustifolium* (Onagraceae). Heredity **74**: 206–215.

Husband, B.C., and Schemske, D.W. 1997. Effects of inbreeding depression in diploid and tetraploid populations of *Epilobium angustifolium*: implications for the genetic basis of inbreeding depression. Evolution **51**: 737–746.

Keating, R.C., Hoch, P.C., and Raven, P.H. 1982. Perennation in *Epilobium* (Onagraceae) and its relation to classification and ecology. Syst. Bot. **7**: 379–404.

Lesuisse, D., Berjonneau, J., Ciot, C., Devaux, P., Doucet, B., Gourvest, J.F., Khemis, B., Lang, C., Legrand, R., Lowinski, M., Maquin, P., Parent, A., Schoot, B., and Teutsch, G. 1996. Determination of oenothein B as the active 5-alpha-reductase-inhibiting principle of the folk medicine *Epilobium parviflorum*. J. Nat. Prod. **59**: 490–492.

Myerscough, P.J., and Whitehead, F.H. 1980. Biological flora of the British Isles. *Epilobium angustifolium* L. (*Chamaenerion angustifolium* (L.) Scop.). J. Ecol. **68**: 1047–1074.

Michaud, J.P. 1990. Observations on nectar secretion in fireweed, *Epilobium angustifolium* L. (Onagraceae). J. Apicult. Res. **29**(3): 132–137.

Mosquin, T. 1966. A new taxonomy for *Epilobium angustifolium* L. Brittonia **18**: 167–188.

Mosquin, T. 1967. Evidence for autopolyploidy in *Epilobium angustifolium* (Onagraceae). Evolution **21**: 713–719.

Mosquin, T., and Small, E. 1971. An example of parallel evolution in *Epilobium* (Onagraceae). Evolution **25**: 678–682.

Praglowsk, J., Nowicke, J.W., Skvarla, J.J., Hoch, P.C., Raven, P.H., and Takahashi, M. 1994. Onagraceae Juss. Circaeeae DC., Hauyeae Raimann, Epilobieae Spach. World Pollen Spore Flora **19**: 1–38.

Purcell, N.J. 1976. *Epilobium parviflorum* Schreb. (Onagraceae) established in North America. Rhodora **78**: 785–787.

Raven, P.H. 1976. Generic and sectional delimitation in Onagraceae, tribe Epilobieae. Ann. MO. Bot. Gard. **63**: 326–340.

Shacklette, H.T. 1964. Flower variation of *Epilobium angustifolium* L. growing over uranium deposits. Can. Field-Nat. **78**: 32–42.

Solbreck, C., and Andersson, D. 1987. Vertical distribution of fireweed, *Epilobium angustifolium*, seeds in the air. Can. J. Bot. **65**: 2177–2178.

Winder, R.S., and Watson, A.K.A. 1994. Potential microbial control for fireweed (*Epilobium angustifolium*). Phytoprotection **75**(1): 19–33.

World Wide Web Links

(Warning. The quality of information on the internet varies from excellent to erroneous and highly misleading. The links below were chosen because they were the most informative sites located at the time of our internet search. Since medicinal plants are the subject, information on medicinal usage is often given. Such information may be flawed, and in any event should not be substituted for professional medical guidance.)

Canadian Willowherb Product Information:
http://www.fytokem.com/willow2.html

Canadian Willowherb Extract Products:
http://www.fytokem.com/extracts.html

A modern herbal by M. Grieve:
http://www.botanical.com/botanicaL/mgmh/w/wilher23.html

Herbaceous plants of the northwoods:
http://www.rook.org/earL/bwca/nature/herbs/epilobiuman.html

Wildflower slide show - fireweed [color photograph]:
http://www.afternet.com/~tnr/wildflower/flowershow.html

Inbreeding and inbreeding depression in fireweed (*Epilobium angustifolium*):
http://www.uoguelph.ca/botany/evollab/inbreed.htm

Fire effects information system [excellent!]:
http://svinet2.fs.fed.us/database/feis/plants/forb/epiang/

Hamamelis virginiana (witch hazel)

Hamamelis virginiana L. Witch Hazel

English Common Names

Witch hazel (also witch-hazel, witchhazel), Virginian witch hazel, common witch hazel. Other names sometimes applied (inappropriately, since most are better used for other plants) include: hazel nut, snapping hazel, spotted alder, striped alder, tobacco wood, and winter bloom.

French Common Names

Hamamélis de Virginie, café du diable.

Morphology

Witch hazel is a deciduous several-stemmed shrub or small tree, 1–5 (very rarely as much as 10) m in height, with smooth, brown, thin, scaly bark and numerous long, flexible, forking, branches. The branches zigzag at the leaf nodes, and this has been interpreted as a way to separate the leaves to achieve maximum exposure to the limited sunlight under the canopies of taller trees. The distinctive hazel-like leaves (i.e., like those of *Corylus* or true hazel species) are wavy or scalloped on the sides, 5–15 cm long, and have an asymmetrical base. The species can spread to some extent by suckering, but reproduces mostly by seeds. It is often shallow-rooted. In well-grown specimens the trunk may achieve a diameter of 10 (very rarely as much as 30) cm. The largest trees are found in the southern portion of the range.

Witch hazel is very unusual in that it flowers in late fall, often after the first frosts. Fragrant, yellow flowers in small axillary clusters appear as the foliage yellows, and the flowers persist after the leaves have fallen. Witch hazel is the only tree in the woods of North America which has ripe fruit, flowers, and the following year's leaf buds on the branch at the same time. The flowers have four twisted, strap-like petals 1.5–2 cm long, which can curl up as if to protect the flower from the cold when the temperatures drop, and unfurl when temperatures rise and pollinators are available. The flowers often survive several frosts. Witch hazel produces a very attractive flush of flowers, which are conspicuous because most deciduous species have lost or are losing their foliage, and (at least in northern woods) virtually all other plants are not in flower, or are well past their peak flowering period. In the northern part of the range, flowering occurs from October to as late as early December, while in the South, blossoms may be present as late as March. The fruit ripens in the following summer, maturing into paired, 2-horned, fuzzy, brown, woody capsules, 1–1.5 cm long, each generally producing a single oblong, hard seed (or sometimes two seeds). The seeds are black and shiny on the outside, white, oily, and farinaceous on the inside, and although quite small, they are edible like the related hazelnuts and filberts (*Corylus* species).

Classification and Geography

The genus *Hamamelis* consists of about six species of deciduous shrubs or small trees. *Hamamelis virginiana* extends from Minnesota, northern Michigan, southern Ontario, southern Quebec, New Brunswick, and southern Nova Scotia, southwards to Texas and central Florida. A number of varieties have been described from this extensive region based on characteristics of leaves, but their taxonomic status requires more study. In its northern range, the leaves are larger, the petals are bright yellow, and the plants are usually shrubs. In South Carolina, Georgia, and Florida, the leaves are usually smaller, the petals are distinctly pale yellow, and the plants sometimes reach the proportions of small trees (these have been called *H. virginiana* var. *parvifolia* Nutt.). The other North American species, *H. vernalis* Sarg. (Ozark witch hazel) occurs from southeastern Missouri through Arkansas to southeastern Oklahoma. On the Ozark Plateau, where *H. virginiana* and *H. vernalis* occur together, the petals of *H. virginiana* are frequently reddish at the base, suggesting that hybridization has occurred there between the two species.

The genus *Hamamelis* exhibits two interesting biogeographical patterns: (1) the eastern Asian/eastern North American disjunctions from the Arcto-Tertiary Forest, which encircled the Northern Hemisphere 15–20 million years ago, like *Panax* (ginseng) and *Hydrastis* (goldenseal); (2) the disjunction of temperate eastern North American elements into the high elevation temperate regions of Mexico. The Mexican disjunctions are believed to be remnants of a more recent and continuous Pleistocene distribution.

Ecology

Witch hazel grows in dry to mesic woods of eastern deciduous forests, usually among mixed hardwoods. In the middle part of its range witch hazel sometimes forms a solid understorey in second-

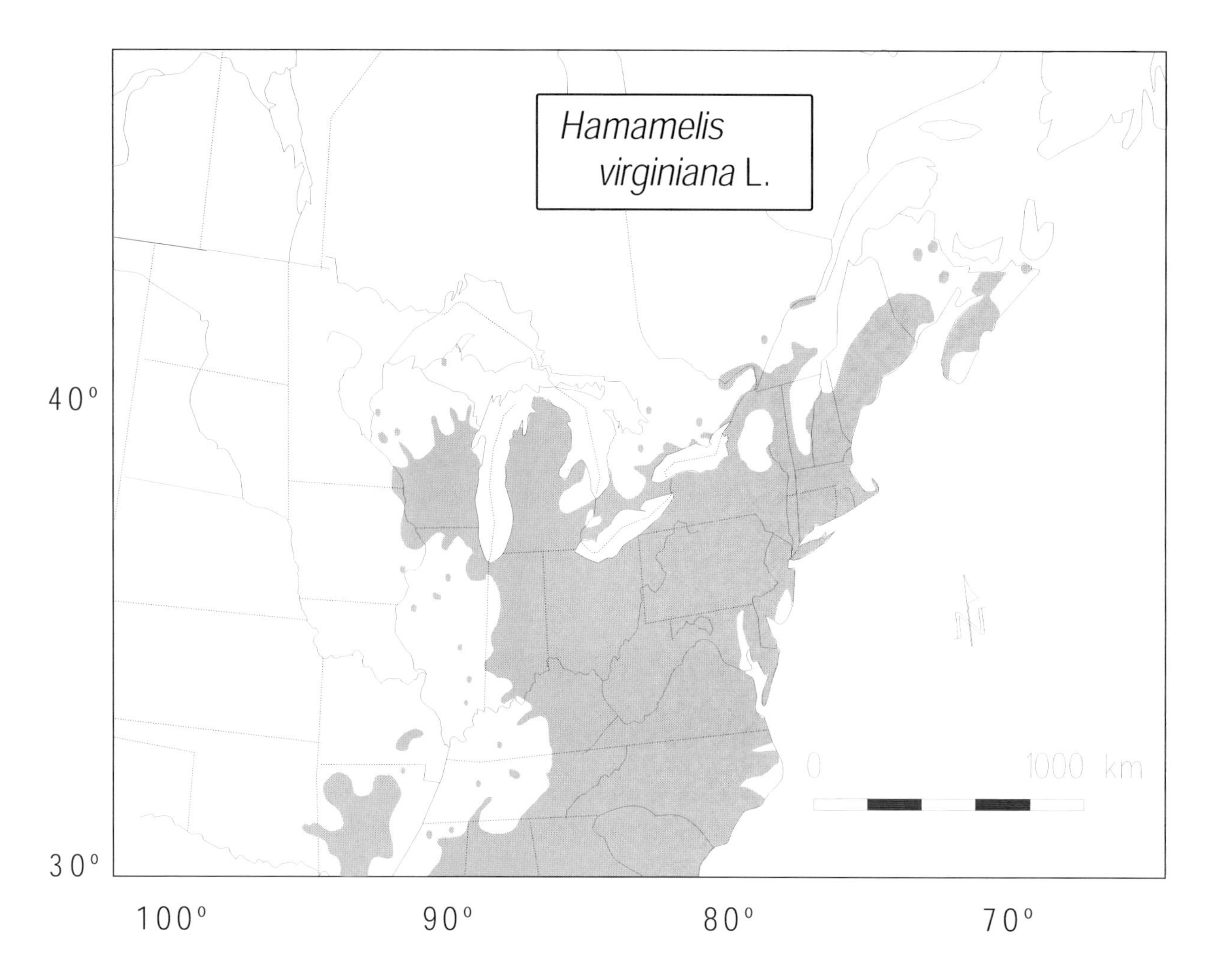

growth and old-growth forests. In the northern part of its range, it is often found as scattered, small colonies. Witch hazel is shade-tolerant and grows well as an understorey species, preferring deep, rich soils. Even shade-tolerant species often benefit from an increase in light, and witch hazel was found to respond to canopy gaps in a central Pennsylvanian oak forest by increasing sexual and vegetative reproduction. It may be found on hills or in stony places, rocky slopes, on the banks of streams, along ravines, trails and forest edges. In the western and southern areas of its range, it is distributed mostly in moist cool valleys and flats, northern and eastern slopes, coves, benches and ravines. In the northern area of its range, it occurs on drier and warmer sites of slopes and hilltops. The species appears to tolerate both acidic and alkaline substrates, but in Canada is mostly associated with sandy, slightly acid substrates.

Staminodes (sterile stamens) in the flowers secrete small amounts of nectar, serving to attract pollinating insects. Because of the late season, cold often limits the availability of pollinators. Experiments have shown that witch hazel can self-pollinate effectively, so that it is not dependent on the unreliable pollinator pool. However, when the weather is favorable, a large variety of insects may be available to cross-pollinate the flowers, and witch hazel is one of the few woodland plants serving nectar-foraging insects in late fall and early winter. The flowers are clearly adapted to pollination by a range of insect species. Although pollination occurs in the fall, fertilization is delayed until the following spring because of pollen and ovule dormancy.

Ripe seeds are dispersed in late autumn, simultaneous with flowering. The seeds are disseminated by mechanical expulsion from the dehiscent capsule. Seeds may be shot to a distance of 10 m (claims of 15 m have been made, although in most cases less than 5 m is achieved), and this has given rise to the name "snapping hazel." An audible pop accompanies the explosive discharge. Birds are thought to have a limited role in dispersing the seeds. The mammals that eat the seeds are likely more important (see non-medicinal uses). The seeds germinate the second year after dispersal. A study in Michigan revealed that successful seed production was irregular, with large numbers of seeds in the occasional good fruiting years related to satiation of host-specific beetles that eat the seeds.

Medicinal Uses

Witch hazel is one of the most popular of medicinal plants, and has been much in demand for centuries. Poultices and infusions of the leaves and (to a much greater extent) the bark have long been used externally to treat wounds and bleeding, including every kind of abrasion, as well as menstrual and hemorrhoidal bleeding. This medical knowledge was first acquired by North American Indians, then by colonists, followed by Europeans. In early times, witch hazel was also employed to treat tumors and inflammations, especially of the eye, and as a liniment. Extracts of witch hazel were also used internally to treat diarrhea. Most of these usages have persisted to the present. About the middle of the 19th century, a product prepared by steam-distillation of the dormant twigs, to which alcohol was added, became extremely popular under the name "hamamelis water." This was intended for external treatment of various skin conditions, and is still marketed today. Alcoholic extracts are popular in Europe for treating varicose veins, and the effectiveness of these extracts in constricting veins has been demonstrated. Modern medicinal uses today also include treatment of inflammation of the gums and mucous membranes of the mouth. The most common present usage is in soothing skin lotions. Witch hazel is employed in toilet water, aftershave lotions, mouth washes, skin cosmetics and the like, and ointments to treat sunburn, chapping, insect stings and bites. Long before such brands as "Obsession," "Passion" and "Old Spice," witch hazel was used as an aftershave. There is some indication of value for treating aging or wrinking of skin, an application with considerable market potential. As with most medicinal plants, usage in Europe considerably exceeds that in North America. Nevertheless, more than a dozen preparations with witch hazel are marketed in Canada.

Hamamelis virginiana (witch hazel)

After an 85-year absence, witch hazel was recently relisted in the US Pharmacopoeia (USP XXIII 1995: 1637).

Corylus avellana (hazel or hazelnut of Europe) is rarely used to adulterate witch hazel, and occasionally it is claimed that the two species have similar medicinal properties.

Toxicity

Witch hazel herbal preparations are often sold in health food stores, for consumption as a bitter tea. Internal consumption should be done cautiously, as the plant has minor amounts of toxic chemicals (such as eugenol, acetaldehyde, and the carcinogen safrole), and an internal dose of as little as a gram can cause nausea, vomiting and constipation. In rare cases, liver damage has been attributed to consumption of witch hazel. External use should also be carried out cautiously, as a concentrated tincture can be sufficiently astringent as to disfigure skin, and contact dermatitis is possible in susceptible individuals. Despite some potential toxicity and misgivings by some that its medicinal value is limited, witch hazel has a long history of popularity.

Chemistry

The medicinal value of witch hazel appears primarily due to its astringency, which seems mostly related to the high tannin content of the plant. The leaves can contain up to 10% tannin, and the bark has up to 3%. Tannins are astringent because they fix proteins, and while this is not helpful to the proteins (which are denatured) it can be helpful to healing of broken or irritated skin by creating a protective covering or constricting the area of injured tissue that is exposed. The numerous personal care products containing witch hazel that are applied to the skin are presumably useful because of the pronounced styptic qualities of the plant. There is some evidence that not just tannins, but other astringent agents are present, and that flavonoids may also play a curative role. Hamamelis water is traditionally prepared as a steam extract (alcohol is subsequently added), and this has very little tannin content, but still considered to be astringent (the astringency of hamamelis water has been attributed simply to the alcohol content).

Non-medicinal Uses

There are a number of garden forms of witch hazel, although hybrids of the Asian species are more popular as ornamental cultivars. Unlike the northeastern North American witch hazel, the Asian species and the Ozark species are all late winter-flowering (February–March). They also have leaves that turn red or orange in the fall instead of yellow, and are consequently more often cultivated than our native species. Propagation by both seeds and cuttings is possible, but the seeds are dormant for a period and the cuttings require a few months under mist. Ornamental cultivars are propagated by grafting onto seedling understock. Utilization of suckers is also a means of propagation, and species with a relatively strong tendency to sucker, such as *H. vernalis*, and races of *H. virginiana* that are more prone to suckering, are potentially useful in this regard.

Many animals have been reported to eat the fruits of witch hazel, including ruffed grouse, northern bobwhite, ring-necked pheasant, white-tailed deer, beaver, cottontail rabbit, and black bear.

Agricultural and Commercial Aspects

A small amount of witch hazel is harvested from plants cultivated in Europe, but most of the world's supply is obtained from wild plants in the eastern United States. The state of Connecticut is a principal supplier of material for production of aqueous witch hazel which is made from twigs collected in autumn, winter and early spring. The witch hazel in cosmetic products comes from stripped leaves and bark collected in summer and early fall in the southern Appalachians. Witch hazel production is a substantial industry. In some years more than a million gallons of hamamelis water has been produced. Given the growing popularity of medicinal plants, it is unlikely that witch hazel will become obsolete. To improve production efficiency, more information is needed on patterns of variation in chemical composition and the influences of ecological factors. With the growing trend to protect wild plants from overharvesting, cultivation of this medicinal crop appears to have considerable promise.

Myths, Legends, Tales, Folklore, and Interesting Facts

- The genus name *Hamamelis* was the Greek name used by Hippocrates for the medlar, *Mespilus germanica* L., a small Eurasian tree of the rose family, with fruits resembling a crab apple that are used to make preserves.
- The origin of the "witch" in witch hazel has been attributed to an Old English term for pliant branches (which are characteristic of the plant). Nevertheless, witch hazel is often associated with witchcraft, an apparent misunderstanding of how the name originated. Historical analysis has shown that the name witch hazel was likely originally applied to English elms with flexible Y-shaped forked branches that were used as the source of divining rods, and the name became transferred by colonists to *H. virginiana* which has similar branches. Divining rods were used to search for water and ores, especially by charlatans (recommended technique: find a branch with forks pointing north and south; twirl it between the fingers and thumbs of the two hands, and point the base of the Y downwards; find a location where the base is attracted by water or minerals, especially gold). Those who dowsed for water by this technique were called "water witches."
- The Menominee Indians (whose former range included northern Wisconsin and adjacent upper Michigan, through which runs the Menominee River) used witch hazel seeds as sacred beads in medicine ceremonies.
- The largest known tree of *H. virginiana*, with a height of 10.6 m and a trunk diameter of 0.4 m, was recorded from Bedford, Virginia in 1994.

Selected References

Anonymous. 1991. Drug therapy of hemorrhoids. Proven results of therapy with a hamamelis containing hemorrhoid ointment. [Results of a meeting of experts. Dresden, 30 August 1991.] Fortschr. Med. Suppl. **116**: 1–11. [In German.]

Berry, E.W. 1920. The geological history of the sweet gum and witch hazel. Plant World **22**: 345–354.

Boerner, R.E.J. 1985. Foliar nutrient dynamics, growth, and nutrient use efficiency of *Hamamelis virginiana* in three forest microsites. Can. J. Bot. **63**: 1476–1481.

Bradford, J.L., and Marsh, D.L. 1977. Comparative studies of the witch hazels, *Hamamelis virginiana* L., and *H. vernalis* Sarg. Proc. Arkansas Acad. Sci. **31**: 29–31.

Brinkman, K.A. 1974. *Hamamelis virginiana* L., witch-hazel. Agric. Handb. U.S. Dep. Agric. **450**: 443–444.

Britton, N.L. 1905. Hamamelidaceae. N. Am. Flora **22**: 187.

Brown, G.E. 1974. Growing witch hazels. J. R. Hortic. Soc. **99**(1): 15–19.

Busher, P.E. 1996. Food caching behaviour of beavers (*Castor canadensis*): selection and use of woody species. Am. Midl. Nat. **135**: 343–348.

Chandler, R.F. 1989. Yarrow. Can. Pharm. J. **122**: 41-43.

Connor, S. 1995. Mystical, medicinal witch hazel. Arnoldia (Jamaica Plain) **55**(3): 20–21.

Darbyshire, S.J., and Dickson, H.L. 1980. Witch-hazel in the Ottawa area. Trail & Landscape **14**: 158–160.

De Steven, D. 1982. Seed production and seed predation in a temperate forest shrub (witch-hazel, *Hamamelis virginiana*). J. Ecol. **70**: 437–443.

De Steven, D. 1983. Floral ecology of witch-hazel (*Hamamelis virginiana*). Mich. Bot. **22**: 163–171.

De Steven, D. 1983. Reproductive consequences of insect seed predation in *Hamamelis virginiana*. Ecology **64**: 89–98.

Dickison, W.C. 1989. Comparisons of primitive Rosidae and Hamamelidae. *In* Evolution, systematics, and fossil history of the Hamamelidae, vol. 1: Introduction and 'lower' Hamamelidae. *Edited by* P.R. Crane and S. Blackmore. Systematics Association Special volume No. 40A. Clarendon Press, Oxford. pp. 47–73.

Dirr, M.A. 1983. Witch hazels deserve a spot in the landscape. Growth and flowering, cultivars available, *Hamamelis*. Am. Nurseryman **157**(5): 53–56, 58, 60–63.

Dirr, R.J. 1994. *Hamamelis* and other witch hazel plants. E. Ulmer, Stuttgart (Hohenheim). 156 pp. [In German.]

Dressler, R.L. 1954. Some floristic relationships between Mexico and the United States. Rhodora **56**: 81–96.

Endress, P.K. 1989. Aspects of evolutionary differentiation of the Hamamelidaceae and lower Hamamelididae. Plant Syst. Evol. **162**: 193–211.

Erdelmeier, C.A., Cinatl, J., Jr, Rabenau, H., Doerr, H.W., Biber, A., and Koch, E. 1996. Antiviral and antiphlogistic activities of *Hamamelis virginiana* bark. Planta Med. **62**: 241–245.

Ernst, W.R. 1963. The genera of Hamamelidaceae and Platanaceae in the southeastern United States. J. Arnold Arbor. **44**: 193–210.

Friedrich, H, and Kruger, N. 1974. New investigations on the tannin of *Hamamelis*. I. the tannin of the bark of *Hamamelis virginiana*. Plant Med. **25**: 138–148. [In German.]

Friedrich, H., and Kruger, N. 1974. New investigations on the tannin of *Hamamelis*. II. The tannin of the leaves of *Hamamelis virginiana*. Plant Med. **26**: 327–332. [In German.]

Friedrich, H, and Kruger, N. 1974. New investigations on the tannin of *Hamamelis*. III. Comparison of different species and seasonal variations. Plant Med. **26**: 333–337. [In German.]

Fulling, E.H. 1953. American witch hazel — history, nomenclature and modern utilization. Econ. Bot. **7**: 359–381.

Gaut, P.C., and Roberts, J.N. 1984. *Hamamelis* seed germination. Comb. Proc. Int. Plant Propag. Soc. (Boulder) **34**: 334–342.

Gleason, H. 1922. The witch hazels. J. N.Y. Bot. Gard. **23**: 17–19.

Granlund, H. 1994. Contact allergy to witch hazel. Contact Dermatitis **31**(3): 195.

Haberland, C., and Kolodziej, H. 1994. Novel galloylhamameloses from *Hamamelis virginiana*. Planta Med. **60**: 464–466.

Halm, I. 1978. *Hamamelis virginiana* L., drug plant. Herba Hung. (Budapest) **17**(3): 97–102. [In Hungarian.]

Hartisch, C., and Kolodziej, H. 1996. Galloylhamameloses and proanthocyanidins from *Hamamelis virginiana*. Phytochemistry **42**: 191–198.

Hicks, D.J., and Hustin, D.L. 1989. Response of *Hamamelis virginiana* L. to canopy gaps in a Pennsylvania oak forest. Am. Midl. Nat. **121**: 200–204.

Hohn, T.C. 1993. Bewitched. Am. Nurseryman **177**(2): 64–73.

Joustra, M.K., and Verhoeven, P.A.W. 1984. Rooting hardwood cuttings of certain woody perennial species [*Cornus* spp., *Hamamelis* spp., *Acer*, *Ligustrum*, *Prunus*, *Corylus*, *Corylapsis*]. Plant Propagat. (Boulder) **30**(2): 3–4.

Khalvashi, T.K., and Fomenko, K.P. 1979. Effect of inorganic fertilizers on the growth and development of *Hamamelis virginiana*, drug plant. Rastit. Resur. (Leningrad) **15**: 98–106. [In Russian.]

Korting, H.C., Schafer-Korting, M., Klovekorn, W., Klovekorn, G., Martin,C., and Laux, P. 1995. Comparative efficacy of hamamelis distillate and hydrocortisone cream in atopic eczema. Eur. J. Clin. Pharmacol. **48**: 461–465.

Korting, H.C., Schaefer-Korting, M, Hart, H,Laux, P., and Schmid, M. 1993. Anti-inflammatory activity of hamamelis distillate applied topically to the skin: influence of vehicle and dose. Eur. J. Clin Pharmacol. **44**: 315–318.

Lamb, J.G.D. 1976. The propagation of understocks for *Hamamelis* [*Hamamelis vernalis*, *Hamamelis virginiana*, *Hamamelis mollis*]. Comb. Proc. Annu. Meet. Int. Plant Propag. Soc. **26**: 127–130.

Li, H.-L. 1952. Floristic relationships between eastern Asia and eastern North America. Trans. Am. Phil. Soc. N.S. **42**: 371–429.

Marquard, R.D., Davis, E.P., and Stowe, E.L. 1997. Genetic diversity among witchhazel cultivars based on randomly amplified polymorphic DNA markers. J. Am. Soc. Hort. Sci. **122**: 529–535.

Masaki, H., Atsumi, T., and Sakurai, H. 1994. Hamamelitannin as a new potent active oxygen scavenger. Phytochemistry **37**: 337–343.

Masaki, H., Atsumi, T., and Sakurai, H. 1995. Protective activity of hamamelitannin on cell damage induced by superoxide anion radicals in murine dermal fibroblasts. Biol. Pharmaceut. Bull. **18**: 59–63.

Masaki, H., Atsumi, T., and Sakurai, H. 1995. Protective activity of hamamelitannin on cell damage of murine skin fibroblasts induced by UVB irradiation. J. Dermatol. Sci. **10**(1): 25–34.

Masaki, H, Sakaki, S., Atsumi, T., and Sakurai, H. 1995. Active-oxygen scavenging activity of plant extracts. Biol. Pharmaceut. Bull. **18**: 162–166.

Messerschmidt, W. 1971. On the knowledge of steam distillates in leaves of *Hamamelis virginiana* L. 4. characterization of leaf drug and distillate. Dtsch. Apoth. ztg. **111**: 299–301. [In German.]

Meyer, F.G. 1997. Hamamelidaceae. *In* Flora of North America north of Mexico, Vol. 3. *Edited by* Flora of North America Editorial Committee. Oxford University Press, New York, NY. pp. 362-365.

O'Gorman, M.V. 1979. Witch hazel *Hamamelis*, varieties. Garden, N.Y. **3**(1): 30–31.

Steyermark, J.A. 1934. *Hamamelis virginiana* in Missouri. Rhodora **36**: 97–100.

Steyermark, J. 1956. Eastern witch hazel. MO Bot. Gard. Bull. **44**: 99–101.

Vennat, B., Pourrat, H., Pouget,, M.P., Gross, D., and Pourrat, A. 1988. Tannins from *Hamamelis virginiana*: identification of proanthocyanidins and hamamelitannin quantification in leaf, bark, and stem extracts. Plant Med. **54**: 454–457.

Weaver, R.E. 1976. The witch hazel family (Hamamelidaceae). Arnoldia (Jamaica Plain) **36**(3): 69–109.

Wood, G.W. 1974. Witch-hazel, *Hamamelis virginiana* L. USDA For. Serv. Gen. Tech. Rep. NE. For. Exp. Stn. **9**: 154–157. [provides information on forest ranges.]

World Wide Web Links

(Warning. The quality of information on the internet varies from excellent to erroneous and highly misleading. The links below were chosen because they were the most informative sites located at the time of our internet search. Since medicinal plants are the subject, information on medicinal usage is often given. Such information may be flawed, and in any event should not be substituted for professional medical guidance.)

A modern herbal by M. Grieve:
http://www.botanical.com/botanicaL/mgmh/w/withaz27.html

Windsor's world of fine and exotic woods:
http://www.windsorplywood.com/worldofwoods/northamerican/WitchHazel.html

Hamamelis insects:
http://www.msue.msu.edu/msue/imp/mod03/01700456.html

Witch hazel, *Hamamelis virginiana* [horticultural information]:
http://www.mpelectric.com/treebook/fact38.html

Trees of the Maritime Forest - *Hamamelis virginiana*:
http://russell4.hort.ncsu.edu/maritime/Hamamvi.htm

Brooklyn Botanical Garden, *Hamamelis virginiana*:
http://www.bbg.org/NYMF/encyclopedia/ham/ham0000.htm

Wholesale nursery growers providing *Hamamelis virginiana*, witch hazel:
http://www.growit.com/plants/growers/SN/2990.htm

Witch hazel, a charming fall bloomer (Wisconsin):
http://www.wnrmag.com/stories/1997/oct97/witchaze.htm

Native plant profile #9: witch hazel:
http://www.agnr.umd.edu/users/mg/natwitch.htm

Painting of Mary Vaux Walcott [color painting]:
http://chili.rt66.com/hbmoore/Images/Walcott/Hamamelis_virginiana.jpg

Hierochloe odorata (sweet grass)

Hierochloe odorata (L.) Beauv. Sweet Grass

The genus name *Hierochloe* was coined from the Greek *hieros*, sacred, and *chloë*, grass, a reference to the use of *H. odorata* in parts of Europe as a strewing herb on porches of churches, especially on saints' days. Reflecting the original Greek, the name is often rendered "*Hierochloë*." However, the Code of Botanical Nomenclature prohibits diacritical signs in Latin names.

English Common Names

Sweet grass, sweetgrass, sweet holygrass, Indian sweet grass, vanilla grass, seneca grass.

The name "sweet grass" is mostly used for species of *Hierochloe*, but is sometimes applied to other species, such as *Glyceria septentrionalis* Hitchc.

French Common Names

Foin d'odeur, herbe sainte.

In French, "foin d'odeur" is also applied to the introduced grass *Anthoxanthum odoratum* L.

Morphology

Sweet grass is a semi-erect perennial, the open panicles reaching heights of (10-)25–60(-100) cm. The creeping rhizomes are slender, and numerous, and shallow feeding roots arise from these as well as the base of the culm. Shiny green vegetative blades, (2-)3–6(-8) mm wide, 20–80 cm long, originate individually from the rhizomes (a form named *H. nashii* Kaczmarek is reputed to produce especially long leaf blades). Sweet grass is notable for its very early flowering, very shortly after spring growth begins. This is possible because the buds for the inflorescences were developed in the preceding autumn. Seeds are ripe in early summer, but seed set is sometimes low or absent, and some colonies produce seeds through apomixis rather than sexually.

Often confused with sweet grass, the European sweet vernal grass (*Anthoxanthum odoratum*) also has a vanilla-like fragrance but differs in its softly hairy foliage and spike-like inflorescence.

Classification and Geography

Hierochloe odorata is sometimes included in other genera, notably *Savastana* or *Torresia*.

Sweet grass is a rather variable polyploid complex, with diploids (with 14 chromosomes), tetraploids (28 chromosomes), octoploids (56 chromosomes), and other cytological phenomena. Some of the chromosome races have been recognized as separate species.

Recent taxonomic study has alleged that there are two very similar species in North America: *H. odorata* ssp. *odorata* in eastern coastal North America (Labrador to New Jersey), and *H. hirta* (Schrank) Borbás ssp. *arctica* G. Weim. in most of northern North America (north of 40° excepting eastern Northwest Territories, Nunavut and the arctic islands). However, the two taxa are difficult to separate and they are best placed together under the name *H. odorata*, pending additional study.

Ecology

Sweet grass is typically found in prairies, moist meadows, or along shorelines, and also in salt marshes, mostly among other herbs and shrubs. In eastern Canada it is spreading along roadsides receiving runoff with deicing salt.

Medicinal Uses

Sweet grass infusions were used by Plains Indians to treat various ailments, including coughs and sore throats, venereal infections, bleeding after childbirth, chapped or windburned skin, and eye irritations. However, the principal medical value of sweet grass is as a spiritually healing plant. Indigenous Peoples have long considered sweet grass to be one of the most important of sacred plants (others are the sages, cedar, alpine fir, juniper, tobacco, and corn). This was the most common sacred plant of Plains Indians, who occupied the region between the Mississippi and the Rockies in the US, and from the Rockies to Manitoba in southern Canada. Sweet grass is a regular component of "medicine pouches," which are typically hung about the neck, the wearer symbolically asking for spiritual protection. Elders often have the responsibility of providing the plants for medicine pouches, and may be given sweet grass as a sign of respect. The plant is in fact symbolic of Native spirituality, a tribute to the Creator or Great Spirit, and an aid to alleviating or removing evil spirits. Sweet grass is frequently braided and burned, customarily in the morning or evening, both in individual and group ceremonies, particularly at pow wows and celebrations. The plant is also burned in an incense (alone or in a mixture) to produce a sacred smoke, the smouldering material passed from person to person as a purification rite and as a symbol of

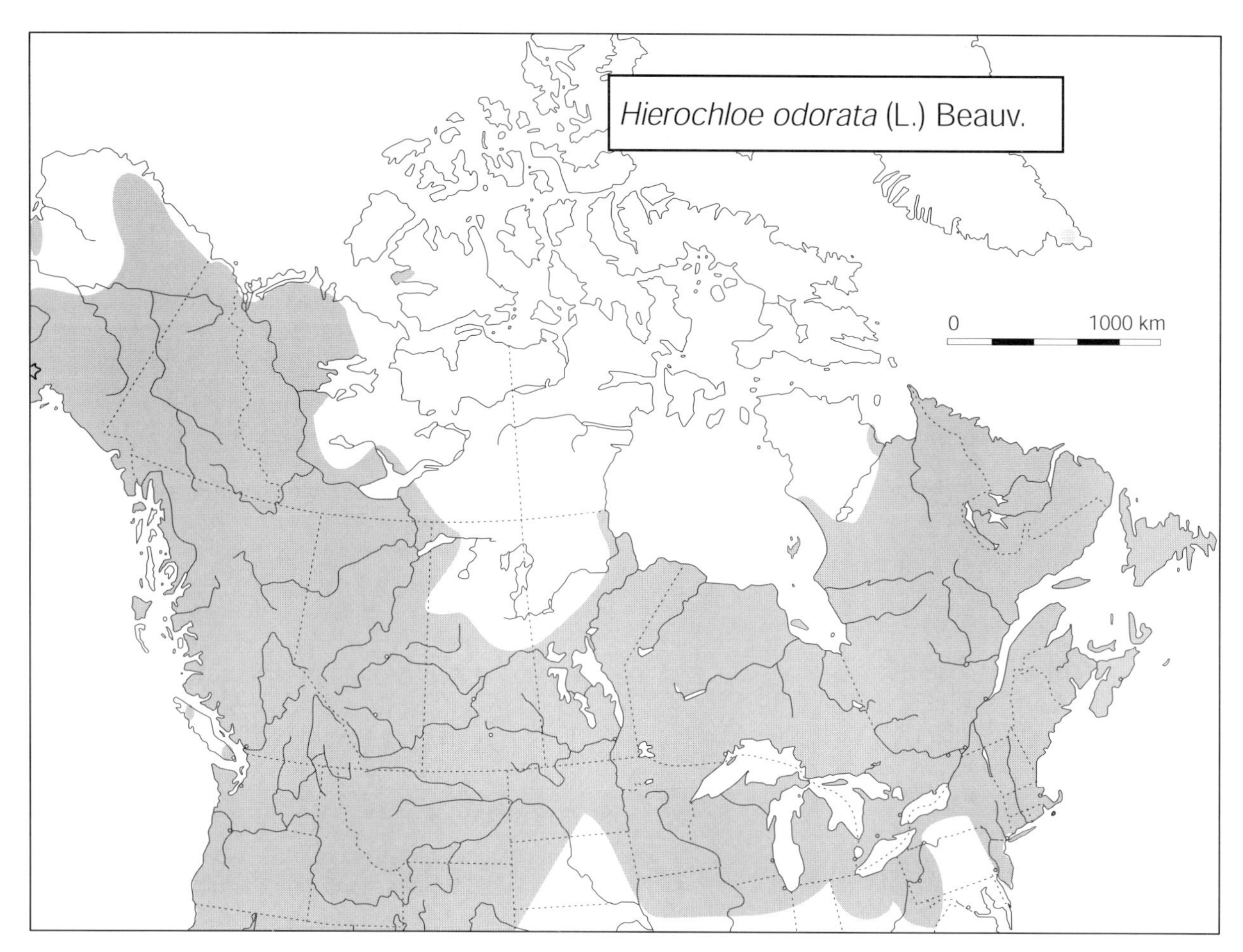

unity. Elders may also conduct a pipe ceremony, in which tobacco (commercial or other plants) is smoked while passing around a smouldering braid of sweet grass. The grass braid requires regular fanning to stay lit.

Toxicity

Coumarin in sweet grass is a flavoring agent (e.g., in the culinary herb sweet woodruff, *Galium odoratum* (L.) Scop.), but is toxic, known in experimental animals to cause liver damage, retard growth, and cause cancer and testicular atrophy. Foraging cattle are believed to have shown toxic reactions to eating large amounts of coumarin-containing plants. In the US coumarin is approved for use as a flavoring agent in alcoholic beverages, but is no longer allowed in food. Since humans do not orally consume much sweet grass, its toxic potential is not of importance. Concern does not seem to be recorded for potential toxicity for livestock.

Chemistry

The odor of sweet grass is due to coumarin.

Non-medicinal Uses

Sweet grass is widely used in weaving and basketwork by native Americans. The leaves of the flowering stalk are too short to be of value. The long vegetative leaves are individually harvested by native people from the middle of July until September, mid-season considered as supplying the best material. The dried leaves are strong and flexible and are often used to prepare handicrafts, largely for the tourist trade. The pleasantly flavored baskets, bowls, trays, and mats are a unique traditional product of Native Peoples, although it may be noted that sweet grass handiwork sold by Indian tribes outside of the distribution range of *H. odorata* may well have been manufactured from other species locally called sweet grass. In southern Quebec, the introduced sweet vernal grass, *Anthoxanthum odoratum*, has also been used by native peoples for basketwork. Cultivated *H. odorata* may produce larger, superior leaves for weaving than wild-growing plants. Some native handicrafts are threatened by recent urbanization and habitat destruction, since this eliminates or reduces the required plant materials.

The essential oil of sweet grass is used in perfumes. The dried leaves have been employed to scent pillows and clothing. On drying, the vanilla fragrance, due to coumarin, becomes more evident. The odor may not be apparent in fresh plants. When properly cured, sweet grass remains pleasantly scented for up to 3 years (some say the scent lasts indefinitely).

The coumarin content results in bitterness, which has been suspected of discouraging its use as a forage by cattle. Although sweet grass is not as palatable as most grasses, it will nevertheless be eaten fairly readily by cattle, and does not transmit its odor to milk. Certainly some humans seem to like the taste. Leaves have been placed in bottles of vodka as a flavorant, and packets of grass have been sold in the US for this purpose. Extract from the plant has been used in Europe to flavor candy, soft drinks, and tobacco.

The aggressively creeping rhizomes of sweet grass are very desirable for stabilizing soil. The tolerance of the species to wet areas makes it especially valuable in controlling erosion by water. The limited height of the plant may be advantageous where appearance is important, since tall-growing herbs often need mowing. Propagating sweet grass as a soil stabilizer is difficult by seed, since so little is produced, and there has been experimentation to establish the plant by rhizomes (underground creeping stems) or stolons (aboveground creeping stems). Experience to date suggests that this species may be one of the best soil stabilizers of moist and moderately saline sites such as road verges. Advantages of vegetative propagation include avoidance of seed shattering and ease of establishment.

Agricultural and Commercial Aspects

There is potential for increased cultivation of sweet grass in support of native culture and handicraft production, as well as for use in soil stabilization. The plant is best raised in a fertile soil enriched with humus, in a moist to wet site. It can be grown in full sun or partial shade. One authority recommended adding a little salt (NaCl) to the soil, since the species is found on the margins of saline as well as fresh water. Plantings may quickly develop into large patches by rhizome reproduction. A one inch plug can develop to a dense 10 square foot mass in 1 year and cover over 50 square feet in 2 years.

Hierochloe odorata (sweet grass)

Myths, Legends, Tales, Folklore, and Interesting Facts

- Aromatic herbs have been used for sacred objectives for millennia, so it is not surprising that Christians in Europe and Native Peoples of North America independently adopted sweet grass for religious purposes.

Selected References

Dore, W.G., and McNeill, J. 1980. Grasses of Ontario. Agricult. Can. Res. Br. Monogr. 26. 566 pp.

Dufault, R.J., Jackson, M., and Salvo, S.K. 1993. Sweetgrass: history, basketry, and constraints to industry growth. *In* New crops. *Edited by* J. Janick and J.E. Simon. Wiley, New York, NY. pp. 442–445. [http://www.hort.purdue.edu/newcrop/proceedings1993/V2-442.html] (N.B. the "sweet grass" referred to in this paper is *Muhlenbergia filipes* Curtis, not *Hierochloe odorata*.)

English, M. 1982. Sweet grass — a sacred herb. The Herbarist **48**: 5–9.

Ferris, C., Callow, R.S., and Gray, A.J. 1992. Mixed first and second division restitution in male meiosis of *Hierochloe odorata* (L.) Beauv. (holy grass). Heredity **69**: 21–31.

Fijalkowski, D., and Wawer, M. 1983 [1986]. *Hierochloe* R. Br. in the Lublin Macroregion (Poland). Ann. Univ. Mariae Curie-Sklodowska Sect. C Biol. **38**(1–21): 109–118. [In Polish.]

Fleurbec. 1985. Plantes sauvages du bord de la mer. Saint-Augustin (Portneuf), Québec. 286 pp.

Hitchcock, A.S. (revised by A. Chase). 1951. Manual of the grassses of the U.S. U.S. Dep. Agricult. Misc. Publ. No. 200, 2nd ed. 1051 pp.

Klebesadel, L.J. 1974. Sweet holygrass, a potentially valuable ally. Agroborealis (Inst. Agricult. Sci., Univ. Alaska) **6**(1): 9–10.

Looman, J. 1982. Prairie grasses identified and described by vegetative characters. Agricult. Can. Publ. 1413. 244 pp.

Ueyama, Y., Arai, T., and Hashimoto, S. 1991. Volatile constituents of ethanol extracts of *Hierochloe odorata* var. *pubescens* Kryl. Flavour Fragrance J. **6**(1): 63–68.

Weimarck, G. 1971. Variation and taxonomy of *Hierochloe* (Gramineae) in the northern hemisphere. Bot. Not. **124**: 129–175.

Wierzchowska-Renke, K., and Sulma,T. 1974. Studies of the hierochloe herb. Evaluation of commercial Herba Hierochloe raw material collected at the Zulawy plantation. Acta Pol. Pharm. **31**: 233–239. [In Polish.]

World Wide Web Links

(Warning. The quality of information on the internet varies from excellent to erroneous and highly misleading. The links below were chosen because they were the most informative sites located at the time of our internet search. Since medicinal plants are the subject, information on medicinal usage is often given. Such information may be flawed, and in any event should not be substituted for professional medical guidance.)

Fire effects information system:
http://svinet2.fs.fed.us/database/feis/plants/graminoid/hieodo/

Sweetgrass (*Hierochloe odorata*) growing and ordering information, Redwood City Seed Co.:
http://www.batnet.com/rwc-seed/sweetgrass.html

Sweet grass ceremony:
http://www.moonchild.com/moonspirit/cere-sweetgrass.html

Growing sweet grass for weaving Chippewa traditional baskets:
http://www.mallardsys.com/savage/sggrow.html

Humulus lupulus (hop)

Humulus lupulus L. Hop

English Common Names

Hop, common hop, English hop, European hop.

"Hop" refers to a hop plant; "hops" refers to the hops of commerce, i.e., the female cones.

French Common Names

Houblon.

Morphology

Hop is a high-climbing, wind-pollinated, perennial vine, sometimes extending more than 10 m. The vines climb by twining, and are assisted in holding onto surfaces by two-hooked hairs that resemble miniature grappling hooks. The annual, above-ground stem is killed by frost each year, re-growth occurring each season from perennial underground rhizomes and from buds on the rootstock (crown). The rootstocks can live for half a century. The perennial crown becomes woody with age, with heavy, rough, dark brown bark. The hop plant is propagated primarily vegetatively rather than by seeds. More than one hundred cultivars have been named, each essentially a clone. Many of these are of hybrid origin, and have been reproduced asexually for centuries. Some plants have male flowers, others have female flowers; occasionally plants bear both kinds. In most natural populations of hop, female plants are about twice as numerous as males. Because the female plants produce the commercially valuable hops (the cones or fruit-clusters), and also are of greater ornamental value than the males, the latter are generally discarded as soon as they can be recognized. However, a few males are usually planted deliberately in England because they increase yield and it is not economical to grow most English cultivars seedless. Male flowers are not organized into "cones," but are in loose inflorescences (panicles). In most hop plantations, males are regarded as a nuisance, since their pollen fertilizes the females, and prevents the formation of the valuable "seedless hops" that are predominantly preferred in commerce. A resin with the valuable brewing constituents is produced in yellow glands on the bracteoles of the cone, as well as on the seeds.

Classification and Geography

Humulus has three species, indigenous in north-temperate areas. Only the perennial *H. lupulus* is native to North America. The Asian annual *H. japonicus* Sieb. & Zucc. is a naturalized weed of eastern North America, including southern Ontario and southwestern Quebec.

There are five varieties of *H. lupulus*: var. *lupulus* of Eurasia; var. *cordifolius* (Miquel) Maximowicz of Japan; var. *neomexicanus* Nelson & Cockerell, the predominant wild hop in the western Cordillera of North America, found from Mexico to British Columbia; var. *pubescens* E. Small, of the midwestern US; and var. *lupuloides* E. Small, of eastern North America, which includes most wild hops from the Prairie Provinces to the maritime provinces. The distribution of the two indigenous Canadian varieties is shown on the map. In much of Canada and the US the European var. *lupulus* is found as an escaped plant from past use in brewing, or as a persisting ornamental around abandoned homesteads. Variety *lupulus* is the ancestor of most brewing cultivars used today. However, both in Japan and in North America, the local wild hops hybridized with the imported European hop to produce unique cultivars.

Ecology

Hop plants frequently occur in moist thickets, slopes, river banks, alluvial woods, or along fences and hedges, often in sandy soils. The plants grow best in rich, alluvial or deep sandy or gravelly loams. Well-drained soils are especially beneficial in areas subject to frost heaving of the roots. Hop is adapted to a wide range of temperate climates. Although it is quite tolerant of low temperatures, good snow cover can reduce winterkill in very cold regions. Hop is somewhat shade tolerant, but prefers full sun.

Medicinal Uses

The high oil and resin content of hops contributed to a reputation, greatly exaggerated, for valuable medicinal properties, and hops have a long history of use in folk medicine. Hop resin is bacteriostatic (against gram-positive organisms) and this factor may lend some credibility to use of hop in former times for treatment of certain types of epidermal sores and irritations, and bacterial infections such as tuberculosis. The hop plant is related to the hemp (marihuana) plant, tempting some to try to smoke the leaves. However, hops are devoid of the mood-altering chemicals found in marihuana. Today, hop extract is used as an aromatic

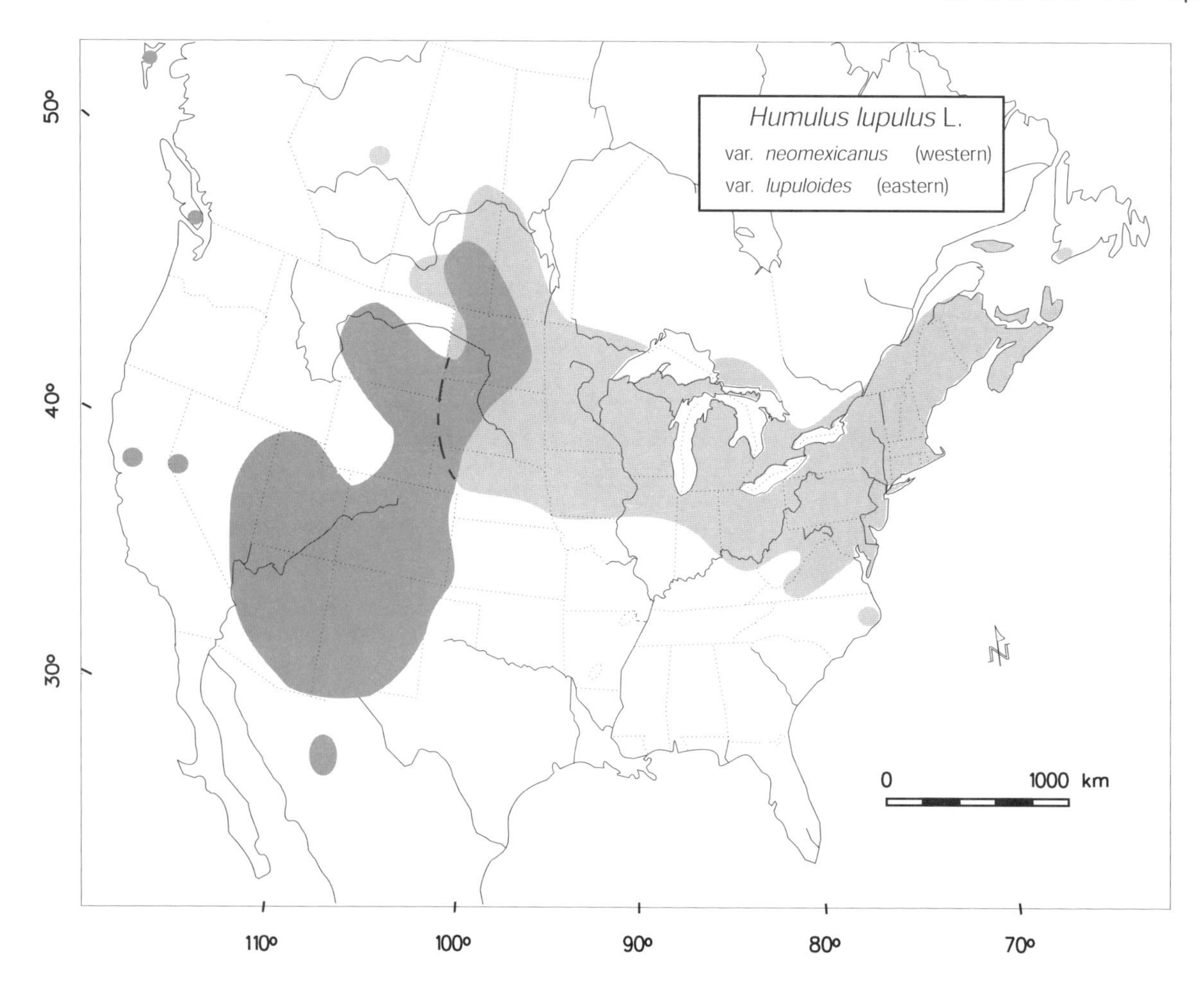

bitter principle in pharmaceutical preparations and in shampoos. Extracts are used in skin creams and lotions in Europe for their alleged skin-softening properties.

There is a long Eurasian tradition of using hops to induce sleep, including putting the cones in pillows, and planting hop beside bedrooms. Remarkably, American Indians independently adopted the soporific use of the plant. The tranquillizing effect commonly alleged in folklore for hops may have a logical basis in a sedative volatile alcohol, dimethylvinyl carbinol, which comprises up to 0.15% of the dried leaves.

Toxicity

Hop plants are not considered toxic, although they have caused dermatitis in as many as one in 30 hop workers.

Chemistry

The commercial value of hops is due to resins which give beer its bitterness, and essential oils which contribute to flavor and aroma. Tannins are also present, which give astringency to preparations made with hops. Over 200 essential oil constituents have been identified, chiefly myrcene, humulene, and caryophyllene, with minor amounts of dipentene, linalool, farnesene and methyl nonyl ketone. The important brewing resins include alpha-acids (α-acids) and the beta-acids (β-acids), also referred to as humulones and lupulones, respectively. Both contribute bitterness to beer, but the α-acids are much more intense than the β-acids. The α-acids are a mixture of chemical analogues, including humulone, cohumulone, and adhumulone; similarly the β-acids are a mixture of lupulone, colupulone, and adlupulone. Brewers have long recognized that North American cultivars have a higher content of alpha acids and produce beer of stronger aroma. Bitter hops are very important commercially, and nearly all owe a great deal to American germplasm. Content of α-acid varies from about 3.5% in traditional European types to as much as 15% in newer bitter varieties. The European forms of hop have a relatively low content of "soft resins" (α- and β-acids collectively), a ratio of α:β approaching one, low cohumulone, moderately low essential oil content, and relatively low myrcene in the essential oil fraction. Native American hop plants are quite high in cohumulone and colupulone content, and have a pungent, unpleasant aroma.

Non-medicinal Uses

Hop has been used through recorded history for various culinary and household purposes, although it is chiefly known as a brewing ingredient. Hop extracts and oil have been used to flavor tobacco, yeast, beverages other than beers, frozen dairy desserts, candy, gelatins, puddings, baked goods, various confections, chewing gums, and condiments. In the past, yeast for bread-making was prepared by culturing wild yeast in a decoction of hops and water. The hops added flavor and apparently prevented the yeast from spoiling by virtue of their antiseptic properties.

Agricultural and Commercial Aspects

The cultivation of hop was not introduced into England until the close of the 15th century. The hop was brought to North America and grown in the early 17th century. The first commercial brewery in Canada was founded in Quebec about 1668 by the Intendant Jean Talon, to control the intemperate use of stronger drink. By the middle of the 19th century, New England and New York produced the bulk of the hops of the New World. However, by the early 20th century, the Pacific Coast became the leading hop-producing area in North America. In the 1920's hop growing in New York was practically wiped out by downy mildew and by Prohibition. Similarly, in eastern Canada commercial hop growing was phased out by the end of the Second World War, but became established in British Columbia.

About two dozen countries, including Canada, raise substantial commercial crops of hops. Germany is the largest producer, followed by the US. Other centers of hop production include Russia, China, England, the former Czechoslovakia, and the former Yugoslavia. About 114,000 tonnes are produced worldwide annually, on more than 90 000 ha.

Hop is a good example of a crop that has been substantially improved recently through incorporation of wild germplasm. A wild hop from Manitoba contributed to the improvement of many standard brewing varieties. Indeed, the improved cultivars account for an ever-increasing proportion of production. The contribution of wild hop germplasm was recently valued at almost $90,000,000 annually in North America.

Humulus lupulus (hop)

The use of hop shoots as a vegetable is an interesting possibility. Young shoots (6–10 cm long) are often consumed as a pot-herb, like asparagus. These spears can be boiled for 2–3 min, and then boiled in a change of water until tender. When steamed for 5 min and served with melted butter or cheese sauce, the shoots taste much like asparagus. In hop-producing areas of Europe, blanched hop spears are often served in fine restaurants. Hop farmers generally have surplus rhizomes from which the spears can be harvested, but because of their desire to maintain possession of unique hop strains, they may not be willing to sell them. In any event, there is a need to select strains that produce tasty shoots rather than good brews.

Given the substantial cultivation and availability of hop for food and flavoring purposes, growing the plant specifically as a pharmaceutical crop seems unwarranted.

Myths, Legends, Tales, Folklore, and Interesting Facts

- St. Hildegard (1098–1179, also known as Hildegard of Bingen and Hildegardis de Pinguia), was an abbess who established a convent and a Benedictine nunnery near the Rhine River. She was one of the most remarkable women the world has ever known, becoming an adviser to popes, kings, and various dignitaries. She wrote on nature and medicine, and was a mystical and spiritual visionary who interpreted Oriental, Judeo-Christian and Greek philosophy. St. Hildegard is often credited with being the first person to popularize the use of hops in brewing.
- In England about 1500, after learning of how well hops preserved beer in continental Europe, British brewers started adding hops to ale (sweet beer made without hops), turning it into bitter beer. Henry VIII (1491–1547), responding to a petition to ban hop, described as "a wicked weed that would endanger the people," outlawed the use of hop by brewers. His son, Edward VI (1537–1553), rescinded the ban in 1552.
- Charles Darwin entertained himself while sick in bed in 1882 by studying a hop plant growing on his window-sill. He noted that the tip of the stem completed a revolution in 2 hours.
- The patent office of the US once granted a patent to a man who claimed to have "invented" the hop's habit of winding from left to right (i.e., circling clockwise, viewed so that the twining stem is growing towards the observer).

Selected References

Barth, H.J., Klinke, C., and Schmidt, C. 1994. The hop atlas - the history and geography of the cultivated plant. Joh. Barth & Sohn, Nurenberg, Germany. 383 pp.

Brady, J.L., Scott., N.S., and Thomas, M.R. 1996. DNA typing of hops (*Humulus lupulus*) through application of RAPD and microsatellite marker sequences converted to sequence tagged sites (STS). Euphytica **91**: 277–284.

Bravo, L., Cabo, J., Fraile, A., Jimenez, J., and Villar, A. 1974. Pharmacodynamic study of the lupulus (*Humulus lupulus* L.) tranquilizing action. Boll. Chim. Farm. **113**: 310–315. [In Spanish.]

De Keukeleire, D., Milligan., S.R., De Cooman, L., and Heyerick, A. 1997. The oestrogenic activity of hops (*Humulus lupulus* L.) Pharmaceut. Pharmacol. Lett. **7**(2–3): 83–86.

Eckel, A., and Fritz, D. 1990. Forcing of hop shoots as a vegetable. 1. Comparison of various hop cultivars. Gartenbauwissenschaft **55**: 34–36. [In German.]

Eckel, A., and Fritz, D. 1990. Forcing hop shoots (*Humulus lupulus* L.) as a vegetable: II. Comparison of different propagation methods and set sizes. Gartenbauwissenschaft **55**(2): 90–92. [In German.]

Edwardson, J.R. 1952. Hops - their botany, history, production and utilization. Econ. Bot. **6**: 160–175.

Guelz, P.G., Mueller, E., Herrmann, T., and Loesel, P. 1993. Epicuticular leaf waxes of the hop (*Humulus lupulus*). Chemical composition and surface structures. Z. Naturforsch. Sect. C, Biosci. **48**: 689–696.

Hampton, R.O. 1988. Health status (virus) of native north American *Humulus lupulus* in the natural habitat. J. Phytopathol. (Berlin) **123**: 353–361.

Haunold, A. 1993. Agronomic and quality characteristics of native North American hops. Am. Soc. Brew. Chem. J. **51**: 133–137.

Katsiotis, S.T., Langezaal, C.R., and Scheffer, J.J.C. 1990. Composition of the essential oils from leaves of various *Humulus lupulus* L. cultivars. Flavour Fragrance J. **5**(2): 97–100.

Katsiotis, S.T., Langezaal, C.R., Scheffer, J.J.C., and Verpoorte, R. 1989. Comparative study of the essential oils from hops of various *Humulus lupulus* L. cultivars. Flavour Fragrance J. **4**(4): 187–192.

Kral, D., Zupanec, J., Vasilj, D., Kralj, S., and Psenicnik, J. 1991. Variability of essential oils of hops, *Humulus lupulus* L. J. Inst. Brewing **97**: 197–206.

Langezaal, C.R., Chandra, A., and Scheffer, J.J. 1992. Antimicrobial screening of essential oils and extracts of some *Humulus lupulus* L. cultivars. Pharm. Weekbl. Sci. **14**: 353–356.

Miller, N.G. 1970. The genera of the Cannabaceae in the southeastern United States. J. Arnold Arbor. **51**: 185–203.

Mizobuchi, S., and Sato, Y. 1985. Antifungal activities of hop (*Humulus lupulus*) bitter resins and related compounds. Agric. Biol. Chem. **49**: 399–404.

Munro, D.B., and Small, E. 1997. Vegetables of Canada. NRC Research Press, Ottawa, ON. 417 pp. [Chapter on *Humulus*: pages 219–225.]

Neve, R.A. 1991. Hops. Chapman and Hall, London, U.K. 266 pp.

Oliveira, A.M.M., and Pais, M.S. 1988. Glandular trichomes of *Humulus lupulus* cultivar Brewer's Gold: Ontogeny and histochemical characterization of the secretion. Nord. J. Bot. **8**: 349–359.

Oliveira, A.M.M., and Pais, M.S. 1990. Glandular trichomes of *Humulus lupulus* cultivar Brewer's Gold (hops): ultrastructural aspects of peltate trichomes. J. Submiscrosc. Cytol. Pathol. **22**: 241–248.

Parker, J.S., and Clark, M.S. 1991. Dosage sex-chromosome systems in plants. Plant Sci. (Limerick) **80**: 79–92.

Pillay, M., and Kenny, S.T. 1994. Chloroplast DNA differences between cultivated hop, *Humulus lupulus* and the related species *H. japonicus*. Theor. Appl. Genet. **89**: 372–378.

Pillay, M., and Kenny, S.T. 1996. Structure and inheritance of ribosomal DNA variants in cultivated and

wild hop, *Humulus lupulus* L. Theor. Appl. Genet. **93**: 333–340.
Polley, A., Seigner, E., and Ganal, M.W. 1997. Identification of sex in hop (*Humulus lupulus*) using molecular markers. Genome **40**: 357–361.
Simpson, W.J., and Smith, A.R.W. 1992. Factors affecting antibacterial activity of hop compounds and their derivatives. J. Appl. Bacteriol. **72**: 327–334.
Small, E. 1978. A numerical and nomenclatural analysis of morpho-geographic taxa of *Humulus*. Syst. Bot. **3**: 37–76.
Small, E. 1980. The relationships of hop cultivars and wild variants of *Humulus lupulus*. Can. J. Bot. **58**: 676–686.
Small, E. 1981. A numerical analysis of morpho-geographic groups of cultivars of *Humulus lupulus* L. based on samples of hops. Can. J. Bot. **59**: 311–324.
Small, E. 1997. Cannabaceae. *In* Flora of North America, north of Mexico, vol. 3. *Edited by* Flora North America Editorial Committee. Oxford University Press, New York, NY. pp. 381–387.
Small, E. 1997. Culinary Herbs. NRC Research Press, Ottawa, ON. 710 pp. [Chapter on *Humulus*: pages 283–289.]
Stevens, R. 1967. The chemistry of hop constituents. Chem. Rev. **67**: 19–71.
Suominen, J. 1994. The northernmost finds in the world of native *Humulus lupulus*. Aquilo Ser Bot. **33**: 121–129. [In Finnish.]
Takahashi, T., Ohsawa, M., Shimakoshi, S., Kishi, H., Kawahara, M., and Yoshikawa, N. 1993. Development cytology of the resin glands of hop (*Humulus lupulus* L.). J. Hortic. Sci. **68**: 797–801.
Tsuchiya,Y., Araki, S., Takashio, M., and Tamaki, T. 1997. Identification of hop varieties using specific primers derived from RAPD markers. J. Ferment. Bioeng. **84**(2) : 103–107.
Whittington, G., and Gordon, A.D. 1987. The differentiation of the pollen of *Cannabis sativa* L. from that of *Humulus lupulus* L. Pollen Spores **29**: 111–120.
Yamanaka, T. 1994. *Humulus lupulus* var. *cordifolius* found in Shikoku. J. Jpn. Bot. **69**: 179 [In Japanese.]

World Wide Web Links

(Warning. The quality of information on the internet varies from excellent to erroneous and highly misleading. The links below were chosen because they were the most informative sites located at the time of our internet search. Since medicinal plants are the subject, information on medicinal usage is often given. Such information may be flawed, and in any event should not be substituted for professional medical guidance.)

Humulus lupulus, J.A. Duke, 1983, Handbook of energy crops, unpublished:
http://www.hort.purdue.edu/newcrop/duke_energy/Humulus_lupulus

Comparative data of thirteen U.S. grown hop varieties:
http://www.john-i-haas.com/variety.htm

Hop growers of America [has several links]:
http://www.usahops.org/

A comparison of the relative merits of leaf hops, hop pellets and hop extracts in brewing by Dr. G.K. Lewis:
http://www.hopunion.com/articles/relative.shtml#Introduction

United States Department of Agriculture, National Agricultural Statistics Service - hops:
http://www.usda.gov/nass/aggraphs/hops.htm

Hops demonstration project (Rutgers University):
http://wwwrce.rutgers.edu/burlington/hops.htm

HopTech's web site:
http://www.hoptech.com/home.htm

Frequently asked questions about hops:
http://hbd.org/brewery/library/mashtun/hop.faq.html

Norm Pyle's hops FAQ [a large document]:
http://www.realbeer.com/hops/FAQ.html

A hops growing primer:
http://hbd.org/brewery/library/HopGrow.html

Grow your own hops:
http://byo.com/byo/Back/grow.html

About hops:
http://www.stpats.com/hopsinfo.htm

HandiLinks [has excellent links]:
http://www.ahandyguide.com/cat1/h/h255.htm

Growing hops - in the home garden:
http://www.oda.state.or.us/hop/extcr104.html

The University of Vermont hops web pages [has excellent links]:
http://www.uvm.edu/~pass/perry/hops.html

The National Hop Association of England:
http://www.breworld.com/nha/

Oregon Hop Commission:
http://www.oda.state.or.us/hop/ohc.html

Beer hops may help prevent cancer:
http://www.junkscience.com/news/thisbud.htm

Home-grown hops, part 2: controlling diseases and pests:
http://tiedhouse.com/rbp/authors/moen/hgrown2.html

Health centre - herb monographs, *Humulus lupulus*:
http://www.healthcentre.org.uk/hc/alternatives/herbal_monographs/hops.htm

Hydrastis canadensis (goldenseal)

Hydrastis canadensis L. Goldenseal

English Common Names

Goldenseal, golden-seal, golden seal, yellow root (yellow-root), orange-root, yellow puccoon, jaundiceroot, yelloweye, yellow paint, Indian turmeric, Indian dye, Indian plant, wild turmeric, tumeric root, ground raspberry, eye-root (eyeroot), eye-balm (eyebalm), warnera, Ohio curcuma, wild curcuma.

The name "golden seal" is evidently derived from the golden or yellow color of the rhizome, and the cup-like depressions remaining on it after the annual flower stems falls away; these scars look like old wax seals.

French Common Names

Sceau d'or, hydraste, hydraste du Canada.

Morphology

In early spring, goldenseal plants emerge from overwintering buds on the perennial rootstock. Generally there are two winter buds at the base of each stem. The rough-hairy, herbaceous, perennial plants reach 20–50 cm in height, and have 1–3 palmately lobed leaves up to 25 cm in diameter with 5–7 doubly serrate lobes. Single inconspicuous greenish-white flowers with the numerous stamens and pistils characteristic of the buttercup family appear in April or May. The most evident parts of the flowers are the white filaments. The sepals and petals are small and fall soon after the flower opens. The leaves are conspicuous and resemble those of Canadian waterleaf (*Hydrophyllum canadense* L.), but differ in being palmate. The raspberry-like fruit, considered inedible, is a distinctive cluster of scarlet, basally-fused berries which ripen in July and August. The stem and leaves usually die down soon after the fruit ripens. The fruit has 10–30 seeds, which are 2–5 mm long, shiny and dark brown or black with a small keel. Seed viability is rather unpredictable, and seeds that are not maintained moist die. Plants develop very slowly from seed: during the first year, most seedlings merely develop cotyledons; production of a true leaf is delayed until the second year, and two leaves and a flower are produced in the third. The horizontal or oblique-growing, sub-cylindrical, knotty, strong-smelling rhizome, 4–7 cm long and 0.5–2 cm wide, when fresh is yellow-brown outside, bright yellow internally, and contains a bright yellow juice. The rhizome has many fibrous rootlets. The so-called root (i.e., the rhizome) is sometimes confused with the yellow roots of other plants such as goldthread (*Coptis trifolia* (L.) Salisb.), yellow root (*Xanthorhiza simplicissima* Marsh.) and wood poppy (*Stylophorum diphyllum* (Michx.) Nutt.).

Classification and Geography

Hydrastis is usually classified with the buttercups (Ranunculaceae), but it is very distinctive and is sometimes placed in its own family (Hydrastidaceae).

Hydrastis canadensis is the only species in the genus. The alleged species "*Hydrastis jezoensis* Sieb. ex Miquel" reported from northeastern Asia likely belongs to another genus. *Hydrastis canadensis* is distributed from southern New England west through southern Ontario to southern Wisconsin, south to Arkansas and northern Georgia. In the US, it has been recorded in 27 states. Goldenseal belongs to the group of plants that occurred in the ancient arctotertiary forest which encircled the northern hemisphere 15–20 million years ago. Its nearest relative, *Glaucidium palmatum* Sieb. & Zucc., occurs in Japan. Goldenseal was only abundant in a limited area in the central portion of its total range (darker shading on accompanying map) in the states of Indiana, Kentucky, Ohio, and West Virginia. Collecting for the drug trade during and after the late 1800s had the effect of drastically reducing natural populations, but habitat destruction also played a role in its decline. In Canada it occurs only in southern Ontario and has been given the official status of a "threatened" species with priority 1 ranking (the highest) for protection in this country.

Ecology

Goldenseal produces colonies in shady, open deciduous woods and the edges of woodland, in the nutrient-rich, hardwood mesic forests of North America. Both seeds and rhizome division are important to natural reproduction of the species. It grows best in soils that are rich, moist, loamy, and with good drainage. In the wild, it usually is provided with a natural mulch of leaf mold or forest litter. Natural drainage is often furnished by location on hillsides. Optimum growth occurs at 75–80% shade. Expansion of agricultural land, timber harvest, building of roads, and particularly wildcrafting, all contribute to the destruction of wild populations. Goldenseal is listed in Appendix II of

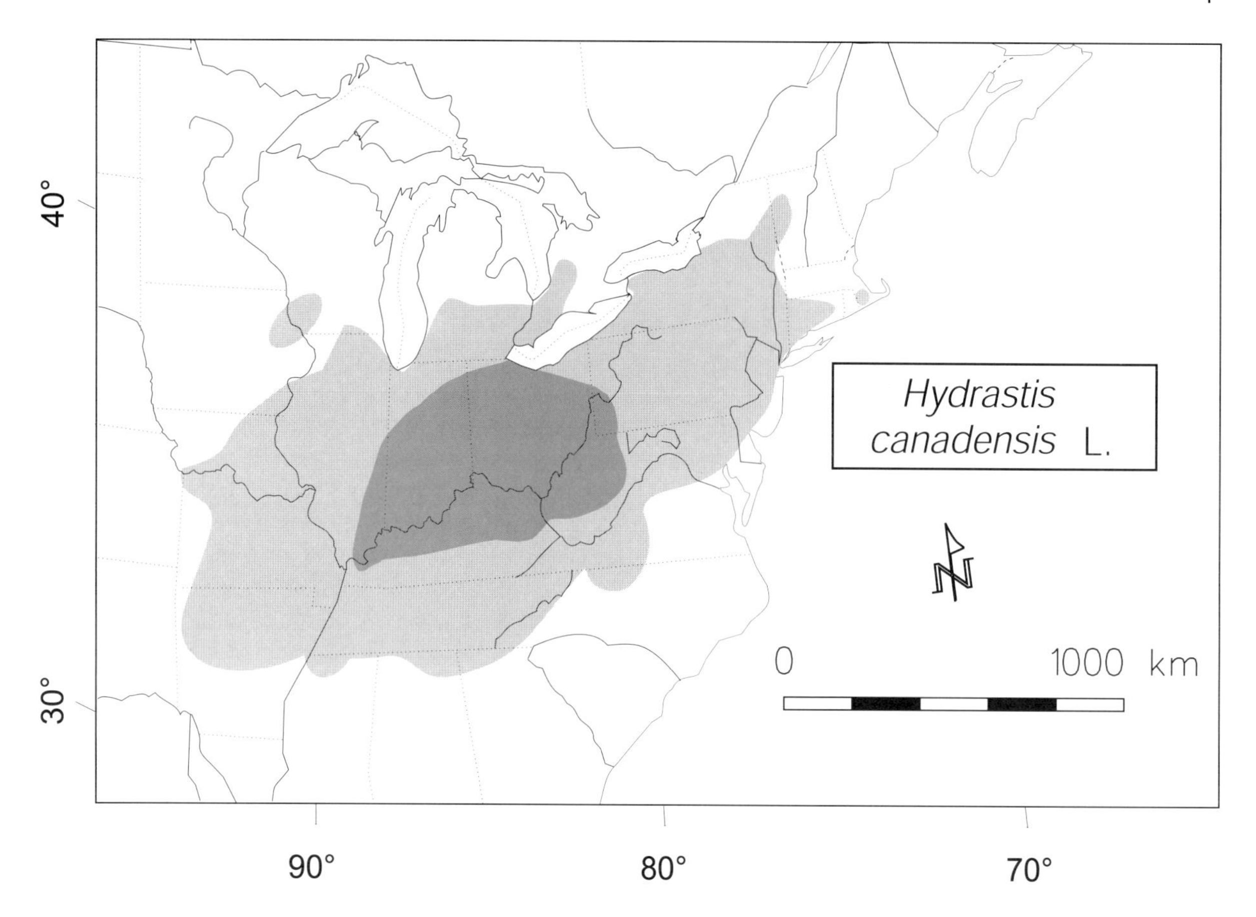

the Convention of International Trade in Endangered Species (CITES), which plays a significant role in controlling international trade in endangered flora and fauna.

Medicinal Uses

The rhizome is generally the source of medical preparations, although occasionally the leaves are also harvested for medicinal use. Maximum concentration of alkaloids in the rhizome is developed in the fall.

The indigenous people of eastern North America used goldenseal to treat various kinds of illness, especially those requiring antimicrobial action. Conditions treated included skin diseases, ulcers, gonorrhea, eye ailments, and cancers. However, goldenseal appears not to have been a medicinal plant of major importance until the mid-1800s when methods of refining the alkaloids hydrastine and berberine were developed. The properties of the alkaloids were conveniently obtained by preparing chemical compounds such as the readily soluble salt, hydrastine hydrochlorate. The primary use of goldenseal in the mid- to late 1800s was in the treatment of digestive disorders, inflammation of mucous tissues and skin diseases, but it soon developed a reputation as a general tonic. The astringent properties (due to hydrastine, as noted below), affect mucosal surfaces both internally and externally, providing a rationale for use of goldenseal to treat mouth and gum disorders, eye afflictions, infected wounds and inflamed skin conditions. These uses were considered the chief virtues of goldenseal in the 19th and early 20th centuries.

In modern medicine, goldenseal alkaloids find some approved applications. Hydrastine and berberine are particularly pharmacologically active, affecting circulation, uterine functions, and the central nervous system. Hydrastine constricts peripheral vessels, decreases blood pressure and stimulates involuntary muscles. Berberine inhibits synthesis of DNA and proteins, and oxidation of glucose. It is used for various digestive and skin problems. Recently berberine was found to be active against the protozoan responsible for Chagas' disease, a major health problem in Central and South America. Goldenseal alkaloids also have some antitumor activity. The wide use of goldenseal to treat traveler's diarrhea, food poisoning, giardia, and cholera could be explained by the various antibiotic properties of the herb (effective against some bacteria, protozoans, and fungi).

In various commercial formulations, goldenseal is used to treat nasal congestion, mouth sores, eye infections, ringworm, hemorrhoids, acne, and as a surface antiseptic. It has a reputation for boosting the immune system. As noted below, self-medication with goldenseal preparations is not recommended. Nevertheless goldenseal is now a component of hundreds of commercial medicinal formulations, sold in national chain drugstores, department stores, convenience stores, health food stores, and mail-order businesses. It is particularly in demand by consumers with severe chronic diseases, especially patients with AIDS.

Toxicity

Use of goldenseal in modern medicine is limited because of its toxicity. Dangerous levels may need to be administered to achieve therapeutically useful effects. Even external use can produce ulceration. Taken orally in excess dosage, goldenseal can produce convulsions like those produced by strychnine, and may lead to paralysis, respiratory failure and death. Eating fresh plant material can ulcerate and inflame the mucous membranes of the mouth. Since goldenseal can promote miscarriage, it should not be taken during pregnancy. People with cardiac problems have been advised to avoid goldenseal as it can raise blood pressure. Goldenseal is a good example of how medicines intergrade into poisons. Its drugs are potent, but pose a serious risk, thus requiring professional guidance and research.

This herb was listed in a 1995 Health Canada document (see "Herbs used as non-medicinal ingredients in nonprescription drugs for human use," http://www.hc-sc.gc.ca/hpb-dgps/therapeut/drhtmeng/policy.html) as a herb that is unacceptable as a nonprescription drug product for oral use.

Chemistry

The rhizome of goldenseal is a source of the medicinal alkaloids hydrastine (1.5–4%), berberine (0.5–6%), and berberastine (2–3%), with lesser amounts of canadine and some minor alkaloids.

Hydrastis canadensis (goldenseal)

Non-medicinal Uses

Indians of eastern North America used goldenseal as a yellow dye for fabrics and a stain for skin. The juice imparts a yellow color to skin and clothing and, mixed with indigo, produces green-colored dyes. Indians also mixed goldenseal with bear grease for use as an insect repellent.

Agricultural and Commercial Aspects

The prominent use of goldenseal in proprietary medicines of past times has been resurrected in a growing popularity of the "health food" industry. Goldenseal has become one of the top selling herbs of North America. Currently there are approximately 40 over-the-counter drugs containing goldenseal or its active ingredients on the Canadian market as elixirs, tablets, capsules or suppositories. It is also sold as an ingredient of some herbal teas. When goldenseal tea was rumored (incorrectly) to prevent detection of morphine in urine samples, it also became popular with drug users for preventing the detection of marihuana and cocaine. It has even been used on drugged race-horses.

Like ginseng and May-apple, a large part of the supply of goldenseal came from mountainous regions of Kentucky and West Virginia, where it was collected by people whose economy was based largely on the virgin forests of the steep mountain slopes and deep valleys. In the late 1850s it was valued at $2.20/kg, but this value decreased as the market became adequately supplied. In the late 1800s 63 500 – 68 000 kg were collected annually, most of this being used in North America, with only 680 kg exported to Europe. About 550 dry roots are required to make one kg. Since it was collected mostly in the Ohio Valley region, Cincinnati became the major source of supply.

Although much of the current supply is from wild-collected plants, goldenseal has become a widely cultivated plant. Because increasing demand poses a threat of extirpation in some areas, it is possible that most of the future supply will come from cultivated plants. It has been cultivated in Arkansas, Michigan, North Carolina, Oregon, Tennessee, Washington, and Wisconsin, the annual production from cultivated plants exceeding several tonnes. Some ginseng growers have found economic benefit in growing some goldenseal as well because it has similar environmental requirements allowing the use of similar equipment. It may be a little easier to grow than ginseng because it tolerates slightly higher light intensities and is less subject to diseases and pests. Because diseases make it difficult or impossible to grow ginseng in the same woodland location as consecutive plantings, goldenseal has excellent potential as a rotation crop for ginseng. The growing popular market for goldenseal products suggests a potential for use of the plant as a diversification crop in southern Canada.

Myths, Legends, Tales, Folklore, and Interesting Facts

- After the American civil war, goldenseal was an ingredient of many patent medicines, notably in "Dr. Pierce's Golden Medical Discovery." Unlike ginseng, which was collected entirely for the export market, goldenseal was consumed in the US, and came to acquire much of ginseng's reputation as a panacea and longevity tonic. Hence one popular name for goldenseal was "poor man's ginseng."
- It has been estimated that more than 95% of the above-ground biomass of goldenseal is produced within the first month of the growing season.
- In 1997, the World Wildlife Fund included goldenseal as one of the "10 most wanted" species in the world (10 of the most threatened species in demand for international trade).

Selected References

Ahluwalia, S.S. 1997. Goldenseal — American gold. Walden House, Bronx, NY. 22 pp.

Bergner, P. 1997. The healing power of echinacea, goldenseal, and other immune system herbs. Prima, Rocklin, CA. 322 pp.

Blecher, M.B., and Douglass, K. 1997. Gold in goldenseal. Hosp. Health Netw. **71**(20): 50–52.

Bowers, H. 1891. A contribution to the life history of *Hydrastis canadensis*. Bot. Gaz. **16**: 73–82.

Caille, G., LeClerc, D., and Mockle, J.A. 1970. Dosage spectrophoto-fluoromètrique des alcaloïdes berbérine, tétrahydroberbérine, hydrastine et application à l'extrait sec et à la teinture d'*Hydrastis candensis* L. Can. J. Pharm. Sci. **5**(2): 55–58.

Carlquist, S. 1995. Wood and bark anatomy of Ranunculaceae (including *Hydrastis*) and Glaucidiaceae. Aliso **14**(2): 65–84.

Catling, P.M., and Sinclair, A. 1998. The history of the golden seal. Recovery - an Endangered Species Newsletter (Canadian Wildlife Service) 1998(Spring): 12.

Cavin, J.C., Krassner, S.M., and Rodriguez, E. 1987. Plant-derived alkaloids active against *Trypanosoma cruzi*. J. Ethnopharmacol. **19**: 89–94.

Combie, J., Nugent, T.E., and Tobin, T. 1982. Inability of goldenseal to interfere with the detection of morphine in urine. J. Equine Vet. Sci. **2**: 16–21.

Creasey, W.A. 1979. Biochemical effects of berberine. Biochem. Pharm. **28**: 1081–1084.

Davis, J.M. 1995. Advances in Goldenseal Cultivation. N.C. State Univ., Agricult. Ext. Serv. Hortic. Info. Leafl. 131 (revised 6/96). 5 pp. [available on internet at http://www.ces.ncsu.edu/depts/hort/hiL/hil-131.html]

Davis, J.M. 1998. Goldenseal. *In* Richters second commercial herb growing conference — transcripts. *Edited by* R. Berzins, H. Snell and C. Richter. Richters, Goodwood, ON. pp. 133–143.

Dyke, S.F., and Tiley, E.P. 1975. The synthesis of berberastine. Tetrahedron **31**: 561–568.

Eichenberger, M.D., and Parker, G.R. 1976. Goldenseal (*Hydrastis canadensis* L.), distribution, phenology and biomass in an oak-hickory forest. Ohio J. Sci. **76**(5): 204–210.

Ford, B.A. 1997. *Hydrastis*. *In* Flora of North America north of Mexico, Vol. 3. *Edited by* Flora of North

America Editorial Committee. Oxford University Press, New York, NY. pp. 87-88.

Foster, S. 1991. Goldenseal, *Hydrastis canadensis*. American Botanical Council, Austin, TX. Bot. Ser. No. **309**: 1–8.

Galeffi, C., Cometa, M.F., Tomassini, L., and Nicoletti, M. 1997. Canadinic acid: An alkaloid from *Hydrastis canadensis*. Planta Med. **63**: 194.

Genest, K., and Hughes, W. 1969. Natural products in Canadian pharmaceuticals IV. Can. J. Pharm. Sci. **4**: 41–45.

Gleye, J., Ahond, A., and Stanislas, E. 1974. La Canadaline: nouvel alacaloïde d'*Hydrastis canadensis*. Phytochemistry **13**: 675–676.

Haage, L.J., and Ballard, L.J. 1989. A growers guide to goldenseal. Nature's Cathedral, Norway, IA.

Hamon, N.W. 1990. Herbal medicine: goldenseal. Can. Pharm. J. **123**: 508–510.

Hardacre, J., Henderson, V.G., Collins, F.B., Andersen, E.L., Harris, V.M., Fewster, B., Beck, R., Bowman, D., and Donzelot, E.L. 1962. The wildcrafters goldenseal manual. Wildcrafters Publications, Rockville, IN.

Henkel, A., and Klugh, G.F. 1908. The cultivation and handling of goldenseal. U.S. Dep. Agric., Bur. Plant Ind. Circ. No. 6. 19 pp.

Hobbs, C. 1990. Goldenseal in early American medical botany. Pharmacy in History **32**(2): 79–82.

Holland, H.L., Jeffs, P.W., Capps, and MacLean, D.B. 1979. The biosynthesis of protoberberine and related isoquinoline alkaloids. Can. J. Chem. **57**: 1588–1597.

Hoot, S.B. 1991. The phylogeny of the Ranunculaceae based on epidermal microcharacters and macromorphology. Syst. Bot. **16**: 741–755.

Kelly, J. 1977. Herb collector's manual and marketing guide: ginseng growers and collectors handbook: a valuable guide for growers of ginseng and golden seal, medicinal herb and root collectors, containing olde tyme herbe recipes and outdoor money-making ideas. 5th ed. Wildcrafters, Looneyville, WV. 97 pp.

Konsler, T.R. 1987. Woodland production of ginseng and goldenseal. Stn. Bull. Purdue Univ. Agric. Exp. Sta., West Lafayette, IN **518**: 175–178.

Li, T.S.C., and Oliver, A. 1995. Specialty Crops infosheet: goldenseal. Province of British Columbia, Ministry of Agriculture, Fisheries and Food. 2 pp.

Liu, C.X., Xiao, P.G., and Liu, G.S. 1991. Studies on plant resources, pharmacology and clinical treatment with berberine. Phytother. Res. **5**: 228–230.

Lloyd, J.U. 1912. The cultivation of *Hydrastis*. J. Am. Pharm. Assoc. **1**: 5–12.

Lloyd, J.U., and Lloyd, C.G. 1884–1885. Drugs and Medicines of North America, Vol. 1, Ranunculaceae. Clarke, Cincinnati, OH. 304 pp.

Lloyd, J.U., and Lloyd, C.G. 1908. *Hydrastis canadensis*. Bull. Lloyd Library of Botany, Pharmacy and Materia Medica 10. 184 pp. [Reproduction of portion of Lloyd and Lloyd 1884–1885, cited above.]

Messana, I., La Bua, R., and Galeffi, C. 1980. The alkaloids of *Hydrastis canadensis* L. (Ranunculaceae). Two new alkaloids: hydrastidine and isohydrastine. Gaz. Chim. Ital. **110**: 539–543.

Sack, R.B., and Froehlich, J.L. 1982. Berberine inhibits intestinal secretory response of *Vibrio cholerae* toxins and *E. coli* enterotoxins. Infect. Immunol. **35**: 471–475.

Sievers, A.F. 1949. Goldenseal under cultivation. U.S. Dep. Agric. Farmers' Bull. 613. 14 pp. [Revision of Van Fleet, W., 1916, 14 pp.]

Shideman, F.E. 1950. A review of the pharmacology and therapeutics of *Hydrastis* and its alkaloids, hydrastine, berberine and canadine. Bull. Nat. Form. Comm. [United Kingdom] **18**(102): 3–19.

Shipley, N. 1956. The hidden harvest. Can. Geogr. J. **52**: 178–181.

Tobe, H., and Keating, R.C. 1985. The morphology and anatomy of *Hydrastis* (Ranunculales): systematic re-evaluation of the genus. Bot. Mag. (Tokyo) **98**: 291–316.

Veninga, L., and Zaricor, B. 1976. Goldenseal/etc: a pharmacognosy of wild herbs. Ruka Publications, Santa Cruz, CA.

Whetzel, H.H. 1918. The *Botrytis* blight of golden seal. Phytopathology. **8**: 73–76.

World Wide Web Links

(Warning. The quality of information on the internet varies from excellent to erroneous and highly misleading. The links below were chosen because they were the most informative sites located at the time of our internet search. Since medicinal plants are the subject, information on medicinal usage is often given. Such information may be flawed, and in any event should not be substituted for professional medical guidance.)

Commercial production of ginseng and goldenseal:
http://www.hort.purdue.edu/newcrop/NewCropsNews/94–4–1/ginseng.html

Purdue University Centre for New Crops and Plant Products [several informative documents]:
http://www.hort.purdue.edu/newcrop/nexus/Hydrastis_canadensis_nex.html

Monograph from J.E. Simon, A.F. Chadwick and L.E. Craker:
http://www.hort.purdue.edu/newcrop/med-aro/factsheets/GOLDENSEAL.html

The information on this page compiled by Elizabeth Burch N.D., the eclectic physician:
http://www.eclecticphysician.com/hydrastis.shtml

AGIS ethnobotany database:
http://probe.nal.usda.gov:8300/cgi-bin/webace?db=ethnobotdb&class=Taxon&object=Hydrastis+canadensis

AGIS phytochemical database, phytochemicals of *Hydrastis canadensis*:
http://probe.nal.usda.gov:8300/cgi-bin/table-maker?db=phytochemdb&definition+file=chems-in-taxon&arg1=Hydrastis+canadensis

Botanical collections. UConn pharmacy garden collection. *Hydrastis canadensis* [not much information but has good links]:
http://florawww.eeb.uconn.edu/acc_num\971052.htm

Canterbury Farms herb of the month [has several good links]:
http://www.nwgardening.com/goldenseal1.html

Goldenseal: are there substitutes?:
http://www.frontierherb.com/herbs/notes/herbs.notes.no9.html

Goldenseal, *Hydrastis canadensis,* Agriculture & Agri-Food Canada, Southern Crop Protection & Food Research Centre:
http://res.agr.ca/lond/pmrc/study/newcrops/goldseal.html

Goldenseal masking of drug test from fiction to fallacy: an historical anomaly:
http://www.frontierherb.com/herbs/notes/herbs.notes.no7.html

Advances in goldenseal cultivation:
http://www.ces.ncsu.edu/depts/hort/hiL/hil-131.html

Wildlife notes: goldenseal – plant in peril?:
http://troy2.fsl.wvnet.edu/wildlife/gseal.htm

Sources of goldenseal seeds, plants or roots:
http://www.ces.ncsu.edu/depts/hort/hiL/hil-123.html

Laminariales (kelp species)

Laminariales Kelp

English Common Names

Kelp.

"Kelp" commonly refers to seaweeds of the brown algal order Laminariales, which have large flat, leaflike fronds (kelp is used both as singular and plural).

French Common Names

Varech (= kelp).

Morphology

Many of the Laminariales are huge, tough, leathery plants. While rather variable, most kelp have a root-like holdfast, a flexible stipe (stem-like portion), and blades (flattened expansions often resembling leaves). The life cycle of the kelp involves alternation of a large spore-producing stage (the sporophyte) and a microscopic gamete-producing stage (the gametophyte). *Macrocystis* is perhaps the most impressive of the kelp, sometimes developing weights of over 200 kg. The longest known *Macrocystis* measured 47 m.

Classification and Geography

Most of the kelp are found in the Pacific Ocean, with a restricted number in the Atlantic.

In Canada, several species of kelp are considered economically important: *Laminaria saccharina* and *L. longicruris* (regarded as belonging to the same species by some authors), are distributed along the entire Canadian Atlantic coast and south to New York, through the eastern Arctic, and the entire Canadian Pacific coast to Washington. *Laminaria setchellii* and *L. groenlandica* have limited economic importance in British Columbia. *Nereocystis leutkeana* presently has limited economic importance in British Columbia but because of its large biomass has considerable potential. *Macrocystis integrifolia* is probably the single most economically important kelp in Canada (based on dollar value, not biomass harvested).

Another brown algal order, the Fucales, also contains algae sometimes marketed pharmacologically as kelp (and also often as "fucus"), but these are best referred to as fucoids. *Fucus vesiculosus* (bladderwrack), frequent on the rocky coasts of the Atlantic and Pacific Oceans, and *Ascophyllum nodosum* (knotted wrack) of the coasts of the North Sea, the western part of the Baltic Sea, and the east coast of Canada, are the species usually encountered in commerce. The pharmacological properties of these fucoids are comparable to those of the true kelp.

Ecology

Members of the Laminariales are well-adjusted for cold waters and overcast skies. Kelp usually dominate the lower intertidal and upper subtidal floras. Kelp forests constitute important marine ecosystems in coastal areas of much of the world, and can support important commercial and recreational fisheries. They vastly increase the surface area upon which other marine organisms can locate, thus increasing diversity. Kelp are fast-growing despite their often cool environment, and provide food and habitat for numerous species of marine organisms. Sea urchins are very significant herbivores of kelp, but their numbers are controlled through predation by lobsters, crabs and sea otters. Widespread destruction of kelp beds in shallow water on both the East and West Coasts of Canada has been attributed to the unchecked overfishing of lobsters and crabs in the East, and the early extirpation of sea otters in the West to supply fur for the European market during the 1700s. Sea otters were recently reintroduced from Alaska to northwestern Vancouver Island and are extending their range along the B.C. coast. The reintroduction has led to a decline in sea urchin populations and a resulting revival in barrier kelp, which was once so extensive that it provided a safe navigation route of sheltered water between the offshore beds and the land.

Medicinal Uses

Seaweed products were used in traditional Chinese herbal medicine as early as the sixteenth century, and the kelp have also been employed in western folk medicine. Until very recently, however, there has been limited evidence of pharmacological value, aside from the effects of compensating for iodine deficiency. Kelp tablets and powders have become popular herbal preparations in North America, and claims have been made that kelp products are useful in treating a variety of ailments. The therapeutic properties of kelp have been attributed particularly to content of trace minerals, especially iodine, which is typically 20,000 times as concentrated in the seaweed by comparison with its aquatic habitat. However, the level of iodine has

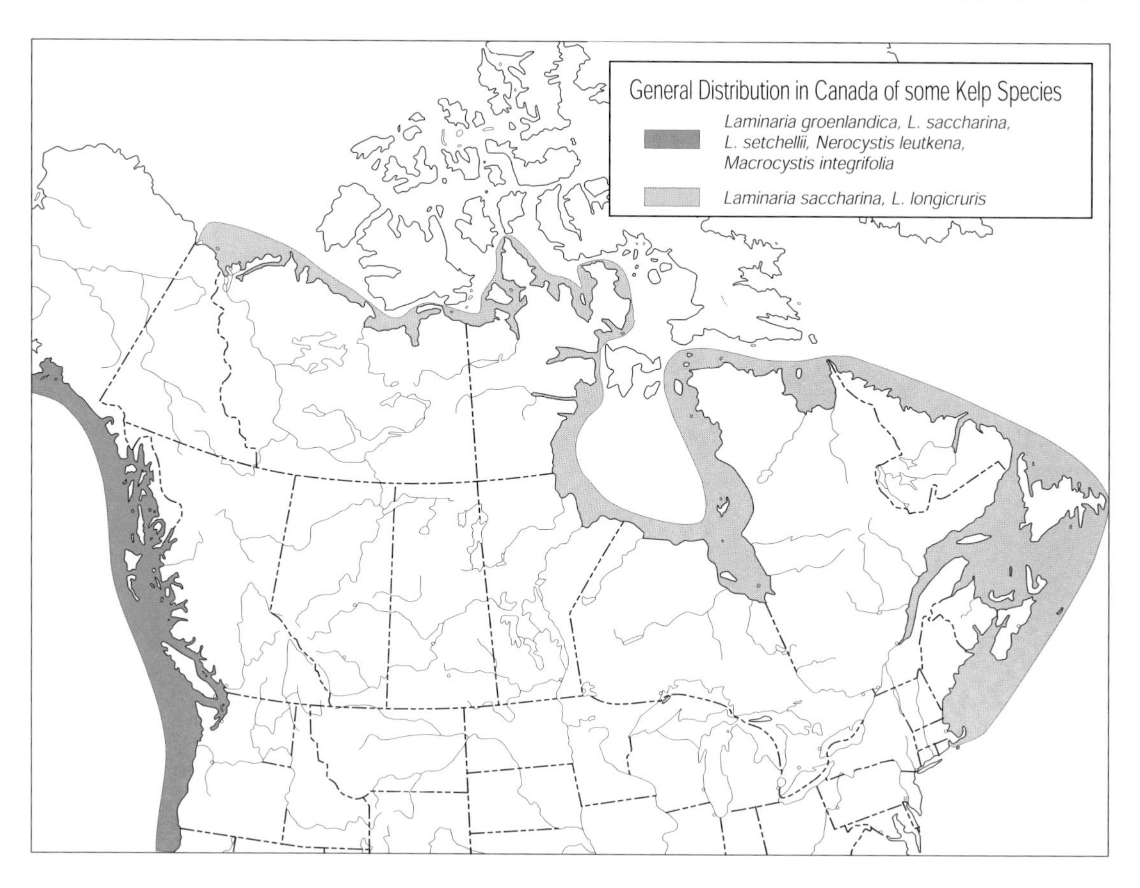

been found to vary considerably among algal species and even within a species, so that commercially available products may deliver different dosages. The very high iodine content of brown algae led to their use in goiter medicines, but the variability of concentrations and the varying absorption conditions for bound and unbound iodine in the plant has made such algal therapy obsolete. Required daily intake of iodine in adults is only 150 micrograms, and the thyroid gland normally does not make use of excess iodine, and indeed very large doses of iodine can induce or intensify hyperthyroidism. Sixty percent of all edible salt in the world is now iodized, reducing iodine deficiency disorders, the world's leading cause of preventable mental retardation until recently. Before 1990, approximately 40 million children were born each year at risk of mental impairment due to iodine deficiency in their mothers' diets. By 1997 that figure was close to 28 million. Kelp are currently marketed as a weight-reducing agent, supposed to be a result of increased production of thyroid hormones that increase metabolism and so remove deposited fats. While beneficial effects on obesity have been claimed, the medical community considers such therapy as potentially dangerous and highly inadvisable.

The polyphenols present in the plants have antibiotic activity, which may be useful for external applications. Algin from kelp is a useful bulk demulcent (soothing) laxative. Fucans from kelp and fucoids are sulfated polysaccharides, also known as kelp slimes. Recent research has suggested that these may be of use as antithrombin, anticancer and anticoagulant agents. Of these, the anticancer activity has proven most exciting, with preliminary results indicating that fucan is a very potent antitumor agent in cancer therapy. Calcium and magnesium, both desirable as nutritional electrolytes, have been found to be very high in some commercial kelp preparations.

Another interesting medical application is based on the ability of the dried stipes of *Laminaria* to expand their original circumference 3–5 times on wetting. These dried stipes have been used to produce non-instrument mechanical dilation of the cervical canal during birth and gynecological treatment.

Toxicity

As noted above, self-medication with kelp in order to regulate iodine levels is hazardous because of the different concentrations of iodine in commercial preparations. Kelp have large amounts of sodium, and so should be avoided in salt-restricted diets. Although kelp are thought to inhibit heavy metal absorption in humans, kelp growing in polluted waters may accumulate very high levels of heavy metals such as strontium and cadmium. Toxic levels of lead can develop. Most dangerously, kelp accumulates arsenic, although in non-polluted waters this should not pose a risk. Manufacturers of kelp products should of course ensure that toxic levels of any of these are not present.

Chemistry

The brown algae have considerable amounts of the high molecular weight polysaccharide algin, which comprises almost half the dry weight of some species. Algin produces viscous colloidal solutions or gels in water, like carageenans and agars commercially obtained from red algae. All of these "hydrocolloids" are primary cell wall structural constituents of brown and red algae. Unlike cellulose in land plants, which provides rigidity to withstand gravity, the hydrocolloids provide seaweeds with flexibility to withstand currents and waves. While primarily useful for a variety of industrial and food uses, the algins also are used medicinally, particularly as a carrier for pharmaceuticals.

Non-medicinal Uses

The larger marine algae or seaweeds have a diversity of uses, like many terrestrial plants. Many brown algae have been extensively used as agricultural fertilizers, especially as a source of potash (which supplies potassium). As a soil amendment, seaweeds also tend to be rich in micronutrients and nitrogen, but are low in phosphate. They have the advantage of being free of terrestrial weeds and fungi. Just as there has developed a certain mystique about the health benefits of brown algae for humans, so too are some convinced that algae have particular nutritional benefits for land plants. This has led to considerable research on plant growth factors in seaweeds (possibly plant hormones, particularly cytokinins), and the marketing of seaweed extracts for use as plant growth stimulants. The hope has also been expressed that alginic acid (marketed as Nomozan) may inoculate some crops against viruses.

Algin from kelp and other algae is used in the manufacture of more than 300 commercial products, and is particularly valued for its ability to suspend agents in food, cosmetics, and a variety of commercial liquid mixtures. Illustrating food applications, algin stops ice crystal formation in ice cream, acts as an emulsifier in salad dressings, sherbets, and cheeses, clarifies and stabilizes the foam of beer, is a filler in candy bars, and a thickener in gravies and puddings. Non-edible products using alginates include plastics, paints, adhesives and rubber tires. Algin can be used as a fiber in fire-resistant clothes and in audio speakers. Alginates are also used in the paper and welding industries. *Ascophyllum nodosum* of Canada's east coast is harvested for its algin content.

Kelp carbohydrates are not normally digested by humans, and three quarters of the dry bulk of the most edible brown seaweeds is indigestible. Nevertheless kelp are used directly as vegetables and condiments, principally in China and Japan. Like other seaweeds, kelp are also harvested as dried fodder for terrestrial livestock in coastal areas (despite the limited digestibility), and sometimes grown as forage for cultivated aquatic animals, such as abalone. A special roe on kelp, greatly valued

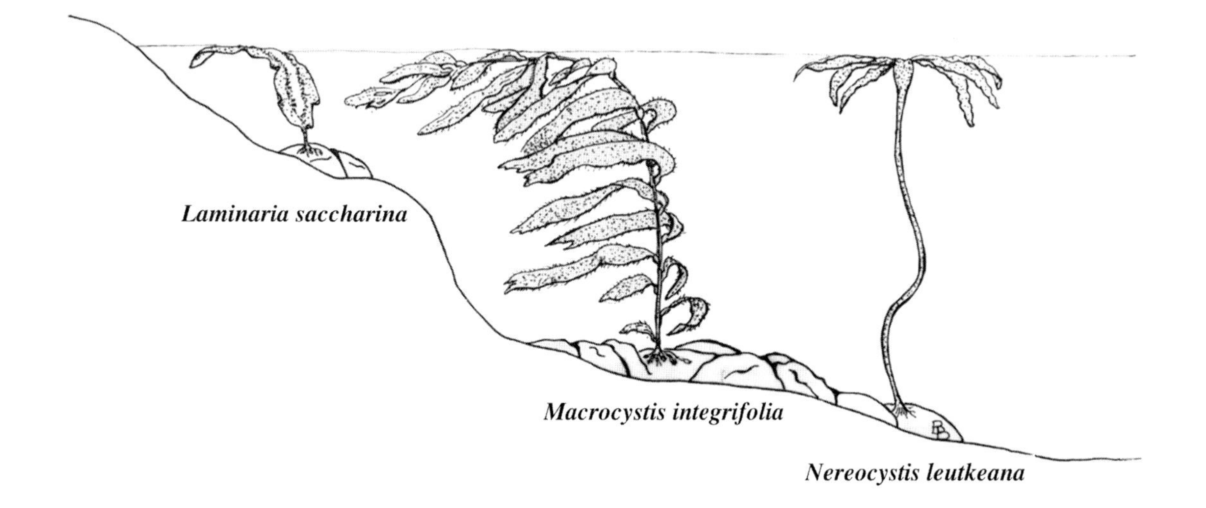

in sushi bars of Japan, has been produced in British Columbia by herring spawning in penned kelp enclosures. The industry is largely managed by native people and the harvest is valued at over 20 million dollars annually.

A recent development is the use of kelp as a substrate for biogas (methane) production, a technology that was developed during the OPEC crisis by General Electric of the United States. The value of this technology in reducing the high energy costs in the Canadian Arctic is currently being explored by Canadian companies.

Agricultural and Commercial Aspects

Uncontrolled harvesting has led to reductions of wild supplies of some algae, but seaweed farms are well established in parts of the world, reducing the pressure on natural stands. "Polyculture" (combining fish farming and seaweed culture) is a clever way of using seaweeds to metabolize by-products of the fish culture. China is a major seaweed producer, growing over 2.5 million tonnes of *Laminaria japonica* annually. Japan's primary aquaculture seaweed is nori (mostly *Porphyra yezoensis*, belonging to the red algae, not the brown), with an estimated value of US $1.5 billion annually (by contrast, the edible seaweed market in the US is currently valued at only about $30 million annually). Canada's development of a cultivated seaweed industry lags far behind that of several countries with low labor costs and a warm climate that allows year-round cultivation. However, there has been some cultivation of *Laminaria saccharina* and *L. groenlandica* on the Pacific coast, for the Oriental and health food markets.

Although Canada currently provides less than 2% of the world's seaweed resources, there is considerable potential for increasing commerce, particularly on the species-rich Pacific coast. In fact, with 20 kelp species, coastal British Columbia is one of the world's major regions of kelp diversity, and it has been estimated that 650,000 tonnes of wild kelp currently grow along the B.C. coastline. Current Canadian regulations permit 100,000 tonnes to be harvested annually under strict conservation guidelines, but less than 1% of this amount was harvested in 1996.

Canada has the longest coastline of any nation, suggesting that greater use should be made of the cold-water marine plant resources. Canada's marine environment may remain much less polluted than elsewhere, providing an attractive situation for algae and algal products intended for human consumption. This potentially very valuable natural resource deserves the best protection possible.

Myths, Legends, Tales, Folklore, and Interesting Facts

- The Ainu in northern Japan collected *Laminaria* for food as early as the eighth century A.D., and during a later period only the privileged classes of Japan were allowed to eat this sea vegetable.
- British Columbia coastal Indians stretched kelp stipes to make fishing lines and used the hollow bulbs and stipe bases as bottles.
- An old sailor's superstition held that winds could be called by circling a strand of kelp over one's head and whistling.
- According to another old superstition, to ensure a steady flow of money into the home, fill a jar with whiskey, place some kelp in it, cap tightly, and place in a kitchen window.

Selected References

Cheng, T.-H. 1969. Production of kelp - a major aspect of China's exploitation of the sea. Econ. Bot. **23**: 215–236.

Chida, K., and Yamamoto, I. 1987. Antitumor activity of a crude fucoidan fraction prepared from the roots of kelp (*Laminaria* species). Kitasato Arch. Exp. Med. **60**: 33–39.

Druehl, L.D. 1983. The integrated productivity of a *Macrocystis integrifolia* plant. Can. J. Bot. **62**: 230–235.

Druehl, L.D. 1988. Cultivated edible kelp. *In* Algae and human affairs. *Edited by* C.A. Lembi and J.R. Waaland. Cambridge University Press. pp. 119–134.

Druehl, L.D., Baird, R., Lindwall, A., Lloyd, K.E., and Pakula, S. 1988. Longline cultivation of some Laminariaceae in British Columbia, Canada. Aquacult. Fish. Manage. **19**: 253–264.

Fleurbec. 1985. Plantes sauvages du bord de la mer. Guide d'identification Fleurbec. Saint-Augustin (Portneuf), QC. 286 pp.

Gendron, L. 1989. Seasonal growth of the kelp *Laminaria longicruris* in Baie des Chaleurs, Quebec (Canada), in relation to nutrient and light availability. Bot. Mar. **32**: 345–354.

Harrell, B.L., and Rudolph, A.H. 1976. Letter: kelp diet: A cause of acneiform eruption. Arch. Dermatol. **112**: 560.

Lamella, M., Anca, J., Villar, R.,Otero, J., and Calleja, J.M. 1989. Hypoglycemic activity of several seaweed extracts. J. Ethnopharmacol. **27**: 35–44.

Lopez-Mosquera, M.E., and Pazos, P. 1997. Effects of seaweed on potato yields and soil chemistry. Biol. Agricul. Horticult. **14**(3):199–205.

Mautner, H.G. 1954. The chemistry of brown algae. Econ. Bot. **8**: 174–182.

Metting, B., Rayburn, W.R., and Reynaud, P.A. 1988. Algae and agriculture. *In* Algae and human affairs. *Edited by* C.A. Lembi and J.R. Waaland. Cambridge University Press. 335–370.

Nardella, A., Chaubet, F., Boisson-Vidal, C., Blondin, C., Durand, P., and Jozefonvicz, J. 1996. Anticoagulant low molecular weight fucans produced by radical process and ion exchange chromatography of high molecular weight fucans extracted from the brown seaweed *Ascophyllum nodosum*. Carbohydrate Res. **289**: 201–208.

Neish, I.C. 1976. Role of mariculture in the Canadian seaweed industry. J. Fish Res. Board Can. **33** (special issue 4, pt. 2): 1007–1014.

Nguyen, M.T., and Hoffman, D.R. 1995. Anaphylaxis to *Laminaria*. J. Allergy Clin. Immunol. **95**(1 part 1): 138–139.

Nishino, T., Ura, H., and Nagumo, T. 1995. The relationship between the sulfate content and the antithrombin activity of an α (1→2)-fucoidan purified from a comercial fucoidan fraction. Bot. Mar. **38**: 187–193.

North, W.J. 1976. Aquacultural techniques for creating and restoring beds of giant kelp, *Macrocystis* spp. J. Fish Res. Board Can. **33** (special issue 4, pt. 2): 1015–1023.

Petrell, R.J., Mazhari-Tabrizi, K., Harrison, P.J., and Druehl, L.D. 1993. Mathematical model of *Laminaria* production near a British Columbian salmon sea cage farm. J. Appl. Phycol. **5**: 1–14.

Prescott, G.W. 1968. The algae: a review. Houghton Mifflin Company, Boston. MA. 436 pp.

Riou, D., Colliec-Jouault, S., Pinczon du Sel, D., Bosch, S., Siavoshian, S., Le Bert, V., Tomasoni, C., Sinquin, C., Durand, P., and Roussakis, C. 1996. Antitumor and antiproliferative effects of a fucan extracted from *Ascophyllum nodosum* against a non-small-cell bronchopulmonary carcinoma line. Anticancer Res. **16**: 1213–1218.

Sanbonsuga, Y., Machiguchi, Y., and Saga, N. 1987. Productivity estimation and evaluation of the cultivation factors in biomass production of *Laminaria*. Bull. Hokkaido Reg. Fish. Res. Lab. **51**: 45–50.

Scagel, R.F. 1967. Guide to common seaweeds of British Columbia. British Columbia Provincial Museum, Department of Recreation and Conservation, Handbook No. 27. Victoria, BC. 330 pp.

Sharp, G.J., and Carter, J.A. 1986. Biomass and population structure of kelp (*Laminaria* spp.) in southwestern Nova Scotia (Canada). Can. Manuscr. Rep. Fish. Aquat. Sci. **1907**: I-IV, 1–42.

Teas, J. 1873. The dietary intake of *Laminaria*, a brown seaweed, and breast cancer prevention. Nutr. Cancer **4**: 217–222.

Tseng, C.L., Lo, J.M., and Huang, C.Y. 1994. Iodine content in *Laminaria*. Radioisotopes **43**(3): 134–136.

Voronova, Y.G., Rekhina, N.I., Nikolaeva, T.A., Tiunova, N.A., Zaikina, I.V., Kobzeva, N. Ya., and Valiente, O. 1991. Extraction of carbohydrates from *Laminaria* and their utilization. J. Appl. Phycol. **3**: 243–246.

Walkiw, O., and Douglas, D.E. 1975. Health food supplements prepared from kelp - a source of elevated urinary arsenic. Clin. Toxicol. **8**: 325–331.

Xia, B., and Abbott, I.A. 1987. Edible seaweeds of China and their place in the Chinese diet. Econ. Bot. **41**: 341–353.

World Wide Web Links

(Warning. The quality of information on the internet varies from excellent to erroneous and highly misleading. The links below were chosen because they were the most informative sites located at the time of our internet search. Since medicinal plants are the subject, information on medicinal usage is often given. Such information may be flawed, and in any event should not be substituted for professional medical guidance.)

Kelp forests [mostly color photographs]:
http://life.bio.sunysb.edu/marinebio/kelpforest.html

Map of extensive kelp beds off the British Columbia coast near Nootka Sound:
http://www.borstad.com/kelp.html

Kelp forests, general and important references:
http://life.bio.sunysb.edu/marinebio/kelp.html

Kelp company listings:
http://www.hua-conex.com/food/kelp.htm

A comparison of two rearing sites of the giant kelp *Macrocystis integrifolia* in Sitka, Alaska (abstract):
http://www.uaf.alaska.edu/seagrant/Pubs_Videos/pubs/AK-SG-90–02.html

Scientists show kelp helps - with a little help from their friends [reports value to sea animals of consuming kelp]:
http://www.gi.alaska.edu/ScienceForum/ASF9/939.html

Oenothera biennis (evening primrose)

Oenothera biennis L. Evening Primrose

English Common Names

(Yellow) evening primrose. Archaic names: tree primrose, scurvish, scabbish (scabish), king's cure-all, nightwillow herb, sundrops, tree primrose, fever plant.

The name evening-primrose is optionally hyphenated. Evening primrose isn't a "primrose," a name best applied to the genus *Primula*. The "evening" in the name relates to the fact that the flowers of many of the 125 species of *Oenothera* open in the evening and release a scent that attracts moths for pollination.

French Common Names

Onagre bisannuelle (commune).

Morphology

Oenothera biennis is a biennial (as the name suggests) or short-lived perennial herb producing strong fleshy roots and a basal rosette of lanceolate leaves in the first year. In the second year the stem grows to 1–2 m tall and develops a spicate inflorescence of 4-parted, yellow, tubular flowers. The fruit is a capsule containing many seeds which mature in the fall. The seeds are very small (ca. 0.5 g/1000), but a single plant can easily produce 150,000. The pollen of many, if not all species of *Oenothera*, is unusual in having protruding apertures and viscin threads.

Classification and Geography

Evening primrose, a native of North America, is found in all provinces of Canada, but is more frequent in the east than the west. The species extends south to Florida and Mexico.

Many texts recognize var. *canescens* T. & G., with dense grayish pubescence, as the predominant plant of western North America, while the eastern plants are referable to var. *biennis*. The classification of the transcontinental *O. biennis* and related species of both North America and Eurasia is, however, very complex. Cytogenetic races of evening primrose are sometimes segregated as distinct species, although these are usually difficult to distinguish morphologically.

Oenothera biennis is a complete translocation complex-heterozygote, with two sets (the "complexes") of seven chromosomes maintained by a system of balanced lethal genes. This type of inheritance is known in a few other genera, but was first described in *O. biennis*, and is the classical example of the phenomenon discussed in evolution and genetics courses. At meiosis, translocations[1] link the chromosomes into a ring of 14, but zig-zag (alternate) separation of the chromosomes generates the original parental sets. Lethal factors kill the pollen carrying one of the sets (so that there is 50% pollen fertility), and ovule lethal factors limit survival to the set of chromosomes complementary to that in the pollen. Self-pollination generates offspring with the two chromosome complements found originally in the maternal plant. The permanent hybrid vigor resulting from the combination of two quite different genomes is thought to explain the success of evening primrose as a colonizing species.

While it is clear that *O. biennis* is the chief *Oenothera* species that has been grown as a medicinal oilseed, related species have also been cultivated, often unknowingly. Other species from which cultivars have been derived include *O. glazioviana* Micheli ("*O. lamarckiana*" of many authors) and *O. parviflora* Micheli.

Ecology

Evening primrose is a frequent weed of roadsides, waste places, and abandoned land, often found in light sandy and gravelly soils. It commonly occurs in association with early successional, biennial and perennial weeds.

---

[1] Translocations are chromosomal abnormalities which occur when chromosomes break and the fragments rejoin to other chromosomes. In "reciprocal translocation" two non-homologous chromosomes break and exchange fragments, and during meiosis chromosome pairing and segregation are abnormal, the chromosomes involved often linking together, leading to odd segregation patterns and reduction of viability. Translocation heterozygotes are the product of reciprocal translocations that link all of the chromosomes together. Evolution of new species by chromosome translocations is very common in plants, less so in animals. In humans, translocations can cause severe diseases; for example, Familial Downs Syndrome, which occurs in 1 of 600 births, may result from translocation of part of chromosome 21 to another chromosome [it may also occur as trisomy (three chromosomes rather than the normal two) of chromosome 21]. One study (http://www.st-elizabeth.edu/~ikessler/hgen/genlec5.html) found that of live human births, about 6% suffered from some kind of chromosomal abnormality; about 2% of live human births have extra or missing sex chromosomes, about 1.4% have a third non-sex chromosome (i.e. trisomy, mentioned above), and about 2% have translocations.

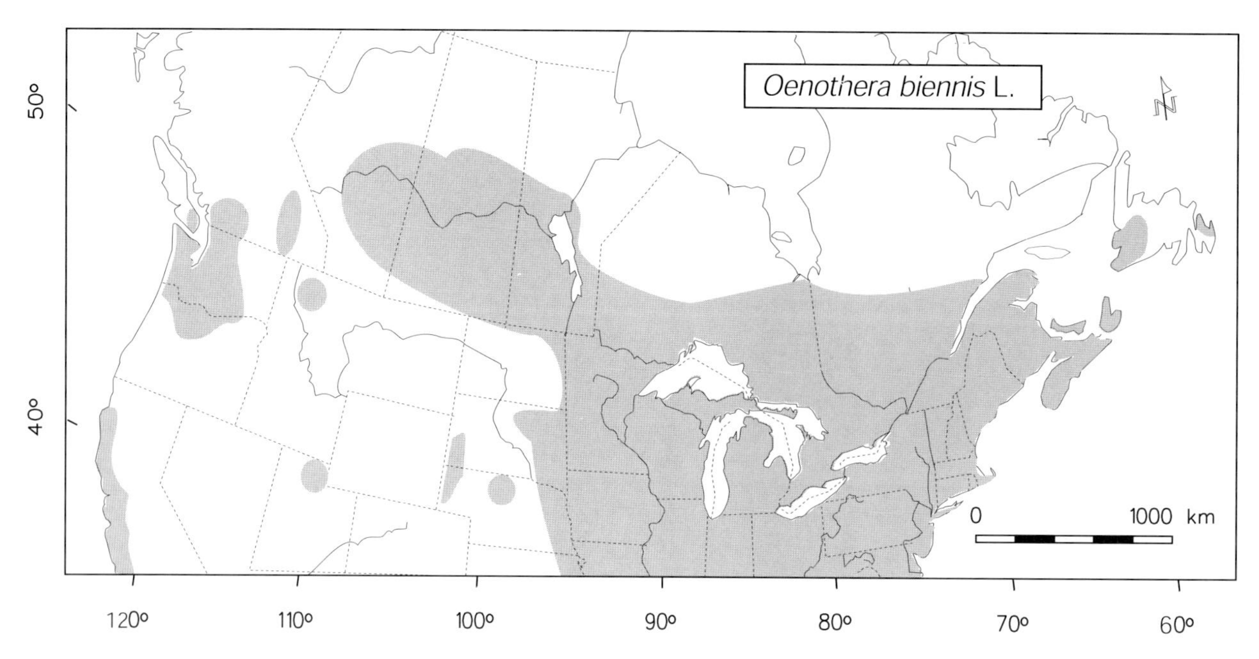

Medicinal Uses

Evening primrose extracts were used medicinally by both Indians and early settlers. In Europe during the early 1600s it was called "King's Cure-all." An infusion of the whole plant was thought to counter asthmatic cough, gastro-intestinal disorders, and whooping cough, and to reduce pain. Poultices were used to treat bruises and wounds

Evening primrose has attracted great interest for its seed oil, used medicinally as a nutritional supplement. The health value of the seed oil resides in an unusual polyunsaturated fatty acid, γ-linolenic acid (gamma-linolenic acid) or simply GLA. The seeds contain 17–25% oil, of which only 7–10% is GLA, although climate and maturity affect oil content and qualitative composition, as well as overall yield. GLA is one of the so-called essential fatty acids needed by humans for maintenance of cell functions. It is a precursor in the biosynthesis of prostaglandins, especially prostaglandin E1, a hormone-like substance that has been clinically shown to regulate metabolic functions in mammals; it affects cholesterol levels, dilates blood vessels, reduces inflammation, and has additional effects. GLA is thought to be important for development of brain tissue and other tissue growth, and nature seems to provide for human infants with high levels of GLA in human milk. GLA is a normal conversion product of linoleic acid, a major constituent of most vegetable oils, so that it would appear that humans should not experience a shortage. Nevertheless, some people, perhaps 10–20% of the population, evidently do not have adequate levels, even when receiving large amount of linoleic acid. The deficiency seems due to lack of an enzyme that metabolizes GLA from linolenic acid, so that there is a deficiency of GLA in the blood. Useful for treating atopic eczema, GLA has therapeutic promise for premenstrual syndrome, diabetes, multiple sclerosis, alcoholism, inflammation, heart disease and stroke. Rubbing GLA into the skin is thought to be an alternative route of assimilation, and so cosmetic preparations sometimes incorporate GLA. Pharmaceutical and food companies are developing GLA-containing supplements and specialty foods for infants, the elderly, and people with health problems.

Toxicity

Side effects of consumption of GLA-fortified foods and supplements have been documented, so that use should be guided by doctors and pharmacists.

Chemistry

Gamma-linolenic acid, the constituent of chief medicinal interest, is discussed above.

Non-medicinal Uses

There are ornamental forms of *Oenothera biennis* with attractive habit and flowers. There are also forms with fleshy edible roots, used as a vegetable, which were more commonly grown in the nineteenth century than today. Evening primrose leaves, shoots, roots, and seed pods were consumed by American Indians as food.

Agricultural and Commercial Aspects

The most significant current economic value of the species lies in its use as a diversification crop. Although GLA has been obtained by fermentation of some yeasts and other fungi, and from currants (*Ribes* species), the chief commercial sources are evening primrose and borage (*Borago officinalis* L.). Companies have engaged in a boastful debate about the comparative efficacy of GLA in their preparations made from evening primrose on the one hand, and from borage on the other. Whether borage or evening primrose is more competitive for GLA production depends on climatic and edaphic factors at a particular location. In Canada, both species are grown. Borage has a higher GLA content, but non-shattering cultivars are not grown in Canada, so that harvest is difficult. Borage is much more suitable for the Canadian prairies, where available cultivars of evening primrose do not overwinter reliably. However it isn't essential to grow evening primrose as a biennial: in Eastern Canada it is often started in greenhouses in mid-winter and transplanted to the field where it is grown as an annual.

As a cultivated plant, evening primrose is tolerant of a variety of soil types and a range of pH, but soils that are prone to crusting after rains and waterlogged soils should be avoided. If planted at too high a density (150 plants m^2) the plants may not bolt.

Evening primrose crops are raised in temperate areas of northern and eastern Europe, North America, and Australasia. US production is centered in North and South Carolina, Texas, and Oregon. Canadian production is centered in Nova Scotia and Ontario. Experimental production in Manitoba has been disappointing. Annual world production of seed has increased at least 20 fold in the last 20 years, and is currently about 4,000 tonnes. Combined US, and Canadian annual production is less than 200 tonnes. In good market years, several hundred ha of evening primrose may be grown in Canada.

Wild evening primrose plants shed their seeds when a pod matures, and since the pods don't mature simultaneously, harvest of seeds is difficult. Nevertheless, seed is gathered from wild plants in northeast China. Most modern evening primrose cultivars have non-shedding pods, which has simplified harvest and reduced seed loss. Crop yields of over 2 tonnes/ha have been recorded in Nova Scotia, although much lower yields are frequent. In Ontario, depending on the rather volatile market and variable production, a hectare may result in a gross financial return of $1,000–2,000.

Oenothera biennis (evening primrose)

The future of evening primrose as a pharmacological crop in Canada is uncertain because of competition from other countries and the unreliability of the present market. Hemp (*Cannabis sativa* L.) is attracting considerable interest as a new crop in Canada, to be grown not only for fiber but also for its high-GLA seed oil. Still another potential source of competition is the possibility that genetic engineers will splice the capacity to produce GLA into crops such as canola (*Brassica* species). Certainly the demand for GLA will continue to grow, and at least from time to time it may be anticipated that evening primrose crops will be grown in Canada on a contracted basis. With respect to climate and native germplasm, Canada is in a good position to develop its share of the evening primrose market.

Myths, Legends, Tales, Folklore, and Interesting Facts

- Some evening primrose seeds have been shown to live to 80 years in the soil.

- Hugo de Vries (1848–1935) was a world-famous student of evolution who, at the beginning of the 20th century, theorized that new species arise by spontaneous changes in individuals called mutations (Charles Darwin had earlier learned that such changes (called "sports" at the time) occur, but did not appreciate their importance for the mechanism of evolution). Unfortunately for de Vries, he chose evening primrose to demonstrate his theory. Later, scientists learned that the odd genetic system of evening primrose was responsible for the generation of altered individuals that de Vries was labelling as mutations and that this system did not occur in many other plants. As a result, his theory was discredited, although de Vries did contribute substantially to evolutionary theory.

Selected References

Baker, J. 1998. Evening primrose. *In* Richters second commercial herb growing conference - transcripts. *Edited by* R. Berzins, H. Snell and C. Richter. Richters, Goodwood, ON. pp. 67–85.

Barthell, J.F., and Knops, J.M.H. 1997. Visitation of evening primrose by carpenter bees: evidence of a "mixed" pollination syndrome. Southwestern Naturalist **42**(1): 86–93.

Baskin, C.C., and Baskin, J.M. 1993. Germination requirements of *Oenothera biennis* seeds during burial under natural seasonal temperature cycles. Can. J. Bot. **72**: 779–782.

Belisle, D. 1991. Potential of evening primrose in Canada. Alternative Crops Notebook **4**: 4–5. [Reprinted from BioOptions, Newsletter of the Center for Alternative Plant & Animal Products **1**(5), 1990.]

Brandle, J.E., Court, W.A., and Roy, R.C. 1993. Heritability of seed yield, oil concentration and oil quality among wild biotypes of Ontario evening primrose. Can. J. Plant Sci. **73**: 1067–1070.

Briggs, C.J. 1986. Evening primrose: La belle de nuit, the king's cureall. Can. Pharm. J. **199**: 248–252.

Budeiri, D., Po, A.L.W., and Dornan, J.C. 1996. Is evening primrose oil of value in the treatment of premenstrual syndrome? Controlled Clin. Trials **17**(1): 60–68.

Court, W.A., Hendel, J.G., and Pocs, R. 1993. Determination of the fatty acids and oil content of evening primrose (*Oenothera biennis* L). Food Res. Int.. 26:181–186.

Cisowski, W., Zielinska-Stasiek, M., Luczkiewicz, M., and Stolyhwo, A. 1993. Fatty acids and triacylglycerols of developing evening primrose (*Oenothera biennis*) seeds. Fitoterapia **64**: 155–162.

Dietrich, W.M., and Wagner, W.L. 1988. Systematics of *Oenothera* section *Oenothera* subsection *Raimannia* and subsection *Nutantigemma* (Onagracceae). Syst. Monogr. **24**: 1–91.

Dietrich, W., Wagner, W.L., and Raven, P.H. 1997. Systematics of *Oenothera* section *Oenothera* subsection *Oenothera* (Onagraceae). American Society of Plant Taxonomists, Ann Arbor, MI. Syst. Bot. Monogr. 50. 234 pp.

Ensminger, P.A., and Ikuma, H. 1987. Photoinduced seed germination of *Oenothera biennis* L. I. General characteristics. Plant Physiol. **85**: 879–884.

Ensminger, P.A., and Ikuma, H. 1987. Photoinduced seed germination of *Oenothera biennis* L. II. Analysis of the photoinduction period. Plant Physiol. **85**: 885–891.

Ensminger, P.A., and Ikuma, H. 1988. Photoinduced seed germination of *Oenothera biennis* L. III. Analysis of the postinduction period by means of temperature. Plant Physiol. **86**: 475–481.

Gates, R.G. 1957. A conspectus of the genus *Oenothera* in eastern North America. Rhodora **59**: 9–17.

Gates, R.G. 1958. Taxonomy and genetics of *Oenothera*. Uitgeverij Dr. W. Junk, Den Haag, Cambridge, MA. 115 pp.

Gregory, D.P. 1963. Hawkmoth pollination in the genus *Oenothera*. Aliso **5**: 357–384.

Gross, K.L. 1985. Effects of irradiance and spectral quality on the germination of *Verbascum thapsus* and *Oenothera biennis* seeds. New Phytol. **101**: 531–542.

Gross, K.L., and Kromer, M.L. 1986. Seed weight effects on growth and reproduction in *Oenothera biennis* L. Bull. Torrey Bot. Club **113**(3): 252–258.

Hall, I.V., Steiner, E., Threadgill, P., and Jones, R.W. 1988. The biology of Canadian weeds. 84. *Oenothera biennis* L. Can. J. Plant Sci. **68**: 163–173.

Hanczakowski, P., Szymczyk, B., and Wolski, T. 1993. The nutritive value of the residues remaining after oil extraction from seeds of evening primrose (*Oenothera biennis* L.). J. Sci. Food Agric. **63**: 375–376.

Horrobin, D.E. 1990. Gamma linolenic acid. Reviews in Contemporary Pharmacology **1**: 1-45.

Hulan, H.W., Hall, I.V., Nash, D.M., and Proudfoot, F.G. 1987. Composition of native evening primrose seeds collected from western Nova Scotia. Crop. Res. Edinburgh (Scottish Academic Press) **27**(1): 1–9.

Kerscher, M.J., and Korting, H.C. 1992. Treatment of atopic eczema with evening primrose oil: rationale and clinical results. Clin. Investig. **70**(2): 167–171.

Kromer, M., and Gross, K.L. 1987. Seed mass, genotype, and density effects on the growth and yield of *Oenothera biennis* L. Oecologia (Berlin) **73**: 207–212.

Lapinskas, P. 1989. Commercial exploitation of alternative crops, with special reference to evening primrose. *In* New crops for food and industry. *Edited by* G.E. Wickens, N. Haq and P. Day. Chapman and Hall, London. pp. 216–221.

Levin, D.A., Howland, G.P., and Steiner, E. 1972. Protein polymorphism and genic heterozygosity in a population of the permanent translocation heterzygote, *Oenothera biennis*. Proc. Nat. Acad. Sci. USA **69**(6): 1475–1477.

Levy, A., Palevitch, D., Ranen, C. 1993. Increasing gamma linolenic acid in evening primrose grown under hot temperatures by breeding early cultivars. Acta Hort. **330**: 219–225.

Loughton, A., Columbus, M.J., and Roy, R.C. 1991. The search for industrial uses of crops in the diversification of agriculture in Ontario. Altern. Crops Notebook **5**: 21–27.

Morrison, K.D., and Reekie, E.G. 1995. Pattern of defoliation and its effect on photosynthetic capacity in *Oenothera biennis*. J. Ecol. **83**: 759–767.

Mukherjee, K.D., and Kiewitt, I. 1987. Formation of gamma linolenic acid in the higher plant evening primrose (*Oenothera biennis* L.). J. Agric. Food. Chem. **35**: 1009–1012.

Munz, P.M. 1965. North American Flora, Series II, Part 5 - Onagraceae. The New York Botanical Garden, NY. 231 pp.

Paccalin, J., Mendy, F., Bernard, M., Delhaye, N., and Spielmann, D. 1983. Rediscovery of an oleaginous plant: *Oenothera biennis*. The importance of gamma-linolenic acid in nutrition. Bull. Acad. Natl. Med. **167**: 923–931. [In French.]

Raven, P.E. 1979. A survey of reproductive biology in Onagraceae. N.Z. J. Bot. **17**: 575–593.

Raven, P.E., Dietrich, W., and Stubbe, W. 1979. An outline of the systematics of *Oenothera* subsect. *Euoenothera* (Onagraceae). Syst. Bot. **4**: 242–252.

Reekie, E.G., and Reekie, J.Y.C. 1991. The effect of reproduction on canopy structure, allocation and growth in *Oenothera biennis*. J. Ecol. **79**: 1061–1071.

Reeleder, R.D. 1994. Factors affecting infection of evening primrose (*Oenothera biennis*) by *Septoria oenotherae*. Can. J. Plant Pathol. **16**: 13–20.

Reeleder, R.D., Monet, S., Roy, R.C., and Court, W.A. 1996. Dieback of evening primrose: Characteristics of *Septoria oenotherae*, its interactions with *Botrytis cinerea*, and use of fungicides to manage disease. Can. J. Plant Pathol. **18**: 261–268.

Rostanski, K. 1982. The species of *Oenothera* L. in Britain. Watsonia **14**: 1–34.

Roy, R.C. 1990. Health food plant may be alternative for tobacco growers. Communication Branch, Agriculture Canada, Ottawa. Agri-Features **2106**: 1–3.

Roy, R.C., White, P.H., More, A.F., Hendel, J.G., Pocs, R., and Court, W.A. 1993. Effect of transplanting date on the fatty acid conposition, oil content and yield of evening primrose (*Oenothera biennis* L.) seed. Can. J. Plant Sci. **74**: 129–131.

Russell, G. 1988. Physiological restraints on the economic viability of the evening primrose crop in eastern Scotland. Crop. Res. (Edinburgh) **28**(1): 25–33.

Simpson, M.J.A. 1994. A description and code of development of evening primrose (*Oenothera* spp.). Ann. Appl. Biol. **125**: 391–397.

Simpson, M.J.A., and Fieldsend, A.F. 1993. Evening primrose: harvest methods and timing. Acta Hort. **331**: 121–128.

Skvarla, J.J., Raven, P.H., Chissoe, W.F., and Sharp, M. 1978. An ultrastructural study of viscin threads in onagraceae pollen. Pollen Spores **20**(1): 5–143.

Skvarla, J.J., Raven, P.H., and Praglowski, J. 1976. Ultrastructural survey of Onagraceae pollen. Reprinted from 'The Evolutionary Significance of the Exine.' *Edited by* I.K. Ferguson and J. Muller. Linn. Soc. Symp. Ser. **1**: 447–479.

Wagner, W.L., Stockhouse, R.E., and Klein, W.K. 1985. The systematics and evolution of the *Oenothera caespitosa* species complex (Onagraceae). Missouri Botanical Gardens, Allen Press, Inc., Lawrence, KS. 103 pp.

Wilson, R. 1989. An alternative crop profile. Crop Development Division, Agriculture Canada. Altern. Crop Notebook **1**: 1–5.

Yaniv, Z., and Perl, M. 1987. The effect of temperature on the fatty acid composition of evening primrose (*Oenothera*) seeds during their development, storage and germination. Acta Hort. **215**: 31–38.

Yaniv, Z., Ranen, C., Levy, A., and Palevitch, D. 1989. Effect of temperature on the fatty acid composition and yield of evening primrose (*Oenothera lamarckiana*) seeds. J. Exp. Bot. **40**: 609–614.

World Wide Web Links

(Warning. The quality of information on the internet varies from excellent to erroneous and highly misleading. The links below were chosen because they were the most informative sites located at the time of our internet search. Since medicinal plants are the subject, information on medicinal usage is often given. Such information may be flawed, and in any event should not be substituted for professional medical guidance.)

Evening primrose, *Oenothera biennis*, Agriculture & Agri-Food Canada, Southern Crop Protection & Food Research Centre:
http://res.agr.ca/lond/pmrc/study/newcrops/eprim.html

Sask Ag & Food: Farmfacts, evening primrose:
http://www.agr.gov.sk.ca/saf/farmfact/sce0190.htm

Paintings of Mary Vaux Walcott [color painting]:
http://chili.rt66.com/hbmoore/Images/Walcott/Oenothera_biennis-2.jpg

Oplopanax horridus (devil's club)

Oplopanax horridus **(J.E. Smith) Miq.** Devil's Club

Synonyms: *Echinopanax horridus* (Sm.) Decne. & Planch. ex H.A.T. Harms; *Fatsia horrida* (Sm.) Benth. & Hook. f.; *Panax horridus* Sm.

The genus name is derived from the Greek *hoplon*, weapon, + *Panax*, i.e., an "armed ginseng," references to the fearsome spines of the plant and to its membership in the same family (Araliaceae) as ginseng. Frequently the genus name has been considered to be a neuter name in Latin, so that the specific epithet ends in "um" (*O. horridum*). However, the Botanical Code of Nomenclature requires that names ending in panax be treated as masculine, so that the name must be spelled *E. horridus*.

English Common Names

Devil's club (sometimes ungrammatically hyphenated as "devil's-club"), Alaskan ginseng.

Devil's club is sometimes confused with devil's claw (*Harpagophytum procumbens* DC.); the latter, an African plant, is commonly marketed in North America as an over-the-counter herbal product, with the information that it has been particularly used as an anti-inflammatory and pain reliever against arthritis and rheumatism. Devil's club should also not be confused with devil's walking-stick (*Aralia spinosa* L.), which similarly has spiny branches but differs in its twice compound leaves with numerous leaflets.

French Common Names

Aralie épineuse.

Morphology

This extremely thorny, deciduous, sweetly-aromatic shrub grows 1–3 (occasionally as high as 5) m, with a stem as thick as 3 cm. The leaves, often described as maple-like, are large, with blades up to 50 cm wide, stalks up to 30 cm long, and 5–9 (-13) palmately arranged lobes with irregularly toothed margins. The stems, leaf stalks and lower veins of the leaves are densely armed with thin stiff spines 5–10 mm long. The spines are very irritating; they can cause festering wounds when imbedded in the skin, and may produce severe allergenic reaction in susceptible people. The root system is shallow, generally without thorns. Roots appear to arise at times from stems that have leaned over to the ground, and such natural layering or suckering spreads the plant vegetatively. The branches often entangle, contributing to the difficulty of walking through a devil's club thicket. Greenish-white flowers, 6 mm long, are developed from May to July (depending on elevation and latitude). The flowers are borne on branching inflorescences up to 25 cm long. In the late summer berries 6–10 mm wide are matured. These are bright red, fleshy, ellipsoid and somewhat compressed. The berries contain two or three seeds, which are believed to be dispersed by animals. The berries persist on the plant through the winter.

Classification and Geography

Oplopanax is a small genus of three species. The Japanese *O. japonicus* (Nakai) Nakai is sometimes considered to be a subspecies of *O. horridus* (ssp. *japonicus* (Nakai) Hult.). The Eurasian *O. elatus* (Nakai) Nakai is closely related.

Devil's club is distributed from Alaska southwards. It is found along the coast through British Columbia and on the west side of the Cascade range through the states of Washington and Oregon. Also, it extends eastward to the Rocky Mountains, including portions of Alberta, Idaho and Montana. Finally, a small enclave is found in northern Michigan and the Thunder Bay district of Ontario, where disjunct populations occur on several islands of northern Lake Superior, including Isle Royale and Passage Island, Michigan, and Porphyry and Slate islands, Ontario.

Ecology

Devil's club typically occurs in moderately well-drained to poorly-drained usually shaded sites. It is found in coniferous woods, especially near streams, and ranges from near sea level to subalpine elevations. It has been recorded on a variety of soils varying from sand and loan to silt, sometimes occurring on very shallow substrates, and generally on acidic soils (pH 3.8 – 6.0). The species is found in both maritime and continental climates. It is a dominant component of understorys of various Pacific Northwest and western boreal forests, often forming dense, nearly impenetrable pure thickets, but also occurring in some understorys with other shrubs and (or) herbaceous plants. The best growth is attained in mature, climax forests. In its eastern disjunct range, devil's club occurs in low rocky woods and wooded ravines, between rock ridges, and on sheltered cliffs.

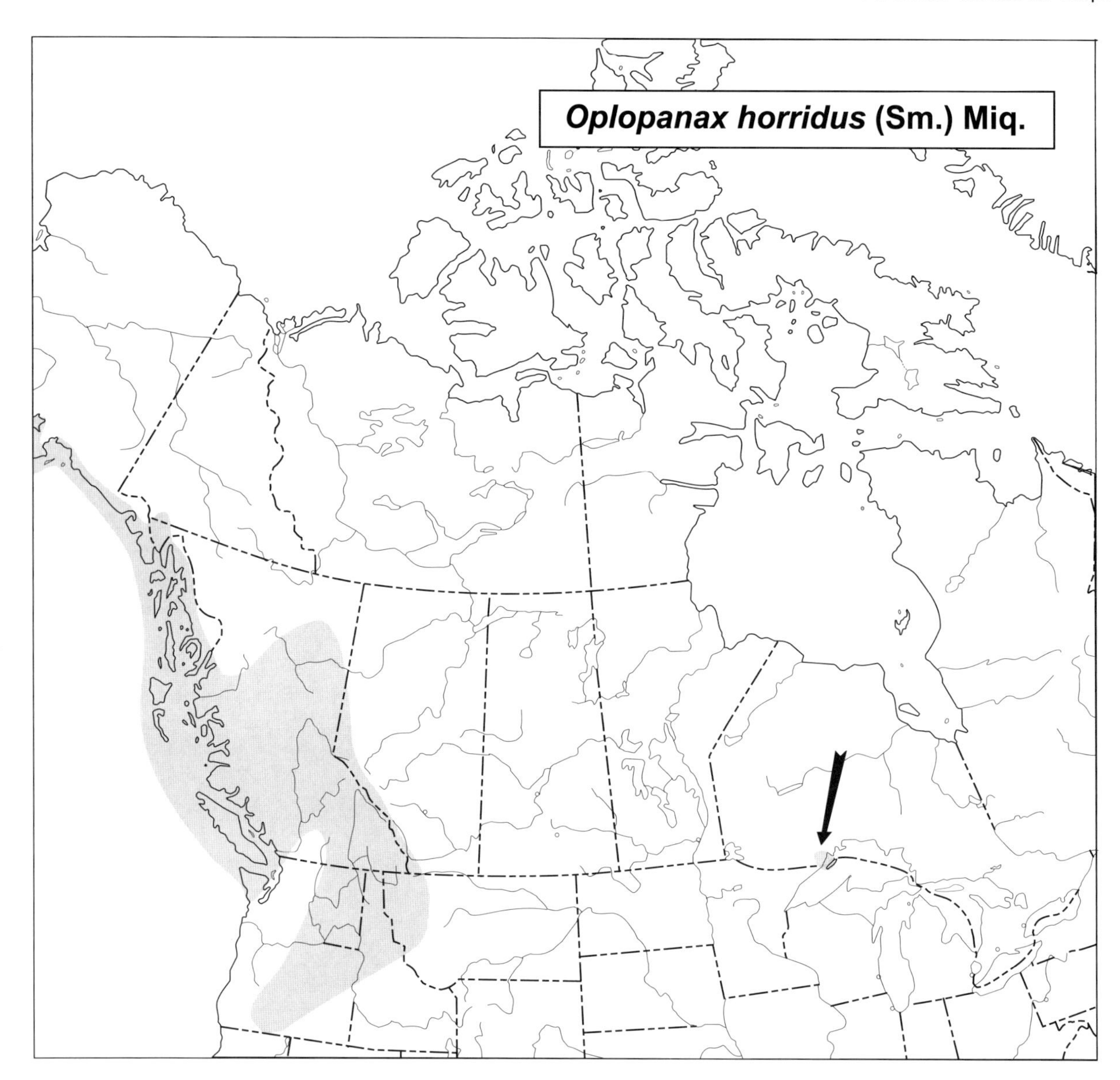

Devil's club receives limited browsing from wild mammals, probably because of its prickly leaves and stems. Black-tailed deer, white-tailed deer, and elk lightly browse in spring and summer, but moose on Isle Royale, Michigan have not been observed browsing the plant. Grizzly and black bears consume devil's club seeds, leaves, and stems.

Medicinal Uses

Devil's club has been used by native peoples in western North America as a medicinal plant since time immemorial. The inner bark and roots were (and continue to be) employed to treat arthritis, rheumatism, stomach and digestive problems, tuberculosis, colds, skin disorders, and many other complaints. In the Northwest Coast region of British Columbia, the bark of devil's club has been the most important of the many barks used medicinally by Indians.

In addition to its use to treat physical ailments directly, devil's club was used by shamans and others in spiritual ceremonies to attain supernatural powers, the plant having a reputation for magical protective properties. Interestingly, Western North American Indians have often ascribed such properties to other prickly or thorny plants. Further, the cathartic properties (causing evacuation of the gut) is also a feature that might be associated with "cleansing" spiritual value. Finally, it has been suggested that the hypoglycemic effect of devil's club might produce a minor lowering of blood sugar and an associated light-headedness that could contribute to a spiritual mood.

The most interesting and potentially useful medicinal property ascribed to devil's club is that it is hypoglycemic (lowering blood sugar), and therefore useful as an anti-diabetic (for controlling diabetes). This has been rather controversial, with some earlier studies not finding hypoglycemic effects, but there is currently a consensus that devil's club is hypoglycemic. Certainly there is a long history of use by native Americans to treat adult-onset diabetes. Insulin therapy is standard treatment for diabetes today, but is not always successful in preventing the complications of diabetes (kidney damage, destruction of the retina and vision, cataracts, arteriosclerosis, neurological dysfunction, and predisposition to gangrene).

Extracts from the inner bark of devil's club have been shown to be antibiotic, specifically against the bacterium genus *Mycobacterium* that causes tuberculosis and other diseases of humans. This validates the use of this plant for treating tuberculosis by Indians of the West Coast.

Toxicity

Apart from allergic reactions to the spines, devil's club is not known to have toxic effects, although diarrhea and weakness have been experienced by those starting to use it. It should be noted that experimental pharmacological work with the plant is limited and so the effects of prolonged use have not been adequately evaluated.

Chemistry

Antibacterial and antifungal activity of extracts of the inner bark has been attributed to several polyynes. The bark has also demonstrated antiviral activity. A sesquiterpene, a sesquiterpene alcohol, and a sesquiterpene ketone have been isolated from the closely related *O. japonicus*, and these same compounds may be present in *O. horridus*. A derivative of the sesquiterpene ketone is used in commercial preparations in Japan to treat coughs and colds.

Non-medicinal Uses

Native peoples of western North America used devil's club wood to construct fishing lures and the charcoal for face paint. Other products produced from the plant included perfume, baby talc, deodorants, and stain for baskets and other materials.

As a horticultural subject, devil's club makes a beautiful but intimidating specimen, that some brave gardeners choose to cultivate. It is particularly suitable as a striking backdrop for shady or partially shady landscapes and water gardens. It can be used to make a formidable hedge to protect areas against human intrusion. The species also has potential as a restoration stream cover plant, preventing or reducing streamside erosion.

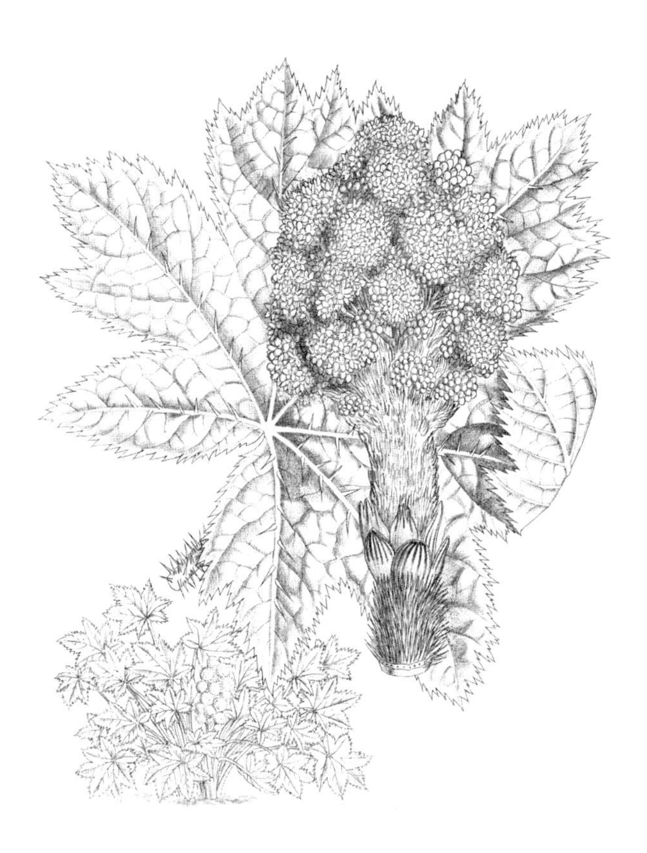

Oplopanax horridus (devil's club)

One account of the plant states that the leaf shoots produced in early spring can be used as a condiment or nibbled raw in very limited amount. However, there is very little use of this plant for human food.

Agricultural and Commercial Aspects

Devil's club is presently not widely known as a herbal medicine. However, it has several characteristics that make it an excellent prospect for penetrating the herbal market. First, it has a well-established folk reputation as possessing a cornucopia of healing actions, especially against some ailments (such as rheumatism, arthritis, stomach and digestive upsets, coughs and colds, and skin disorders) that are chronic and widespread, not consistently curable by conventional western medicine, and often not so serious as to require pharmaceuticals. Second, it is acquiring a reputation as an "adaptogen" (a substance with properties that strengthen the body to withstand a wide variety of stresses and illnesses; see discussion in chapter on *Rhodiola*). Given that it is related to the world's most important adaptogen, ginseng, and even shares in part its scientific name (*Panax*), devil's club clearly has the kind of background that makes it attractive to the herbal market. Thirdly, research has shown that the plant has at least some genuine medicinal virtues. Fourth, the plant has the kind of impressive common name and extraordinary appearance that make medicinal herbs memorable and attractive. In short, devil's club is very appealing in a marketing sense. It certainly deserves both medicinal (pharmacological) and marketing (especially as a source of nutraceutical preparations) research.

Because the natural distribution of devil's club includes the declining old growth forests in the west, and a small disjunct area in the east, increased harvest could erode genetic variation. It is important to ensure the well-being of this quite unique plant of considerable potential medicinal value.

As a result of cultivation for ornamental purposes, a basic horticultural knowledge of the plant is available. Devil's club is most easily propagated from suckers and root cuttings but can be grown from seed. The species is cold-hardy. Although typically found as an understory shrub in partial shade, and moderately shade tolerant, it can be planted in full sun. Devil's club withstands pruning very well. It is drought intolerant and should be provided with a moist situation.

Myths, Legends, Tales, Folklore, and Interesting Facts

- In at least two Northwest Indian cultures, devil's club was associated with bears. The Tlingit thought bears chewed the roots to soothe their battle wounds. The Bella Coola thought that bears ate the unpalatable fruits (known to them as "grizzly's berries") and used the thorny branches as bedding.
- The Haida placed a stick of devil's club under their mattresses or across the top of their doorways to protect against evil spirits.
- Canadian Pacific was incorporated in 1881 to build a transcontinental railway linking Eastern Canada with the Pacific coast. In parts of British Columbia where devil's club grows the search for a route was made extremely difficult because of virtually impenetrable large patches of the shrub, and the railway was rerouted to avoid some of these areas.
- The devil is commemorated in numerous names of plants that have been used medicinally, besides devil's club and devil's claw alluded to earlier: devil's apple (*Datura stramonium* L.), devil's bit (*Liatris squarrulosa* Michx., and *Scabiosa succisa* L.), devil's cherries (*Atropa bella-donna* L), devil's dung (*Ferula assa-foetida* L.), devil's guts (*Cuscuta epithymum* Murr.), devil's nettle (*Achillea millefolium* L.), devil's shoe string (*Tephrosia virginiana* (L.) Pers.), devil's tobacco (*Lobelia tupa* L.), devil's tongue (*Amorphophallus rivieri* Durieu), devil tree (*Alstonia scholaris* Brown), devil's trumpet (*Datura stramonium* L.).

Selected References

Gottesfeld, L.M.J. 1992. The importance of bark products in the aboriginal economies of northwestern British Columbia, Canada. Econ. Bot. **46**: 148–157.

Justice, J.W. 1966. Use of devil's club in Southeast Alaska. Alaska Med. **8**(2): 36–39.

Kobaisy, M., Abramowski, Z., Lermer, L., Saxena, G., Hancock, R.E., Towers, G.H., Doxsee, D., and Stokes, R.W. 1997. Antimycobacterial polyynes of devil's club (*Oplopanax horridus*), a North American native medicinal plant. J. Nat. Prod. **60**: 1210–1213.

Lee, C., and Lee, S.A. 1991. Palynotaxonomic study of the genus *Fatsia* Decne. and Planch., and its relatives (Araliaceae). Korean J. Plant Taxon. **21**: 9–26.

Marquis, R.J., and Voss, E.G. 1981. Distribution of some western North American plants disjunct in the Great Lakes region. Mich. Bot. **20**: 53–82.

McCutcheon, A.R., Stokes, R.W., Thorson, L.M., Ellis, S.M., Hancock, R.E.W., and Towers, G.H.N. 1997. Anti-mycobacterial screening of British Columbian medicinal plants. Int.. J. Pharmacogn. **35**(2): 77–83.

McCutcheon, A.R., Roberts, T.E., Gibbons, E., Ellis, S.M., Babiuk, L.A., Hancock, R.E., and Towers, G.H.N. 1995. Antiviral screening of British Columbian medicinal plants. J. Ethnopharmacol. **49**: 101–110.

Smith, G.W. 1983. Arctic pharmacognosia II. Devil's club, *Oplopanax horridus*. J. Ethnopharmacol. **7**: 313–320.

Takeda, K., Minato, H., and Ishikawa, M. 1966. Studies on sesquiterpenoids XII. Structure and absolute configuration of oplopanone, a new sesquiterpene from *Oplopanax japonicus* (Nakai) Nakai. Tetrahedron, Suppl. No. **7**: 219–225.

Turner, N.J. 1982. Traditional use of devil's-club (*Oplopanax horridus*; Araliaceae) by Native Peoples in western North America. J. Ethnobiol. **2**: 17–38.

World Wide Web Links

(Warning. The quality of information on the internet varies from excellent to erroneous and highly misleading. The links below were chosen because they were the most informative sites located at the time of our internet search. Since medicinal plants are the subject, information on medicinal usage is often given. Such information may be flawed, and in any event should not be substituted for professional medical guidance.)

Fire effects information system [excellent general botanical account!]:
http://svinet2.fs.fed.us/database/feis/plants/shrub/oplhor/

Northwest nation, devil's club or Alaskan ginseng [short personal account]:
http://www.audubon.org/chapter/wa/rainier/nn/d-club.htm

Devil's club in rainforest [color photograph]:
http://www.wainet.com/forest.htm

Mystery plant - devil's club [color photograph and brief text]:
http://www.calgaryzoo.ab.ca/calgzoo/devilclb.htm

Devil's cub (*Oplopanax horridus*) [color photograph and brief text]:
http://www.for.gov.bc.ca/hfp/pubs/standman/bear/Devil.htm

Paintings of Mary Vaux Walcott [color painting]:
http://chili.rt66.com/hbmoore/Images/Walcott/Oplopanax_horridum-2.jpg

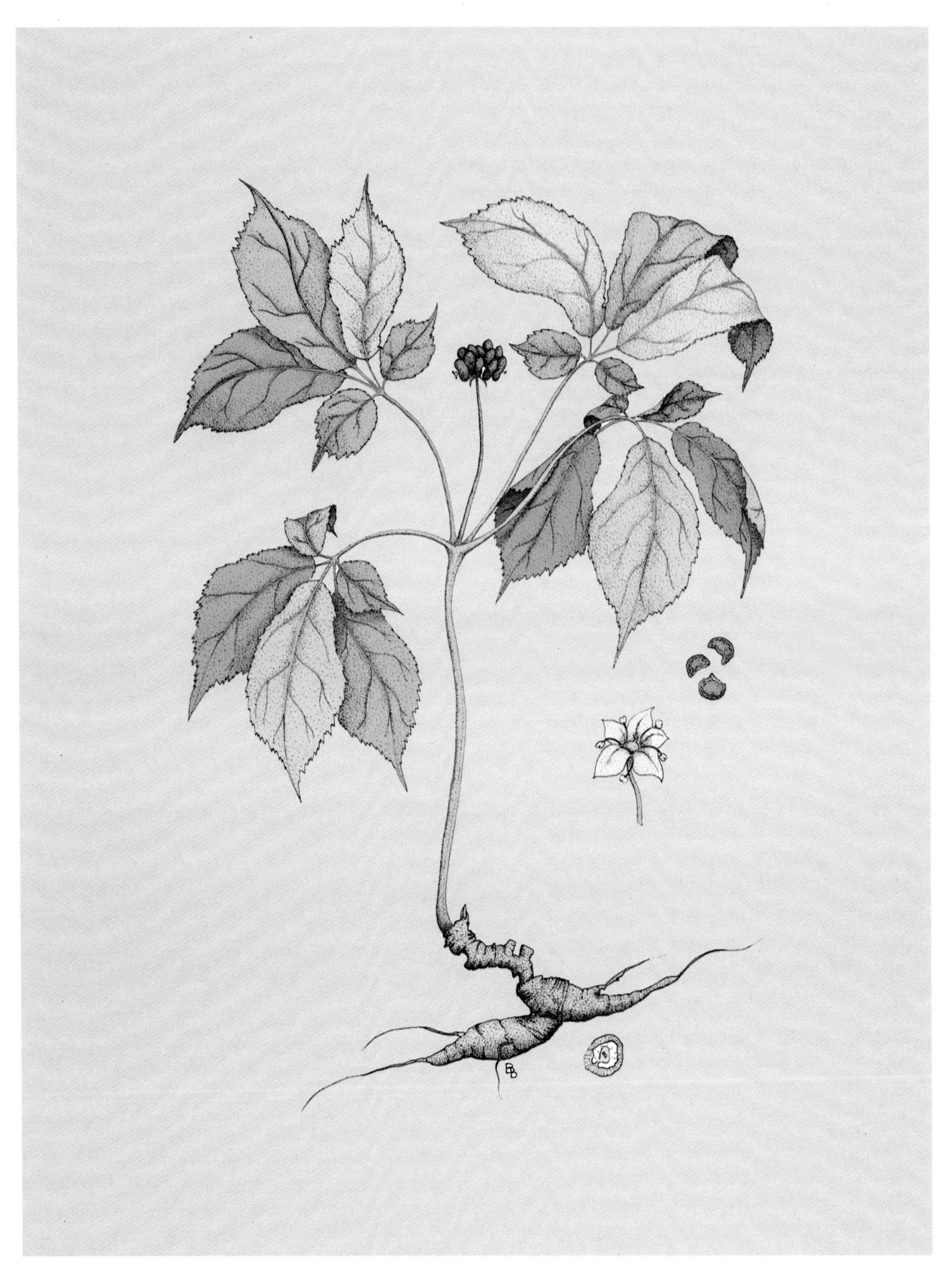

Panax quinquefolius (ginseng)

Panax quinquefolius L. Ginseng

The genus name *Panax* comes from the Greek *pas*, all, and *akos*, cure; or *panakes*, all healing, referring to the medicinal properties of the ginseng plant. Panacea was a goddess in Greek mythology who could heal all diseases, and who found a remedy for maintaining good health.

Frequently one sees the scientific name spelled "*Panax quinquefolium*," in the belief that the genus name *Panax* is neuter, and so requires the second word of the binomial to use the Latin suffix "um." However, the name *Panax* is masculine in both Greek and Latin, and the Code of Botanical Nomenclature requires that the masculine form be used, i.e., *quinquefolius*.

English Common Names

American ginseng, Canadian ginseng, five-fingers, occidental ginseng, sang, seng.

The name ginseng, first applied to Asian ginseng (*P. ginseng*), is derived from the Chinese *ren-shen* (in standard Chinese piryin) [also rendered *jen-sheng*, *jin-hsien*, *shen seng*, and *shinseng*]. This is usually translated as "man-shaped root," but may be better rendered as "man-essence;" the name originates from (a) the fancied resemblance of ginseng root to the human form, or (b) from the belief that the root represents the essence of the earth crystallized in a human form.

French Common Names

Ginseng, ginseng à cinq folioles.

Morphology

The unbranched, erect stems of American ginseng arise from a short rhizome and an elongated tuberous root. The root is 1–3 cm thick and 5–10 cm long, spindle-shaped, and often forked. The roots of older plants become branched and acquire prominent circular wrinkles; they are slightly aromatic, and have a sweetish, somewhat bitter taste. Mature plants are 20–70 cm tall, with a whorl of three or four long-stalked, palmate leaves, each generally with five large leaflets, of which the upper three are larger than the lower two. In midsummer, 6–20 small, yellowish flowers are borne on a short stalk arising from the centre of leaf attachment at the top of the stem. Fruits begin to ripen at the end of July and mature to a deep red color. Reproduction is entirely by seed. Seeds are produced by plants more than 3 years old with up to 150 being generated by a single plant. However, commonly, only a few seeds are matured by wild plants. Once a seedling of American ginseng has become established, its life expectancy is over 20 years, and some plants have been reported to live to at least 60 years of age. The age of a plant can be estimated by the number of stem scars on the short rhizome on top of the root and by leaf number. Wild plants with two palmate leaves ("prongs"), are generally over 3 years old, those with three prongs are mostly more than 6 years old and those with four prongs are mostly over 13 years old.

Asian ginseng is extremely similar in appearance to American ginseng, but the former is reported to have tapered leaflet bases while they are rounded in the latter. However, it has proven difficult to identify many plants using this character, and certain other recommended discriminating characters also appear to be quite variable.

Classification and Geography

Panax is a genus of perennial herbs, with two species in eastern North America and perhaps 5–10 species in Asia. Best known is the eastern Asian species *P. ginseng* C.A. Mey., known as ginseng, Asian ginseng, Oriental ginseng, Chinese ginseng and Korean ginseng, which is the major source of ginseng of commerce. American ginseng occurs from southern Ontario and southwestern Quebec south to Oklahoma, Louisiana, and northern Florida.

In North America, two species could be confused with American ginseng. Dwarf ginseng (*P. trifolius* L.) is a smaller species with stemless leaflets, which does not appear to have the properties of American ginseng, and is not harvested or cultivated. It is nevertheless a very unusual plant being one of the 0.1% of flowering plants that can change their gender from male to female and *vice versa*. Wild sarsaparilla (*Aralia nudicaulis* L.) and other species of *Aralia* are also superficially similar, but have pinnate instead of palmate leaves.

Ecology

Ginseng occurs in colonies of a few to hundreds of plants in rich, shady, deciduous forests, in deep leaf litter. Sites are frequently on northern or northeastern cool rocky slopes, commonly in areas with limestone outcrops in damp but well-drained soils. American ginseng thrives in 75% shade, and even shadier locations in the southern limits of its

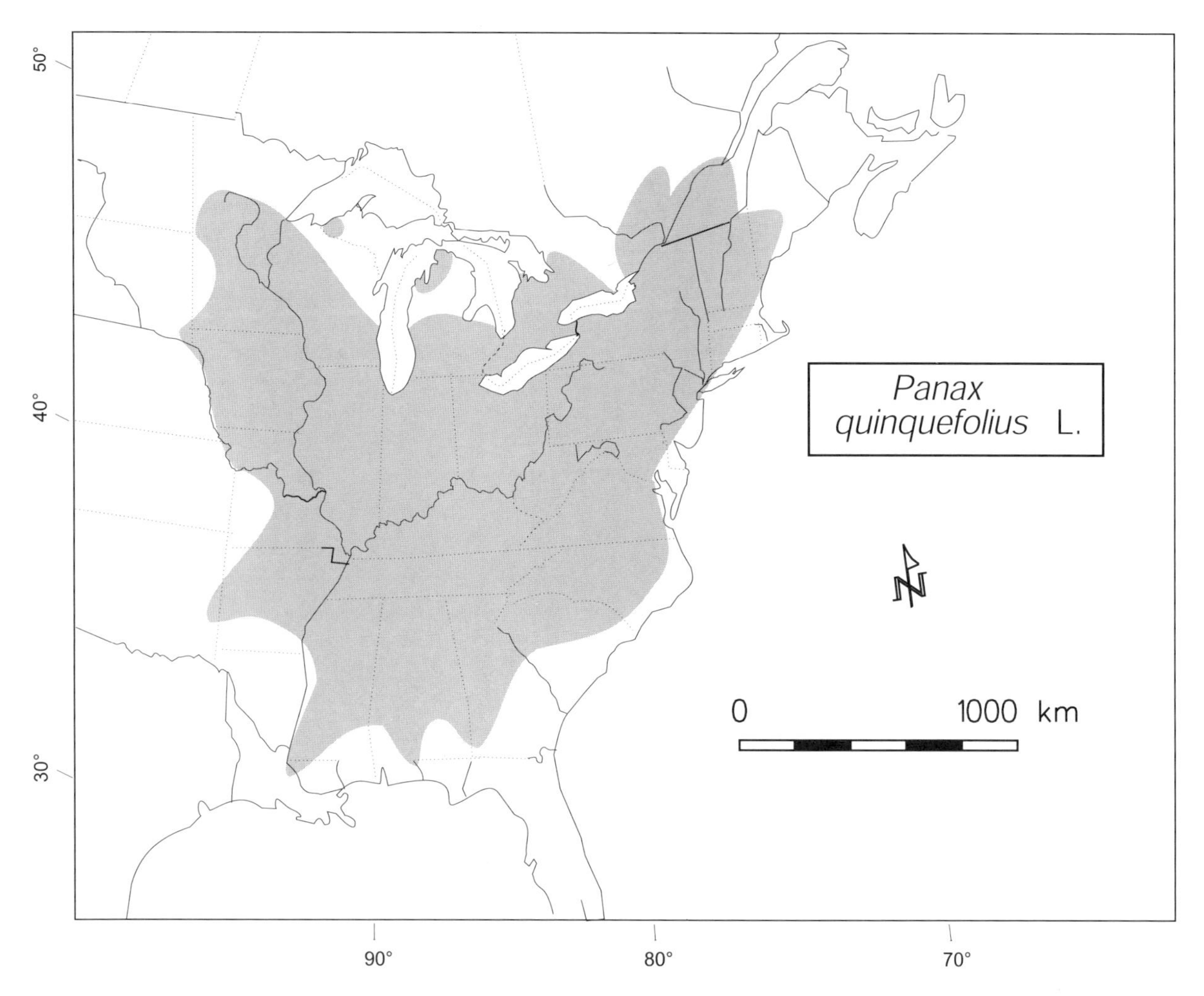

range. Clearcutting of virgin forests and overharvesting have drastically reduced the size of wild populations.

Although American ginseng cannot change its sex like dwarf ginseng, it can modify its gender through varying ratios of flowers with one or two ovules. In general, the larger and older plants that are able to mature more seeds can be viewed as more female. The flowers are adapted to cross-pollination through different maturation times of the male and female parts of the flower. Soon after the petals separate, the anthers mature and release pollen, prior to the stigmatic lobes separating and becoming receptive. Both wild and cultivated plants of American ginseng are visited by a wide variety of insects, but a few species of small bees are the most important pollinators. The attractive fruits are probably dispersed by animals.

Medicinal Uses

Ginseng has been used in Asian medicine for perhaps 5000 years. Wild American ginseng was apparently used by many Indian tribes for increasing the fertility of women, as a tonic to increase mental powers, and to treat headache, cramps, fevers, rheumatism, and cough. However, the extent to which these uses were acquired from visiting Europeans is uncertain.

Ginseng is the world's most widely used medicine, a consequence of its popularity in Asian medicine. Like vitamin C, it is widely used as a preventive medicine, and to maintain good health. It is commonly believed, especially in western countries, that it is an aphrodisiac, has amazing healing properties, provides energy, lowers blood pressure, retards the aging process, cures neurological disorders, and speeds recovery from sickness. Ginseng has been said to enhance digestion, stimulate blood circulation, relieve fatigue and cure blood diseases, and in general have a stimulating tonic effect. Ginseng has the reputation of being the ultimate elixir of life, a symbol of strength and long life, and a source of happiness. An extremely impressive number of ginseng recipes are employed in Asian medicine for various ailments. Particularly in the Orient, ginseng preparations are used medicinally to treat hypotension, hypertension, stress, insomnia, fatigue, depression, arthritis, diabetes, high cholesterol levels, bronchitis, some cancers, anemia, impotence, and premature aging.

There is some good evidence for the therapeutic value of ginseng, but it has been a subject of continuing controversy, with western scientists generally rejecting the claims of eastern medicine that ginseng has manifest benefits in the treatment of numerous illnesses. Much of ginseng medical research has been supported by those with commercial motives, and the design of experiments has often allowed researchers to draw whatever conclusions they wished, but could hundreds of millions of users be wrong?

Toxicity

The incidence of adverse reactions to ginseng is very low. Nevertheless it has been suggested that those with hayfever, asthma, emphysema, and cardiac or blood clotting problems, as well as pregnant women, should limit consumption.

Chemistry

Ginseng's alleged virtues are believed to be due to a large variety of root triterpene saponins called ginsenosides (less frequently panaxosides and panaquilins). These occur in the foliage as well as the roots, but by tradition only the roots are used. Many of these chemicals were assigned different names by American and Asian ginseng researchers, which of course can be confusing.

Non-medicinal Uses

Ginseng is almost exclusively used medicinally, although there is some very minor consumption of the root as a vegetable.

Agricultural and Commercial Aspects

The commercial significance of American ginseng was not discovered until after knowledge of Asian ginseng was disseminated to the West. Asian ginseng was first described in western literature in 1714 by Père Jartoux, a missionary in China. Jartoux conjectured that ginseng would be found in similar habitats in North America, and this information was transmitted by the Jesuits in Paris to their Canadian outposts. In 1704, Michel Sarrazin, the King's Physician to "New France," discovered American ginseng in Quebec, and brought some roots to Paris. However, not until Father Lafitau, a Jesuit priest and missionary among the Iroquois, read Jartoux's paper and found American ginseng near Montreal in 1716, did trade in the New World species begin. By 1718 the Jesuits were shipping dried roots, collected by the Iroquois, to China. They realized in the early 18th century that trade in American ginseng with China was extremely lucrative, and so they attempted to keep this trade secret. However, just why the celibate fathers were taking such an unusual interest in a certain low-growing herb rumored to be an aphrodisiac attracted attention, and eventually the secret leaked out. In the early 1700s, American ginseng became second only to fur as a trading commodity in New France. The practice of gathering ginseng in North America has continued to the present, especially by rural people in southern Appalachia, who harvest the roots for shipment to Asia and Europe. Historically, ginseng (mostly Asian) has fetched absurdly inflated prices - on occasion, thousands of dollars a kilogram for unusually shaped or larger roots.

American ginseng was first cultivated in Canada about 1890, but has been grown commonly only since the 1930s. The major centres of cultivation of American ginseng today are in the Haldimand-Norfolk region of southwestern Ontario, and southern British Columbia, and Wisconsin (mostly Marathon County, where 80% of US ginseng is grown). In the US, American ginseng is also grown in Kentucky, North Carolina, Tennessee and other states. About 545 tonnes were produced in 1991 in Wisconsin. In recent years, dried wild roots from Canada have sold for as much as $200.00/kg, while cultivated roots have sold for about $50.00/kg. In 1991, about 295 tonnes were grown in Ontario, worth an estimated $35 000 000. In British Columbia in 1992, about 109 tonnes of root worth about $13 000 000 were produced, as well as about 25 tonnes of seeds, worth about $5 000 000. Canadian production of ginseng is increasing. About 3/4 of recent production has been shipped to Hong Kong.

American ginseng plants take 5–7 years to mature and are grown from seeds or seedlings (1–3 years old). Loamy soils are best and 75% shade provided by screen or lath is essential. The crop is susceptible to fungus diseases and usually requires fungicide treatments as well as good air circulation. A good yield of roots is about 4 tonnes per hectare. The Ginseng Growers Association of Canada, centered in Simcoe, Ontario, helps Canadian farmers and promotes the market for the Canadian crop.

Ginseng is a major commodity of Asian commerce. The market for ginseng extends throughout the Far East, and there is comparatively minor usage as well in western nations. Five to six million people in the US alone consume ginseng regularly, and perhaps 20 million in western nations have used it. Until recently, ginseng products in North America were largely imported Asian ginseng, but North American-grown American ginseng is rapidly becoming popular as well in North America.

Until recently, export of wild roots from Ontario amounted to 40 000 per year. A Canadian study completed in 1987 found that the rate at which the wild roots were being harvested would likely eliminate the plant over large portions of its Ontario range. The species is also quite rare in Quebec, so that collecting from the wild in Canada has been inadvisable for some time. In 1988, American ginseng was officially listed as "threatened" in Canada by the Committee on the Status of Endangered Wildlife in Canada (COSEWIC). In 1989 export of wild American Ginseng was officially discontinued from Canada pending evaluation of the magnitude of the threat. The Convention on International Trade in Endangered Species (CITES) requires countries to show that exporting indigenous ginseng will not endanger the plant's survival. In the US, collection and sale of wild American ginseng are subject to registration, permits, and an official season. American ginseng provides an encouraging model of how to protect plants that have become seriously reduced in numbers because of over collecting. Measures have included both national protection and expansion of cultivation, thus reducing pressure on natural populations. These natural populations are potentially important sources of genetic variation for the improvement of the crop.

Myths, Legends, Tales, Folklore, and Interesting Facts

- Chinese herbal medicine is based on the concept of *yin* and *yang* forces of Daoist herbal theory. Yang represents masculinity, strength, and heat, and yin by contrast is feminine, mild, and cold. Ginseng is generally considered to represent yang. However, American ginseng is "yin of the yang," and is considered good for the respiratory or digestive system to "reduce heat." The "cooler" American ginseng also is considered more desirable in hot climates. American ginseng supposedly also has more aphrodisiac effect.
- Ginseng has a "contractile root" - a kind of root possessed by many perennial herbaceous species, which is designed to pull the root down into the ground. This is necessary to maintain the growing tip, which regenerates the stem each year, at ground level, where it is protected. Were it not for the contractile root, each season the growing rhizome would extend the growing tip higher into the air. To counteract the vertical growth of the rhizome, the ginseng roots contract yearly at the same rate at which the rhizome grows upward, pulling the plant downward.
- In 1788 Daniel Boone collected about 13.6 tonnes of ginseng roots in what is now West Virginia and Kentucky to sell in Philadelphia. At a typical 18th century price of 25cents/kg (10 cents/pound), this would have fetched about $3000.00, an enormous sum of money in the 1780s. Unfortunately during transportation in a boat, the ginseng was damaged by flooding, and had to be redried. Adding to Colonel Boone's woes, during the delay required to dry the ginseng again the price in Philadelphia fell.
- In 1908, Mr. L.J. Wilson of Pennington Gap, Virginia became totally exasparated at the prospect that the thieves who had robbed his ginseng garden the previous 2 years would return. He set up a series of shotguns with fine trip wires in his garden, with the result that a thief was shot to death and his companion wounded. A coroner's jury exonerated him from criminal blame.
- The largest wild ginseng root collected was found in woods near Benzonia, Michigan by Custer Higgins; it weighed 1.2 kg.
- The fable is told about the man who was arrested for selling a ginseng preparation allegedly adding years to one's life. The charlatan, it was found, had been arrested for the same offence before - in 1870, 1910, 1949, and 1975!

Selected References

Anderson, R.C., Fralish, J.S., Armstrong, J.E., and Benjamin, P.K. 1993. The ecology and biology of *Panax quinquefolium* L. (Araliaceae) in Illinois. Am. Midl. Nat. **129**: 357–372.

Bai, D., Brandle, J., and Reeleder, R. 1997. Genetic diversity in North American ginseng (*Panax quinquefolius* L.) grown in Ontario detected by RAPD analysis. Genome **40**: 111–115.

Bailey, W.G., Skretkowicz, A.L., Sawchuk, A.M., Proctor, J.T.L., Clark, L., and Lefebvre, C.M. 1994. International ginseng conference - Vancouver 1994. Program, abstract and trade show booklet. 78 pp.

Catling, P.M. 1995. Pollen vectors in an American ginseng (*Panax quinquefolius*) crop. Econ. Bot. **49**: 99–102.

Chandler, R.F. 1988. Ginseng — an aphrodisiac. Can. Pharm. J. **122**: 36–38.

Charron, D., and Gagnon, D. 1991. The demography of northern populations of *Panax quinquefolium* (American ginseng). J. Ecol. **79**: 431–445.

Cheung, K.S., Kwan, H.S., But, P.P.H., and Shaw, P.C. 1994. Pharmacognostical identification of American and Oriental ginseng roots by genomic fingerprinting using arbitrarily primed polymerase chain reaction (AP-PCR). J. Ethnopharmacol. **42**: 67–69.

Court, W.A., Reynolds, L.B., and Hende, J.G. 1996. Influence of root age on the concentration of ginsenosides of American ginseng (*Panax quinquefolium*). Can. J. Plant Sci. **76**: 853–855.

Cronin, F. 1982. Elixir or not, ginseng is a lucrative cash crop. Can. Geogr. **102**(6): 60–63.

Curran, D.F. 1983. The complete ginseng grower's manual. D.F. Curran Productions, Missoula, MT. 146 pp.

Curran, D.F., and Curran, P.A. 1985. The ginseng disease and pest reference guide. D.F. Curran Productions, Missoula, MT. 118 pp.

Duc, N.M., Kasai, R., Ohtani, K., Ito, A., Nham, N.T., Yamasaki, K, and Tanaka, O. 1994. Saponins from Vietnamese ginseng, *Panax vietnamensis* Ha et Grushv. collected in central Vietnam. II. Chem. Pharm. Bull. (Tokyo) **42**: 115–122.

Duc, N.M., Kasai, R., Ohtani, K. Ito, A., Nham, N.T., Yamasaki, K, and Tanaka, O. 1994. Saponins from Vietnamese ginseng, *Panax vietnamensis* Ha et Grushv. collected in central Vietnam. III. Chem. Pharm. Bull. (Tokyo) **42**: 634–640.

Duke, J.A. 1989. Ginseng: a concise handbook. Reference Publication, Inc., Algonac, MI. 273 pp.

Gagnon, D. 1993. L'étude démographique du ginseng à cinq folioles et de l'ail des bois. L'Euskarien (La Société Provancher d'Histoire Naturelle du Canada) **1993**(winter): 33-36.

Harrison, H.C., Parke, J.L., Oelke, E.A., Kaminski, A.R., Hudelson, B.D., Martin, L.J., Kelling, K.A., and Binning, L.K. 1992. Ginseng. BioOptions (Newsletter, Center Alternative Plant and Animal Products) **3**(4): 1–4. [Exerpt from Alternative Field Crops Manual published by Univ. Wisconsin-Extension, Center for Alternative Plant and Animal Products, and Minnesota Extension Service.]

Jo, J., Blazich, F.A., and Konsler, T.R. 1988. Postharvest seed maturation of American ginseng: stratification temperatures and delay of stratification. Hortscience **23**(6 part 1): 995–997.

Kelly, J. 1977. Herb collector's manual and marketing guide. Ginseng growers and collectors handbook. U.S. Dept. of Agriculture, Looneyville, WV. 97 pp.

Konsler, T.R. 1986. Effect of stratification temperature and time on rest fulfillment and growth in American ginseng (*Panax quinquefolium*). J. Am. Soc. Hortic. Sci. **111**: 651–654.

Konsler, T.R. 1987. Woodland production of ginseng and goldenseal. Stn. Bull. Purdue Univ. Agric. Exp. Stn. **518**. 175–178.

Konsler, T.R., and Shelton, J.E.1990. Lime and phosphorus effects on American ginseng: I. Growth, soil fertility, and root tissue nutrient status response. J. Am. Soc. Hortic. Sci. **115**: 570–574.

Konsler, T.R., Zito, S.W., Shelton, J.E., and Staba, E.J. 1990. Lime and phosphorus effects on American ginseng: II. Root and leaf ginsenoside content and their relationship. J. Am. Soc. Hortic. Sci. **115**: 575–580.

Lee, J.C., Strik, B.C., and Proctor, J.T.A. 1985. Dormancy and growth of American ginseng (*Panax quinquefolium*) as influenced by temperature. J. Am. Soc. Hortic. Sci. **110**: 319–321.

Lewis, W.H. 1986. Ginseng: a medical enigma. *In* Plants in indigenous medicine and diet. Biobehavioral approaches. *Edited by* N.L. Etkin. Redgrave Publ., Bedford Hills, NY. pp. 290-305.

Lewis, W.H., and Zenger, V.E. 1982. Population dynamics of the American gingseng *Panax quinquefolium* (Araliaceae). Am. J. Bot. **69**: 1483–1490.

Lindsay, K.M., and Cruise, J.E. 1975. Ginseng - native plant now rare. Ont. Nat. **15**(2): 16–19.

Liu, C.X., and Xiao, P.G. 1992. Recent advances on ginseng research in China. J. Ethnopharm.. **36**: 27–38.

Nantel, P., Gagnon, D., and Nault, A. 1996. Population viability analysis of American ginseng and wild leek harvested in stochastic environments. Conserv. Biol. **10**: 608–621.

Oliver, A. 1993. American ginseng culture in the arid climates of British Columbia (abridged version). British Columbia Ministry of Agriculture, Fisheries and Food. 6 pp.

Oliver, A., Van Lierop, B., and Buonassisi, A. 1990. American ginseng culture in the arid climates of British Columbia. British Columbia Ministry of Agriculture and Fisheries. 37 pp.

Persons, W.S. 1986. American ginseng: green gold. Bright Mountain Books, Asheville, NC. 172 pp. [Revised edition 1994, 203 pp.]

Proctor, J.T.A. 1992. The ginseng industry and the Canadian Ginseng Research Foundation. The Grower **42**(10): 1–6.

Proctor, J.T.A., and Bailey, W.G. 1987. Ginseng: industry, botany, and culture. Hort. Rev. **9**: 187–236.

Schlessman, M.A. 1985. Floral biology of American ginseng (*Panax quinquefolium*). Bull. Torrey Bot. Club **112**: 129–133.

Schreiner, J. 1994. Root of success. Can. Geogr. **114**(6): 42–46, 48.

Small, E. 1997. Culinary Herbs. NRC Research Press, Ottawa, ON. 710 pp. [Chapter on *Panax*: pp. 438–456.]

Smith, R.G., Caswell, D., Carriere, A., and Zielke, B. 1996. Variation in the ginsenoside content of American ginseng, *Panax quinquefolius* L. roots. Can. J. Bot. **74**: 1616–1620.

Song, Y., and Xie, C. 1986. A taxonomical study on plants of genus *Panax* in Sichuan (China). J. West China Univ. Med. Sci. **17**: 322–327. [In Chinese.]

Thompson, G.A. 1987. The field cultivation of American ginseng. Stn. Bull. Purdue Univ. Agric. Exp. Sta. **518**: 179–185.

Wen, J., and Zimmer, E.A. 1996. Phylogeny and biogeography of *Panax* L. (the ginseng genus, Araliaceae): inferences from ITS sequences of nuclear ribosomal DNA. Mol. Phylogenet. Evol. 6: 167–177.

Ye, J., Duhui, L., and Shengzhen, G. 1991. Application of the grey system theory in deciding climatic regions suitable to introduce *Panax quinquefolium*. Int. J. Biometeorol. **35**: 55–60.

World Wide Web Links

(Warning. The quality of information on the internet varies from excellent to erroneous and highly misleading. The links below were chosen because they were the most informative sites located at the time of our internet search. Since medicinal plants are the subject, information on medicinal usage is often given. Such information may be flawed, and in any event should not be substituted for professional medical guidance.)

Ginseng: old crop, new directions:
http://www.hort.purdue.edu/newcrop/proceedings1996/v3–565.html

Medical attributes of *Panax* spp. - ginseng:
http://wilkes1.wilkes.edu/~kklemow/Panax.html

Ginseng production in Ontario:
http://www.gov.on.ca/OMAFRA/english/crops/facts/gpak.htm

Ginseng, QB 96–07 ISSN:1052–5378. Ginseng. January 1984-March 1996. Quick Bibliography Series no. QB 96–07 (Updates QB 90–32) 217 Citations in English from the AGRICOLA:
http://www.nal.usda.gov/afsic/AFSIC_pubs/qb9607.htm

Ginseng cost of production of 1 acre of ginseng in Ontario. OMAFRA Ginseng Series:
http://www.gov.on.ca/OMAFRA/english/crops/facts/gincop.htm

Coût de production d'un acre de ginseng en Ontario. Série du MAAARO sur le ginseng:
http://www.gov.on.ca/OMAFRA/french/crops/facts/gincopf.htm

Ginseng books & reference materials:
http://www.ginseng.bc.ca/gi02005.htm

American ginseng. Species at risk in Canada:
http://www.nature.ca/english/ginseng.htm

La rhizoctonie du ginseng:
http://www.gov.on.ca/OMAFRA/french/crops/facts/95–002.htm

Peter Hellyer Ginseng Growers web page:
http://www.simcom.on.ca/~hellyer/index2.html

General information and contacts for ginseng, Ontario:
http://www.gov.on.ca/OMAFRA/english/crops/facts/infcngin.htm

Commercialisation et exportation du ginseng. Série du MAAARO sur le ginseng:
http://www.gov.on.ca/OMAFRA/french/crops/facts/ginmkexf.htm

British Columbia ginseng traders:
http://www.ginsengtraders.com/

Information on ginseng (North Carolina):
http://www2.ncsu.edu/ncsu/cals/hort_sci/comm/ginseng.html

The ginseng page. A comprehensive guide on all aspects ginseng: cultivation, botany, physiology, chemistry, pharmacology, medical uses, economics and business plans:
http://www.imageon.com.au/ginseng/

Questions fréquentes au sujet du ginseng. Dr Richard Reeleder. Contexte: la Production de ginseng en Ontario:
http://res.agr.ca/PUB/lond/pmrc/francais/faq/ginseng.html

Growing ginseng, United States Dept. of Agriculture Farmers Bulletin Number 2201. Reviewed by J. R. Nuss:
http://www.penpages.psu.edu/penpages_reference/29401/2940169.html

Ginseng industry, Alberta Agriculture, Food and Rural Development:
http://www.agric.gov.ab.ca/agdex/100/8883002.html

Growing ginseng, Michigan State University Extension Home Horticulture:
http://www.msue.msu.edu/msue/imp/mod03/03900053.html

Strengthening farming. Farm practices in B.C. A reference guide:
http://www.agf.gov.bc.ca/bcfarmscape/fppa/refguide/commodty/ginseng.htm

General information and contacts for ginseng. OMAFRA Ginseng Series:
http://www.gov.on.ca/OMAFRA/english/crops/facts/infcngin.htm

Market News, Volume 1, Issue 4. Ginseng- proceed with caution:
http://www.agr.ca/pfra/sidcpub/mnv1no4.htm

Market News, Volume 4, Issue 2. Ginseng update:
http://www.agr.ca/pfra/sidcpub/mnv4no2.htm

Ginseng research at PMRC (Delhi):
http://res.agr.ca/lond/pmrc/news/news_595.html#Ginseng Research

Canadian ginseng, development of novel technology and germplasm, Agriculture & Agri-Food Canada, Southern Crop Protection and Food Research Centre [has good links to Canadian ginseng sites]:
http://res.agr.ca/lond/pmrc/study/newcrops/ginseng1.html

Ginseng: a great Canadian product:
http://res.agr.ca/lond/pmrc/study/newcrops/ginseng.html

Ginseng Growers Association of Canada:
http://alpha.nornet.on.ca/ginseng/

Commercial production of ginseng and goldenseal:
http://www.hort.purdue.edu/newcrop/NewCropsNews/94–4–1/ginseng.html

Top herbal products encountered in drug information requests (Part 1) by J.L. Muller and K.A. Clauson [requires registration (free) with Medscape; one of the herbals discussed is ginseng]:
http://www.medscape.com/SCP/DBT/1998/v10.n05/d3287.mulL/d3287.mull-01.html

The search for new pharmaceutical crops: drug discovery and development at the National Cancer Institute [c information on ginseng]:
http://www.hort.purdue.edu/newcrop/proceedings1993/V2–161.html

Ginseng: old crop, new directions, by J.T.A. Proctor:
http://www.hort.purdue.edu/newcrop/proceedings1996/V3-565.html

Podophyllum peltatum (May-apple)

Podophyllum peltatum L. May-apple

English Common Names

May-apple, May apple, mayapple, American mandrake (not to be confused with the European mandrake, *Mandragora officinarum* L., a European plant used throughout history for medicines and potions), mandrake, wild mandrake, Devil's apple, Indian apple, ground-lemon (ground lemon), hog-apple, wild citron, wild lemon, yellow berry, duck's foot, raccoon berry, wild jalap, umbrella plant, vegetable calomel, vegetable mercury.

"Calomel' is mercurous chloride, which was used medicinally in early times; the names "vegetable calomel" and "vegetable mercury" reflect use of May-apple for the same medicinal purposes (*Acorus calamus*, treated in this work, has been referred to as "sweet calomel").

French Common Names

Citron sauvage, citronnier, ipécacuanha de la Caroline, pied de canard, podophylle pelté, podophylle à feuilles peltées, podophylle en bouclier, pomme de mai.

Morphology

An unusual perennial herb up to 60 cm tall, May-apple has one to three umbrella-shaped and deeply lobed leaves. The stems are produced from branched underground rhizomes about 6 mm thick and sometimes as long as a meter. These rhizomes are dark or reddish-brown, subcylindrical, with stem and leaf scars above and root scars below. They are yellowish-white internally, faintly odorous, and bitterly acrid in taste. Generally in non-botanical literature for May-apple, the term "root" is applied collectively to the harvested rhizome and its attached roots. The solitary, waxy-white, nodding flowers are 2 to 4 cm across. Plants with pink flowers and maroon or red fruits (forma *deamii* Raymond) also appear rarely. Flowering occurs in May in the north and the yellow, oval or roundish, mucilaginous, pulpy fruit, 2 to 4 cm across, ripens in July and August, turning from green to yellow. The fruit is disagreeably scented when immature but considered pleasantly fragrant when ripe by most, and with an odd subacid, faintly strawberry flavor. Some people, however, find the ripe fruit nauseous, and as noted below the unripe fruit is poisonous. Rarely, plants may have a cluster of fruits (forma *polycarpum* Clute).

Classification and Geography

May-apple occurs throughout the eastern United States from southern New England to southern Minnesota south to eastern Texas and northern Florida. In Canada it is found throughout much of southern Ontario south of the Canadian Shield and in the eastern townships of Quebec. It is frequent over much of its Ontario range. In Quebec it is rare and occurs mostly along the upper St. Lawrence River. Quebec has designated the species as threatened, and in the province it may not be collected or destroyed, or possessed outside of its natural habitat, under penalty of substantial fines. Most of the few known populations in Quebec may have been introduced by North American Indians. May-apple has persisted at various points where it has been introduced north of the natural range limit shown on the accompanying map, such as in the Gatineau Hills north of Ottawa and in Nova Scotia.

May-apple is one of several eastern North American plants that have very close relatives in eastern Asia. The similarity between the deciduous forests of eastern Asia and eastern North America, along with other evidence, suggests that a temperate deciduous forest, the arcto-tertiary flora, formed a continuous band around the northern hemisphere 15 to 20 million years ago. This band was later fragmented by climatic cooling, uplift of mountains with isolation and rainshadow effects, and continental glaciations, the two major remaining fragments surviving in eastern Asia and eastern North America. At least four Asian species have been recognized. The most important medicinal Asian relative is the Himalayan May-apple, *Podophyllum hexandrum* Royle (= *P. emodi* Wall.), a plant with shiny mottled leaves, erect, conspicuous and often pink flowers, red fruits, and a clumped instead of spreading habit. Another Asian species, *P. pleianthum* Hance, is also used medicinally.

Ecology

May-apple occurs in moist, shady, deciduous, rich woodlands, forest edges, thickets and marshy meadows and ditches.

Moths and bumblebees are the likely pollinators. Canadian researchers have reported that the flowers do not produce nectar and that fruit and seed set is greater where May-apple plants grow close to flowering wood betony (*Pedicularis*

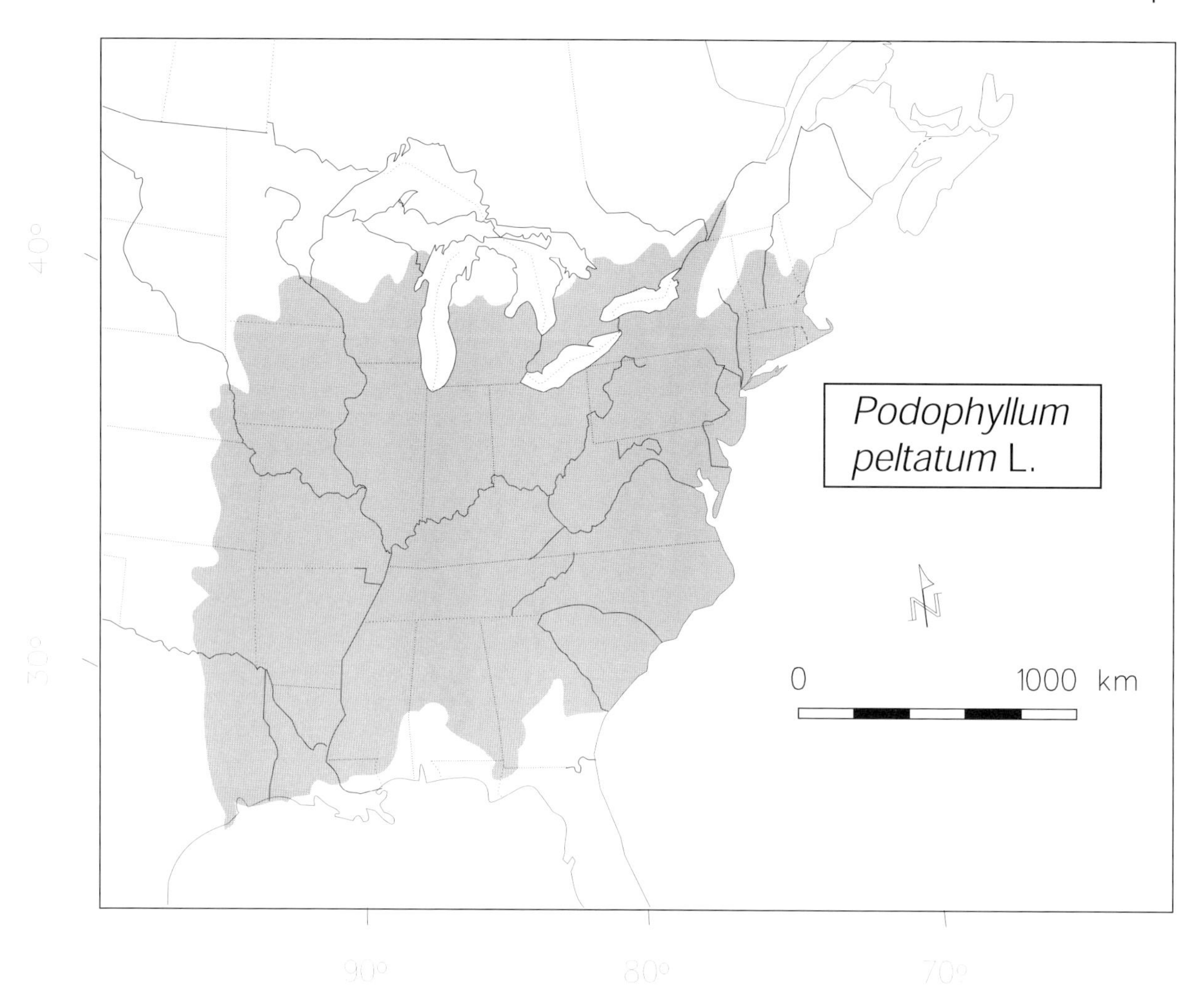

canadensis L.) which produces nectar prolifically. The wood betony is a so-called "magnet species," attracting pollinators and thus facilitating the pollination of flowers of other species growing near it. May-apple fruit is eaten and dispersed by mammals, birds, and Eastern box turtles.

Medicinal Uses

North American Indians used May-apple as a purifying medicine to expel parasitic worms, and in the treatment of certain cancers, such as tumorous skin growths. A brew of powdered May-apple was employed as a laxative (the purgative action is very strong). The powder was also applied as a poultice to get rid of warts. Settlers came to use May-apple for a variety of illnesses, including typhoid fever, cholera, dysentery, hepatitis, rheumatism, kidney, bladder and prostate problems, and venereal diseases. In 1885 Canadian pioneer botanist Catharine Parr Traill reported that most Canadian physicians were using May-apple root in the treatment of complaints of the liver. This use persisted many years, for example in the well known Carter's liver pills.

The root of May-apple is the source of the powdered mixture of resins referred to pharmaceutically as *Resina Podophylli* — resin of podophyllum or simply podophyllum. Highest activity occurs in the rhizomes just after the foliage is shed, and again in the early spring. Since its discovery by Europeans, this poisonous plant has had many medicinal uses, especially internally as a digestive medicine, but also externally in treatment of sores and skin problems. Of particular interest is the fact that both the North American and the Asian May-apple contain anti-cancer agents. At present, extracts of May-apple are used in topical medications for genital warts and some skin cancers.

Toxicity

The roots as well as the unripe fruit, seeds and leaves can be fatally poisonous. Recorded symptoms include vomiting, diarrhea, dizziness, headache, bloating, lowered blood pressure, hallucinations, loss of reflexes, confusion, and stupor. Grazing animals avoid the plant because of its bitterness, but pigs have died after eating May-apple shoots. May-apple should not be used during pregnancy, as it is teratogenic and can cause abortion (even topical application has caused minor fetal anomalies). May-apple extract is sometimes applied to cervical warts, and can cause the development of atypical epithelial cells; a false positive PAP smear has been recorded 6 months after such topical application. Extreme caution is recommended in the handling of May-apple. Fragments of root flicked into the eye during grinding can cause extreme swelling, internal bleeding, severe pain and temporary loss of sight. Even handling the rhizomes can cause dermatitis, and there are recorded cases of death following topical application. May-apple is too toxic to attempt self-medication. May-apple is sold through mail-order nursery catalogues, and the potential danger of handling the plant is not always made clear.

Podophyllum peltatum (May-apple)

Chemistry

Podophyllotoxins are called aliphatic alkaloids, which are functionally and structurally similar to alkaloids, with multiple rings, but lack nitrogen. These active principles are in the unripe fruit, the leaves and stem, and especially in the rhizome and roots. The rhizome and roots contain 3.5 to 6% of a resinous mixture, podophyllum resin or podophyllin. Podophyllin, like the drug colchicine, arrests cell division through an effect on RNA and DNA synthesis during interphase, a property useful in treating cancer. Podophyllin also is responsible for the laxative effects of May-apple. The resin contains lignan glycosides, including about 20% podophyllotoxin, 10% alpha-peltatin and 5% beta-peltatin. These have anticancer and purgative properties. Podophyllotin blocks the release of iodine from the thyroid gland and the release of catecholamines from the adrenal medulla. Synthetic and semisynthetic variations of the podophyllotoxin, known as epipodophyllotoxins, have been employed as anticancer agents; the most active are the drugs Teniposide and Etopiside (also known as Vepeside) from the cultivated Asian species, which are widely used to treat cancers.

Non-medicinal Uses

North American Indians employed May-apple as a poison, and cases of its use for suicide have been recorded. The whole plant was boiled by the Menomini and the liquid then used as an insecticide on recently introduced potato plants. Since it was utilized by Indians, probably for thousands of years, it was possibly cultivated and undoubtedly dispersed by them. A number of sites in Canada near the northern range limit are associated with former Indian habitation. One of the earliest observations of the plant at an Indian site on the northern range limit was that of Champlain, who in 1619 found the Hurons of Cahiagué (northern Simcoe County, Ontario, then one of the largest settlements in Canada with 5000 inhabitants) eating the berries. Champlain noted that they were "plentiful and extremely good to eat." The early settlers made jam and preserves from the ripe fruits, and they cultivated the plant to a limited degree. Although Harvard botanist Asa Gray described the fruit as "eaten by pigs and boys," most current books on edible wild plants report an agreeable taste. It must be remembered that the unripe fruits are not only bitter, but poisonous. The fully ripe fruits are claimed to make excellent jellies and marmalades. Drinks made by admixture with lemonade or Madeira wine are locally popular. The eminent student of medicinal plants James A. Duke coined the following ditty:

> "Toxicologists may deem it an error,
> To imbibe of May-apple Madeira.
> But if the list was complete
> Of the toxins we eat,
> Perhaps we would all die of terror."

Agricultural and Commercial Aspects

In 1990 the market for drugs from May-apple was worth over 100 million dollars. Both the Eurasian and the North American May-apples are an important source of anti-cancer drugs, on a par with Madagascar periwinkle and species of yew. May-apple is currently an ingredient of prescription drugs sold in the United States and is used in at least seven Canadian drug products. The supply is obtained exclusively from material collected in the wild, mostly in Indiana, Kentucky, North Carolina, Tennessee, and Virginia. The roots are collected in the autumn and dried. Several hundred tonnes are collected annually for both domestic and international markets. Since the demand is increasing, harvesting from the wild is laborious, and drug content of wild plants is variable, it has been suggested that high-yielding clones be identified and cultivated. Although seed germination presents problems, the plants are readily grown from root divisions.

Myths, Legends, Tales, Folklore, and Interesting Facts

- May-apple sometimes forms circles of plants (fairy rings) like some mushrooms and ferns.
- E.M. Frieders (cited in World Wide Web Links, below) wrote: "The small edible fruits are lemon-shaped and are yellowish in color (so why may-apple and not maylemon, I ask?)."

Selected References

Bedows, E., and Hatfield, G.M. 1982. An investigation of the antiviral activity of *Podophyllum peltatum*. J. Nat. Prod. **45**: 725–729.

Bennet, R.G., and Grist, W.J. 1985. Nasal papillomas: sucessful treatment with podophyllin. South. Med. J. **78**: 224–225.

Bhadula, S.K., Singh, A., Lata, H., Kuniyal, C.P., and Purohit, A.N. 1996. Genetic resources of *Podophyllum hexandrum* Royle, an endangered medicinal species from Garhwal, Himalaya, India. Plant Genet. Resourc. Newsl. **106**: 26–29.

Chandler, R.F. 1990. Podophyllum. Can. Pharm. J. **123**: 330-331, 333.

Chatterjee, R. 1952. Indian *Podophyllum*. Econ. Bot. **6**: 342–354.

Couillard, L., and Forest, G. 1998. Espèces menacées au Québec - Le podophylle pelté. Gouvernement du Québec, Ministère de l'Environnement et de la Faune, Direction de la conservation et du patrimoine écologique, Québec. 4 pp.

Demaggio, A.E., and Wilson, C.L. 1986. Floral structure and organogenesis in *Podophyllum peltatum* (Berberidaceae). Am. J. Bot. **73**: 21–32.

Duke, J.A. 1983. The marvelous mayapple. Bot. Grower (Newsl.) **1**(1): 3–4.

Ernst, W.R. 1964. The genera of Berberidaceae, Lardizabalaceae, and Menispermaceae in the southeastern United States. J. Arnold Arbor. **45**: 1–35.

Fisher, AA. 1981. Severe systemic and local reactions to topical podophyllum resin. Cutis **28**: 233.

Geber, M.A., De-Kroon, H., and Watson, M.A. 1997. Organ preformation in mayapple as a mechanism for historical effects on demography. J. Ecol. **85**: 211–223.

George, L.O. 1997. *Podophyllum*. *In* Flora of North America north of Mexico, Vol. 3. *Edited by* Flora of North America Editorial Committee. Oxford University Press, New York, NY. pp. 287-288.

Graham, N.A., and Chandler, R.F. 1990. Herbal medicine - *Podophyllum*. Can. Pharm. J. **123**: 330–333.

Jackson, D.E., and Dewick, P.M. 1984. Aryltetralin lignans from *Podophyllum hexandrum* and *Podophyllum peltatum*. J. Phytochem. **23**: 1147–1152.

Kelly, M.G., and Hartwell, J.L. 1954. The biological effects and the chemical composition of podophylloton. A review. J. Nat. Cancer Inst. **14**: 967–1010.

Krochmal, A., Wilkins, L.,Van Lear, D., and Chien, M. 1974. Mayapple, *Podophyllum peltatum* L. USDA Forest Service Research Paper NE-296. U.S. Department of Agriculture, Upper Darby, PA. 9 pp.

Kroon, H., Whigham, D.F., and Watson, M.A. 1991. Developmental ecology of mayapple: effects of rhizome severing, fertilization and timing of shoot senescence. Ecology **5**: 360–368.

Kutney, J.P., Arimoto, M., Hewitt, G.M., Jarvis, T.C., and Sakata, K.1991. Studies with plant cell cultures of *Podophyllum peltatum* L. I. Production of podophyllotoxin, deoxypodophyllotoxin, podophyllotoxin, and 4'-demethylpodophyllotoxin. Heterocycles (Tokyo) **32**: 2305–2309.

Laverty, T.L. 1992. Plant interactions for pollinator visits: a test of the magnet species effect. Oecologia **89**: 502–508.

Laverty, T.M., and Plowright, R.C. 1986. Fruit and seed set in mayapple (*Podophyllum peltatum*): influence of intraspecific factors and local enhancement near *Pedicularis canadensis*. Can. J. Bot. **66**: 173–178.

McFarland, M.F., III, and McFarland, J. 1981. Accidental ingestion of *Podophyllum*. Clin. Toxicol. **18**: 973–977.

Meijer, W. 1974. *Podophyllum peltatum* — May Apple a potential new cash-crop plant of Eastern North America. Econ. Bot. **28**: 68–72.

Miller, R. 1985. Podophyllin. Int. J. Dermatol. **24**: 491–498.

Montgomery, F.H. 1965. Poisonous fruits. Federation of Ontario Naturalists, Edward Gardens, Don Mills, ON. 24 pp.

Rosenstein, G., Rosenstein, H., Freeman, M., and Weston, N. 1976. Podophyllum — a dangerous laxative. Pediatrics **57**: 419–421.

Rust, R.W., and Roth, R.R. 1979. Seed production and seedling establishment in the mayapple, *Podophyllum peltatum* L. Am. Midl. Nat. **105**: 51–60.

Sadowska, A., Wiweger, M., Lata, B., and Obidoska, G. 1997. In vitro propagation of *Podophyllum peltatum* L. by the cultures of embrya and divided embrya. Biol. Plant. **39**: 331–336.

Sohn, J.J., and Policansky, D. 1977. The cost of reproduction in the mayapple *Podophyllum peltatum* (Berberidaceae). Ecology **58**: 1366–1374.

Taylor, C., and Taylor, J. 1964. *Podophyllum peltatum* f. Deamii from Bryan County, Oklahoma. Rhodora **66**: 167.

Von Krogh, G. 1981. Podophyllotoxin for condyloma acuminatum eradication. Acta Derm. Venereol. **98** (Suppl.):1-48

Whisler, S.L., and Snow, A.A. 1992. Potential for the loss of self-incompatibility in pollen-limited populations of mayapple (*Podophyllum peltatum*). Am. J. Bot. **79**: 1273–1278.

World Wide Web Links

(Warning. The quality of information on the internet varies from excellent to erroneous and highly misleading. The links below were chosen because they were the most informative sites located at the time of our internet search. Since medicinal plants are the subject, information on medicinal usage is often given. Such information may be flawed, and in any event should not be substituted for professional medical guidance.)

May apple [mostly color photographs]:
http://cribbage.cit.cornell.edu/flowers/pages/mayapple.html

Poisonous plants of North Carolina - *Podophyllum peltatum*:
http://russell4.hort.ncsu.edu/poison/Podoppe.htm

Hathaway's virtual trail, Mayapple or mandrake:
http://eclipse.dtl.pcs.k12.va.us/vtraiL/a24.htm

Description and natural history of the Mayapple:
http://biotech.chem.indiana.edu/botany/mayhist.html

Wildflower notes, Mayapple (*Podophyllum peltatum*):
http://www.duc.auburn.edu/~deancar/wfnotes/mayap.htm

Minnesota's native Mayapple: the plant alkaloid answer to cancer. Elizabeth M. Frieders, Spring 1996, Minnesota Plant Press 15(3):
http://www.stolaf.edu/depts/biology/mnps/papers/frieders.html

Polygala senega (Seneca snakeroot)

Polygala senega L. Seneca Snakeroot

The genus name *Polygala* means "much milk," and has also been interpreted as a reference by the ancient Greek physician to the Roman Army, Dioscorides, to some low shrub reputed to promote milk secretion. "Senega" is derived from the Seneca tribe of North American Indians, who used the plant as a remedy for snake-bites.

English Common Names

Seneca snakeroot, Seneca-snakeroot, Senega root, Senega snakeroot, rattlesnake root, mountain flax, white snakeroot (a name more generally applied to *Eupatorium rugosum*).

Several unrelated species have the common name "snakeroot," and are sometimes confused with seneca snakeroot.

French Common Names

Polygala sénéca, polygala de Virginie.

Morphology

Seneca snakeroot is a perennial herb. Several erect stems, 10–50 cm tall, arise from the branched root, and bear alternate, lance-shaped leaves. The lowest leaves are small and scalelike.

Spikes of roundish, small, white (frequently tinged with green) flowers appear in June and early July. The short capsules ripen in July and August, and have two blackish, hairy seeds with long white appendages (arillodes). The dandelion-like root is slender, conical, with an enlarged crown, tortuous, yellowish-grey or brownish on the outside and cream-colored internally, with a sweetish odor reminiscent of wintergreen, and a strongly acrid taste. In commerce, two root types are often recognized. The more desirable Northern or Manitoba Seneca snakeroot has larger roots (up to 15 cm long and 12 mm thick), which are dark brown and purplish near the crown. It is collected in Manitoba, Saskatchewan and Minnesota. Southern or small Seneca snakeroot, which comes from South Carolina and Georgia, has roots up to 8 cm long and 7 mm thick, which are brownish-yellow.

Classification and Geography

Polygala senega has a broad range in Canada, extending from the St. John River system of New Brunswick, southwestern Quebec, Ontario, Manitoba and Saskatchewan to southern Alberta. It also occurs throughout much of the eastern and midwestern United States south to Georgia.

Ecology

Habitats include prairies, savannas and periodically dry, often somewhat open woods. Light availability varies from open sky to partial shade. Soils occupied are often rocky, and the plants have been recorded in quite basic soils (the pH as high as 9).

The breeding system of Seneca snakeroot requires study, but related species are pollinated by bees. The seed appendages may be utilized as food by ants which disperse the seed over short distances. Dispersal over longer distances is probably by birds.

Medicinal Uses

Seneca snakeroot was utilized by the Seneca Indians in treatment of rattlesnake bite. Canadian botanist Frère Marie-Victorin suggested that the resemblance of the knotty root crown to a rattlesnake's tail may have contributed to its use by the Seneca as an antidote. Its use, however, was not confined to the Seneca, for it was considered to have several medicinal properties, especially in relation to respiratory ailments, among other tribes including Ojibwa, Huron, Chippewa, Menomini and Iroquois. Walpole Island Ojibwa of southern Ontario used the root for headache, nasal congestion and stomach ache. It was utilized as a general cure by the Chippewa Indians and carried by them on their journeys for general health and safety.

Seneca snakeroot was sent to Europe in the early 1700s and held a regular place in European drug stores during the 1800s for use in treatment of pneumonia.

The root is ground into powder and used in various patent medicines, particularly in cough medicines, as a stimulant expectorant. It is present in some prescription drugs used in the treatment of bronchitis and asthma. Seneca snakeroot is also used in veterinary medicine. Aside from its effects on the respiratory system, it promotes perspiration and urination.

Toxicity

The root is a severe and serious irritant when too much is consumed. It can cause nausea, dizziness, diarrhea and violent vomiting.

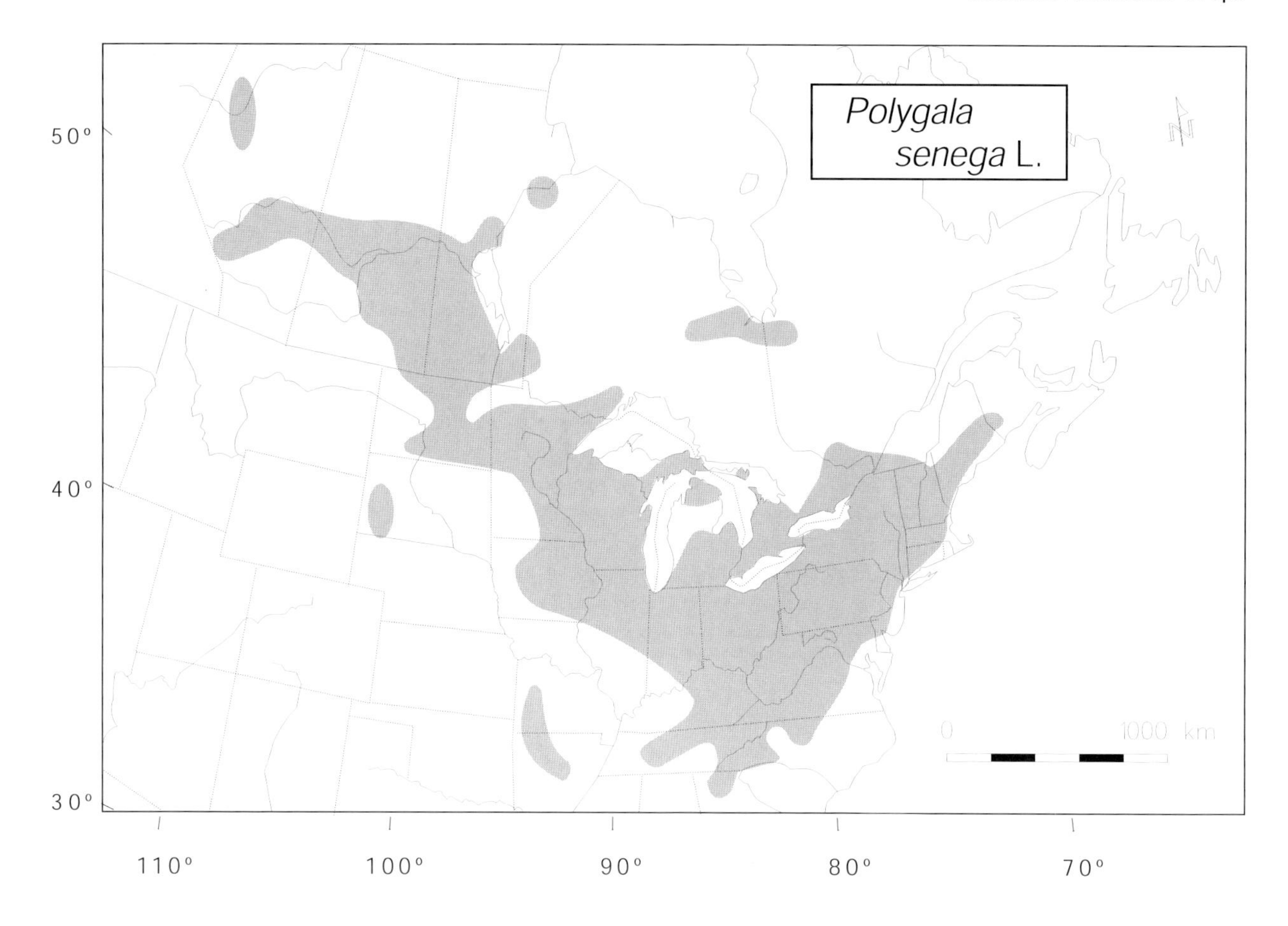

Chemistry

The active root (pharmaceutically referred to as *Senegae Radix*) constituents are triterpinoid saponins (notably senegin). Also recorded are phenolic acids, polygalitol (a sorbitol derivative), methyl salicylate, and sterols.

Non-medicinal Uses

Seneca snakeroot is grown to a minor extent as a garden ornamental.

Agricultural and Commercial Aspects

Seneca snakeroot was once grown to a limited extent in parts of Europe, and is still cultivated in Japan, India and Brazil. Up until the early 1960s Canada was the chief supplier of Seneca snakeroot with exports worth several hundred thousand dollars a year. Most of the exported material originated from plants collected from the wild in Saskatchewan and Manitoba. There has been a relative decline in its value as a Canadian export. Seneca snakeroot is used in about a dozen drug products in Canada.

Other species of *Polygala* are sometimes used to adulterate Seneca snakeroot. *Polygala tenuifolia* Willd., Japanese seneca, from India and Japan, is the most frequently encountered adulterant.

Early experimental work in North America suggested that Seneca snakeroot can be grown in soil suitable for ordinary field crops, and without shade. The plants are set 40 cm apart in rows of at least the same width. The seeds have a reputation as difficult to germinate. Seedlings may be protected with straw during the first winter. About 4 years are required to produce roots of marketable size.

Although little, if at all cultivated in Canada, Seneca snakeroot is still gathered from the wild, especially by indigenous people of Manitoba and Saskatchewan, for sale to commercial pharmaceutical companies. Three quarters of the world's wild supply originates from Manitoba's Interlake District — the land about Lakes Winnipeg, Winnipegosis, and Manitoba. Wild plants are sufficiently plentiful that the wild resource can continue to be harvested, but since collecting mature roots sacrifices the plant, cultivation is a potentially desirable alternative. Seneca snakeroot is under consideration in several provinces as a new or alternative crop, especially to enhance its economic value to indigenous people. The most frequent admixture involves the North American ginseng (*Panax quinquefolius*).

The greatest current interest in seneca snakeroot is in Japan, where it is cultivated on a modest scale. Japanese scientists have recently reported on various experiments with cultivation, and they have also conducted the most recent and comprehensive work on chemical composition of the saponins. In 1993, 72 farmers produced about 10 tonnes of commercial root in Japan, suggesting that there is appreciable potential for the cultivation of seneca snakeroot in Canada.

Myths, Legends, Tales, Folklore, and Interesting Facts

- In respect to ancient religious custom that required some fruits of the forest to be left for the gods, Manitoba Indians experienced in collecting Seneca root leave a portion of the root intact to permit regrowth.
- The floral essence of Seneca snakeroot is thought to stimulate old memories; and also to reduce quarrelling and showing off.
- The Cree and the Chippewa Indians believed that Seneca snakeroot had the power to protect one on long journeys.
- Methyl salicylate in the root of *Polygala senega* appears to contribute to its strange characteristic of smelling and tasting sweet at first and then sour and acid.

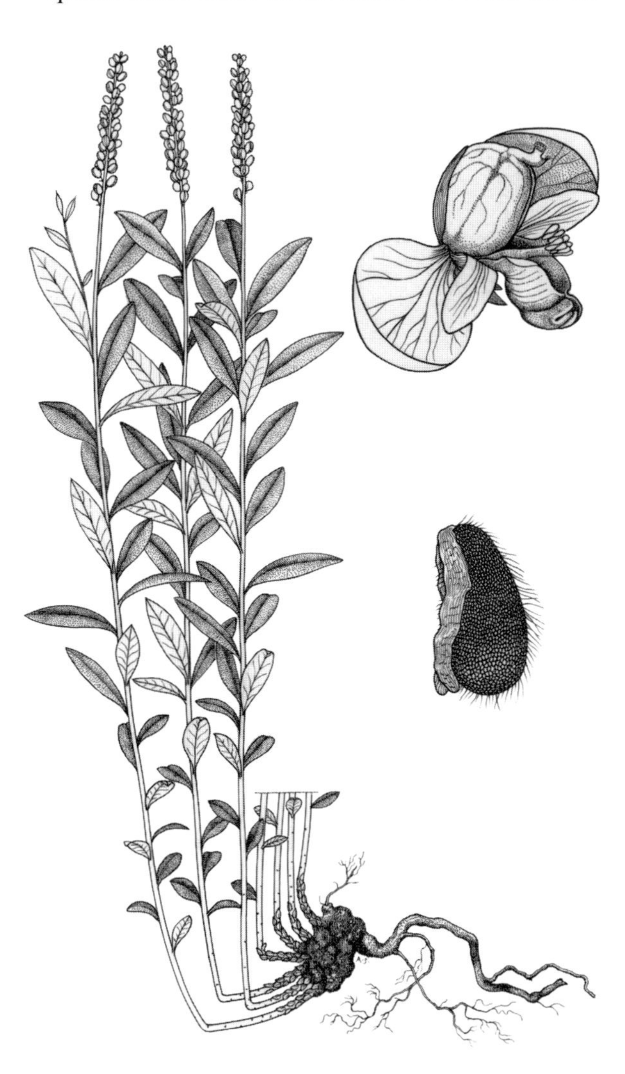

Polygala senega L.

Selected References

Briggs, C.J. 1988. Senega snakeroot — a traditional Canadian herbal medicine. Can. Pharm. J. **121**: 199–201.

Gillett, J.M. 1968. The milkworts of Canada. Canada Dept. of Agriculture, Research Branch Monograph 5. 24 pp.

Harris, G.H. 1891. Root foods of Seneca Indians. Proc. Rochester Acad. Sci. **1**: 106–115.

Hayashi, S., and Kameoka, H. 1995. Volatile compounds of *Polygala senega* L. var. *latifolia* Torrey et Gray roots. Flavour Fragrance J. **10**: 273–280.

Johnson, K.L. 1984. Whorled milkwort (*Polygala verticillata*) and its common relatives Seneca snakeroot and gaywings. Bull. Manit. Nat. Soc. **8**(1): 10.

Kawatani, T., and Ono, T. 1968. Effect of light intensity on the growth and root yield of *Polygala senega* L. var. *latifolia* Torr. et Gray. Eisei Shikenjo Hokoku **86**: 105–107. [In Japanese.]

Miller, N.G. 1971. The Polygalaceae in the southeastern United States. J. Arnold Arbor. **52**: 267–284.

Moes, A. 1966. A parallel study of the chemical composition of *Polygala senega* and of "*Securidaca longepedunculata*" Fres. var. *parvifolia,* a Congolese polygalacea. J. Pharm. Belg. **21**: 347–362. [In French.]

Pelletier, S.W., and Nakamura, S. 1967. A prosapogenin from *Polygala senega* and *Polygala tenuifolia*. Tetrahedron Lett. **52**: 5303–5306.

Saitoh, H., Miyase, T., and Ueno, A. 1993. Senegoses A-E, oligosaccharide multi-esters from *Polygala senega* var. *latifolia* Torr. et Gray. Chem. Pharm. Bull. (Tokyo) **41**: 1127–1131.

Saitoh, H., Miyase, T., and Ueno, A. 1993. Senegoses F-I, oligosaccharide multi-esters from the roots of *Polygala senega* var. *latifolia* Torr. et Gray. Chem. Pharm. Bull. (Tokyo) **41**: 2125–2128.

Saitoh, H., Miyase, T., Ueno, A., Atarashi, K., and Saiki, Y. 1994. Senegoses J-O, oligosaccharide multi-esters from the roots of *Polygala senega* L. Chem. Pharm. Bull. (Tokyo) **42**: 641–645.

Shoji, J., and Tsukitani, Y. 1972. On the structure of senegin - 3 of *Senegae radix*. Chem. Pharm. Bull. (Tokyo) **20**: 424–426.

Takeda, O., Azuma, S., Ikeda, M., Mizukami, H., Ikenaga, T., and Ohashi, H. 1986. Cultivation of *Polygala senega* var. *latifolia*: I. Effects of seeding density and heavy manuring and dense planting cultivation on plant growth, root yield and senegin content. Shoyakugaku Zasshi **40**: 103–107. [In Japanese.]

Takeda, O., Azuma, S., Ikeda, M., Mizukami, H., Ikenaga, T., and Ohashi, H. 1987. Studies on the cultivation of *Polygala senega* var. *latifolia*: III. Effect of cultivation temperature on the growth, root yield and senegin content. Shoyakugaku Zasshi **41**: 121–124. [In Japanese.]

Takeda, O., Azuma, S., Mizukami, H., Ikenaga, T., and Ohashi, H. 1986. Cultivation of *Polygala senega* var. *latifolia*: II. Effect of soil moisture content on the growth and senegin content. Shoyakugaku Zasshi **40**: 434–437. [In Japanese.]

Takiura, K., Yamamoto, M., Murata, H., Takai, H., and Honda, S. 1974. Studies on oligosaccharides. XIII. Oligosaccharides in *Polygala senega* and structures of glycosyl-1,5-anhydro-D-glucitols]. Yakugaku Zasshi **94**: 998–1003. [In Japanese.]

Yoshikawa, M., Murakami,T., Ueno,T., Kadoya, M., Matsuda, H., Yamahara, J., and Murakami, N. 1995. E-senegasaponins A and B, Z-senegasaponins A and B, Z-senegins II and III, new type inhibitors of ethanol absorption in rats from Senegae Radix, the roots of *Polygala senega* L. var. *latifolia* Torrey et Gray. Chem. Pharm. Bull. (Tokyo) **43**: 350–352.

World Wide Web Links

(Warning. The quality of information on the internet varies from excellent to erroneous and highly misleading. The links below were chosen because they were the most informative sites located at the time of our internet search. Since medicinal plants are the subject, information on medicinal usage is often given. Such information may be flawed, and in any event should not be substituted for professional medical guidance.)

Healthlink online resources, seneca snakeroot:
http://www.healthlink.com.au/nat_lib/htm-data/htm-herb/bhp736.htm

A modern herbal by M. Grieve:
http://www.botanical.com/botanicaL/mgmh/s/senega41.html

Genus of the week, *Polygala*:
http://fisher.bio.umb.edu/pages/JFGenus/Jfgen10.htm

Herbs for the Prairies, senega:
http://paridss.usask.ca/specialcrop/tour_senega.html

Constituents of *Polygala senega*:
http://chili.rt66.com/hbmoore/Constituents/Polygala_senega.txt

Snake root- HealthWorld Online:
http://www.healthy.net/library/books/hoffman/materiamedica/snake.htm

Rhamnus purshianus (cascara)

Rhamnus purshianus DC. — Cascara

Rhamnus has been considered both masculine and feminine, so that one encounters both "*R. purshianus*" and "*R. purshiana*." Early botanists favored the masculine version, and the *Code of Botanical Nomenclature* recommends that tradition be followed.

English Common Names

Cascara, cascara sagrada, bayberry, bearberry, bearwood, bitterbark, cascara buckthorn, Californian buckthorn, holybark, Persian bark.

The name "bearberry" is sometimes employed for cascara in British Columbia, but is best reserved for *Arctostaphylos uva-ursi*, treated in this work. "Bayberry" generally refers to *Myrica pensylvanica* Loisel., a shrub of eastern North America. The common name "Californian buckthorn" should not be confused with the scientific name *Rhamnus californica* Esch. (California coffeeberry). The drug (i.e., bark preparation) is known as cascara sagrada (French = écorce sacrée, cascara sagrada), from the Spanish for "sacred bark," or rarely chittem bark. Below, following majority practice, the plant is referred to as cascara while the laxative bark preparation is called cascara sagrada.

French Common Names

Écorce sacrée, cascara, nerprun cascara, nerprun de Pursh.

Morphology

Cascara is a deciduous shrub or tree as tall as 12 m (very rarely to 15 m) and with a trunk diameter sometimes approaching a meter, although usually no more than 50 cm. The ashy-gray to dark brown, often red-tinged bark, which constitutes the economic part of the plant, is thin and smooth, and develops brown to gray scales. The freshly cut interior surface is bright yellow, but darkens rapidly. The leaves are 5–18 cm long, oblong, with 10–12 pairs of prominent parallel veins. Most plants have both bisexual and unisexual flowers, although occasionally a plant bears flowers of only one sex. The berry-like drupes are red when immature, ripening to black or purplish black, sweet, juicy fruits 6–12 mm in diameter. As the plants are frequently cut down for their bark, it is fortunate that stump sprouts arise readily.

Classification and Geography

Rhamnus purshianus occurs in the Pacific Coast region from British Columbia (including Vancouver Island) southward in the coastal ranges and in the Sierra Nevada to southern California; it is also found in the Rocky Mountain region, east to northern Idaho, and northwestern Montana. Overcollecting has eliminated the plant in parts of its range.

Ecology

Rhamnus purshianus is often found in moist areas, by streamsides and in woods, often in lowlands and canyons, and in submontane areas to about 1500 m. It is also distributed along fence rows and roadsides. While cascara is widespread, it is usually not abundant, generally occurring amidst local forest and woodland species. It is very tolerant of shade, and often is an understory species. The fruits are eaten by birds and mammals, which disperse the seeds.

Medicinal Uses

Cascara is mainly useful for its laxative bark, and in this respect is one of the most valuable commercial native pharmacological crops of North America. Pacific Northwest Indians used cascara sagrada for centuries as a traditional remedy for constipation, and white settlers took up this use in the early 1800s.

Cascara sagrada is an example of the maxim that good medicines should taste bad. If taken as a tonic or tea (rather than in capsule form) it is bitter and tends to provoke nausea. It is very useful for habitual constipation, but is also employed for digestive complaints, and in treating hemorrhoids. It is most recommended for disorders in which an easy evacuation of the bowel is desired. Cascara sagrada is not habit-forming, and indeed should only be used on a short-term basis. It is classified as a tonic laxative, strengthening the peristaltic muscles of the intestinal wall so that additional use of a laxative becomes unnecessary.

Toxicity

Cascara sagrada should not be used during pregnancy (cathartics may induce labor) or lactation (the laxative may be transferred to the infant), or in cases of intestinal obstruction. Frequent use may result in loss of water and salts, deposition of pigment in the intestinal mucosa, and red urine. Apparently handling cascara for a prolonged interval can transfer the laxative effects through the skin, although this does not seem to be a significant problem for individuals harvesting the bark.

Chemistry

The cathartic activity of cascara sagrada is due to a mixture of hydroxyanthracene derivatives (particularly anthraquinone glycosides). These excite peristalsis in the colon, and so are useful in treating chronic constipation. The mechanism has been explained as an inhibition of the absorption of electrolytes and water from the large intestine, with a consequent increase in volume of the bowel contents strengthening the dilation pressure and thereby stimulating peristalsis.

Non-medicinal Uses

There are several minor uses for cascara. The wood is locally used for posts, fuel, and for turnery. Cascara honey is considered very tasty, although somewhat laxative. The fruits may be eaten by humans, although a temporary reddish cast to the skin is said to occasionally result. Extracts from cascara have been employed to flavor liqueurs, soft drinks, ice cream and baked goods. Cascara is sometimes grown as an ornamental, principally in the eastern US and Europe. The tree is also planted to provide food and habitat for wildlife, and to control soil erosion.

Agricultural and Commercial Aspects

Today, cascara sagrada accounts for about 20% of the US laxative market, the latter estimated to be worth about US$400 million annually. The overall retail value of cascara bark is of the order of $100 million. Cascara sagrada has been recorded in about 200 drug products sold in Canada, and occurs in more North American drug preparations than any other wild-collected material. It has been called the most widely used cathartic on earth. However, the demand for it appears to have diminished somewhat since the 1960s because of the development of alternative drugs.

Cascara bark is still collected from wild trees throughout the range of the plant, particularly in Washington. It has been cultivated for harvest of the bark in British Columbia, Washington and Oregon, and occasionally in Eurasia, but plantations have not achieved much economic success, and the plant remains largely gathered from the wild. Because it is a secondary host of Oat Crown Rust

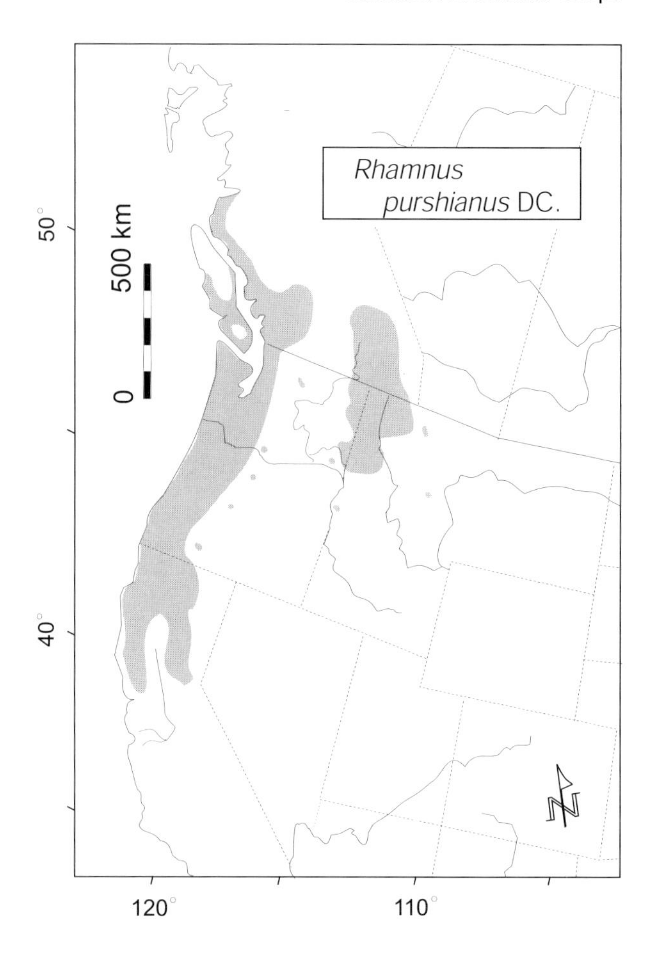

(although not as significant as such species as *R. frangula* L.), local regulations may limit its cultivation. Formerly, harvest involved stripping the bark from the trunk of standing trees, a wasteful and unacceptable procedure that kills the trees and harvests only some of the bark. By contrast, felling the tree and leaving a 30 cm stump cut at an angle (to shed water) allows complete harvesting of bark, including that of the branches, and lets the plant regenerate. The average yield of bark per tree is 4.5 kg, but ranges from about 2 kg for a 7.5 cm diameter tree to 71 kg for a 43 cm diameter tree. The harvested bark is sun-dried, broken into pieces, packed and shipped to dealers. Because fresh bark tends to cause griping and nausea, the bark is aged for a year, or artificially aged by heat (e.g., 1 hour at 100°C). The drug preparation should be stored away from moisture and light.

Overharvesting of wild trees is a continuing problem. In British Columbia, cutting trees on Crown land requires a permit, and provincial legislation also governs how such cutting may be carried out. In the first half of this century, wild trees sometimes supplied annual harvests of over 300 tonnes in B.C. alone, with harvests from throughout the natural range estimated in some years at about 2,000 tonnes. Reliable estimates for recent times are not available. Cultivation may become more economically feasible as natural populations continue to decline. The tree is so readily established that the cascara industry in Canada could easily be strengthened by either cultivation and (or) increasing its frequency in natural habitats to that of former times.

Myths, Legends, Tales, Folklore, and Interesting Facts

- The conquistadores were so impressed by the effectiveness and mildness of *Rhamnus purshianus* that they christened it cascara sagrada, Spanish for holy or sacred bark.
- The name *Rhamnus purshianus* commemorates the botanist Frederick Pursh (1774–1820), born in Germany. The victim of misfortune and ill health, his plant collection was destroyed in a fire, and he died in poverty in Montreal while preparing a book on the plants of Canada.
- The laxative utility of *Rhamnus purshianus* should not be confused with that of the comparatively dangerous *R. catharticus* L. (buckthorn). Buckthorn was once widely employed in Europe, and imported by North American colonists, who reportedly needed an explosive laxative that would "act like dynamite." Settlers commonly planted buckthorn conveniently close to their dwellings, and today it is a commonly naturalized weedy shrub in southeastern Canada and northeastern US.
- The following recommendation for cascara sagrada was encountered on the world wide web: "Sprinkle an infusion around your home before going to court; it will help you win your case."
- Due to its bitter taste, cascara sagrada can be put on finger nails to discourage nail biting.
- Cascara sagrada has been used in amulets for protection against evil and hexes.

Rhamnus purshianus (cascara)

Selected References

Anon. 1968. The chemical assay of cascara dry extract, cascara tablets and cascara bark. Analyst **93**: 749–755.

Anon. 1973. Recommended methods for the evaluation of drugs. The chemical assay of cascara bark and cascara dry extract. Analyst **98**: 830–837.

Bonmassar, E. 1970. Use of the scanning electron microscope in pharmacognosic research. Studies of the bark of some species of *Rhamnus*. Atti. Accad. Med. Lomb. **25**: 175–182. [In Italian.]

Davidson, J. 1949. The cascara tree in British Columbia. British Columbia Ministry of Agriculture.

Dunn, L. 1942. Cascara. Oregon State College, School of Forestry, Corvallis, OR. 10 pp.

Fairbairn, J.W., and Simic, S. 1970. A new dry extract of cascara (*Rhamnus purshiana* DC.) bark. J. Pharm. Pharmacol. **22**: 778–780.

Giavina-Bianchi, P.F., Jr., Castro, F.F., Machado, M.L., and Duarte, A. 1997. Occupational respiratory allergic disease induced by *Passiflora alata* and *Rhamnus purshiana*. Ann. Allergy Asthma Immunol. **79**: 449–454.

Gyanchandani, N.D.,Yamamoto, M., and Nigam, I.C. 1969. Anthraquinone drugs. Thin-layer chromatographic identification of aloes, cascara, senna, and certain synthetic laxatives in pharmaceutical dosage forms. J. Pharm. Sci. **58**: 197–200.

Johnston, M.C., and Johnston, L.V.A. 1978. *Rhamnus*. New York Botanical Garden, New York, NY. 96 pp.

Kinget, R. 1967. Studies of the drugs of anthraquinone principles. XVI. Determination of the structure of anthracene derivatives reduced from the bark of *Rhamnus purshiana* DC. Planta Med. **15**(3): 233–239. [In French.]

Longo, R. 1965. Distinctive chromatographic identification of the bark of *Rhamnus frangula* L., *Rhamnus purshiana* DC., *Rhamnus alpina* L., *Rhamnus fallax* Boiss. Boll. Chim. Farm. **104**: 828–833. [In Italian.]

Longo, R., and Fumagalli, U. 1965. Method of chemical analysis of the anthrone components of cascara sagrada. Boll. Chim. Farm. **104**: 824–827. [In Italian.]

MacDonald, K.C. 1941. Propagation of the cascara tree, a conservation measure. Bulletin 108. British Columbia Ministry of Agriculture, Victoria, BC. 9 pp.

Manitto, P., Monti, D., Speranza, G., Mulinacci, N., Vincieri, F.F., Griffini, A., and Pifferi, G. 1995. Studies on cascara, part 2. structures of cascarosides E and F. J. Nat. Prod. **58**: 419–423.

Melotte, R., and Denoel, A. 1971. Anatomical and micrograph study of the cortex of branches of *Rhamnus staddo* A. Rich. Comparison with cascara and frangula. J. Pharm. Belg. **26**: 31–37. [In French.]

Parke, Davis & Company. 1889. Cascara sagrada. Parke, Davis & Company, Detroit, MI. 71 pp.

Quercia, V. 1976. Separation of anthraquinone compounds of cascara sagrada by means of high-pressure liquid chromatography. Boll. Chim. Farm. **115**: 309–316. [In Italian.]

Radwan, M.A. 1976. Germination of cascara (*Rhamnus purshiana*) seed. Tree Plant Notes U.S. For. Serv. **27**(2): 20–23.

Taylor, R.L., and Taylor, S. 1980. *Rhamnus purshiana* in British Columbia. Davidsonia **11**: 17–23.

Turner, N.J., and Hebda, R.J. 1990. Contemporary use of bark for medicine by two Salishan native elders of southeast Vancouver Island, Canada. J. Ethnopharmacol. **29**: 59–72.

World Wide Web Links

(Warning. The quality of information on the internet varies from excellent to erroneous and highly misleading. The links below were chosen because they were the most informative sites located at the time of our internet search. Since medicinal plants are the subject, information on medicinal usage is often given. Such information may be flawed, and in any event should not be substituted for professional medical guidance.)

Fire effects information system:
http://svinet2.fs.fed.us/database/feis/plants/shrub/arcuva/

A modern herbal by M. Grieve:
http://www.botanical.com/botanicaL/mgmh/c/cassag27.html

Herbal Materia Medica, D. Hoffman, cascara sagrada:
http://www.healthy.net/library/books/hoffman/materiamedica/cascara.htm

Cascara: herb of the week:
http://www.thedance.com/herbs/cascara.htm

Rhamnus purshianus [in German; has an excellent photograph]:
http://apv.ethz.ch/CD-ROM/Familien/Rhamnaceae/Rhamnus%20purshianus/Rhamnus%20purshianus.html

Monograph: *Rhamnus*:
http://www.healthlink.com.au/nat_lib/htm-data/htm-herb/bhp718.htm

Rhamnus diseases, Michigan State University Extension Home Horticulture:
http://www.msue.msu.edu/msue/imp/mod03/01700789.html

Rhamnus insects, Michigan State University Extension Home Horticulture:
http://www.msue.msu.edu/msue/imp/mod03/01700788.html

Rhamnus pesticide recommendations, Michigan State University Extension Home Horticulture:
http://www.msue.msu.edu/msue/imp/modop/00001892.html

Rhodiola rosea (roseroot)

Rhodiola rosea (L.) Scop. Roseroot

Rhodiola, a relatively advanced group of the Crassulaceae (Orpine family), has been interpreted (mostly in Eurasia) as a separate genus, or (particularly in North America) as a section of the genus *Sedum*. Extensive European and Asian literature on this drug plant is found under the name *Rhodiola*. There are about three dozen species of *Rhodiola*, mostly in the mountains of Asia, and their classification is in need of study. When roseroot is considered to be a species of *Sedum*, it is known as *S. rosea* (L.) Scop. (erroneously as *S. roseum*; or sometimes by the invalidly published name *S. rhodiola* DC.). The genus name originates from the Greek *rhodon*, meaning rose, an allusion to the odor of the rootstock.

Rootstock and lower shoot of *Rhodiola rosea* (roseroot)

English Common Names

Roseroot, golden root, Arctic root.

French Common Names

Orpin rose, rhodiole rougeâtre.

Morphology

Roseroot plants are perennial dioecious herbs, 5–40 cm tall. Most are male or female, while occasional plants have both male and female flowers in the same cyme. The stout rootstocks are either erect or horizontal, 0.4–5 cm in diameter, and produce pale brown fibrous roots below and floriferous (or occasionally sterile) ascending or erect stems above. The common name roseroot is a reference to the rootstocks possessing the scent of rose petals. The names "gold root" and "golden root" have been said to reflect the perceived value of the rootstocks, not their color. Stems of male plants are said to separate much more readily from the rhizome than those of the female plants. The leaves on the rhizome are scalelike, 2–7 mm long, and reddish-brown. The stem leaves are 7–40 mm long, fleshy, oval, obovate or oblong, spirally arranged, usually with dentate margins, dark green to glaucous or sometimes reddish. The flowers occur in terminal, corymbose to umbellate cymes, 0.5–7 cm in diameter, with one to over 150 flowers, these up to 7 mm in diameter. Flowering commences in early summer, sometimes before the leaves are fully expanded. The petals may be greenish, yellow, reddish or purple, or various intermediate shades. The fruits are erect, brown follicles, with small, brown or orange-brown winged seeds. In addition to dispersion by seeds, the plants are sometimes disseminated as pieces of rhizome. The rhizomes break or rot readily into pieces which can be transported by water, ice and wind.

Classification and Geography

Rhodiola rosea in the broad sense is an extremely variable circumpolar species of cool temperate and subarctic areas of the northern hemisphere, including North America, Greenland, Iceland, and Eurasia. The European distribution includes northern Europe and most of the mountains of central Europe, southwards to the Pyrenees, central Italy, and Bulgaria. The Asian distribution includes Arctic and alpine regions in the Altai Mountains, Eastern Siberia, Tien-Shan, the Far East, and south to the Himalayan Mountains.

In North America, *Sedum* subgenus *Rhodiola* as defined by R.T. Clausen includes at least five North American taxa of the *S. rosea* complex, and *S. rhodantha* Gray, a species of the southern and central Rocky Mountains, with an elongate inflorescence of pink flowers. Western North American plants with dark red flowers and green leaves have most recently been listed as *S. integrifolium* (Raf.) A. Nelson (a species difficult to separate from *S. rosea* in the narrow sense), including var. *integrifolium*, var. *procerum* Clausen and var. *neomexicanum* (Britt.) Clausen. Dark red-flowered plants with glaucous leaves, isolated in Minnesota and New York, have been listed as *S. integrifolium*

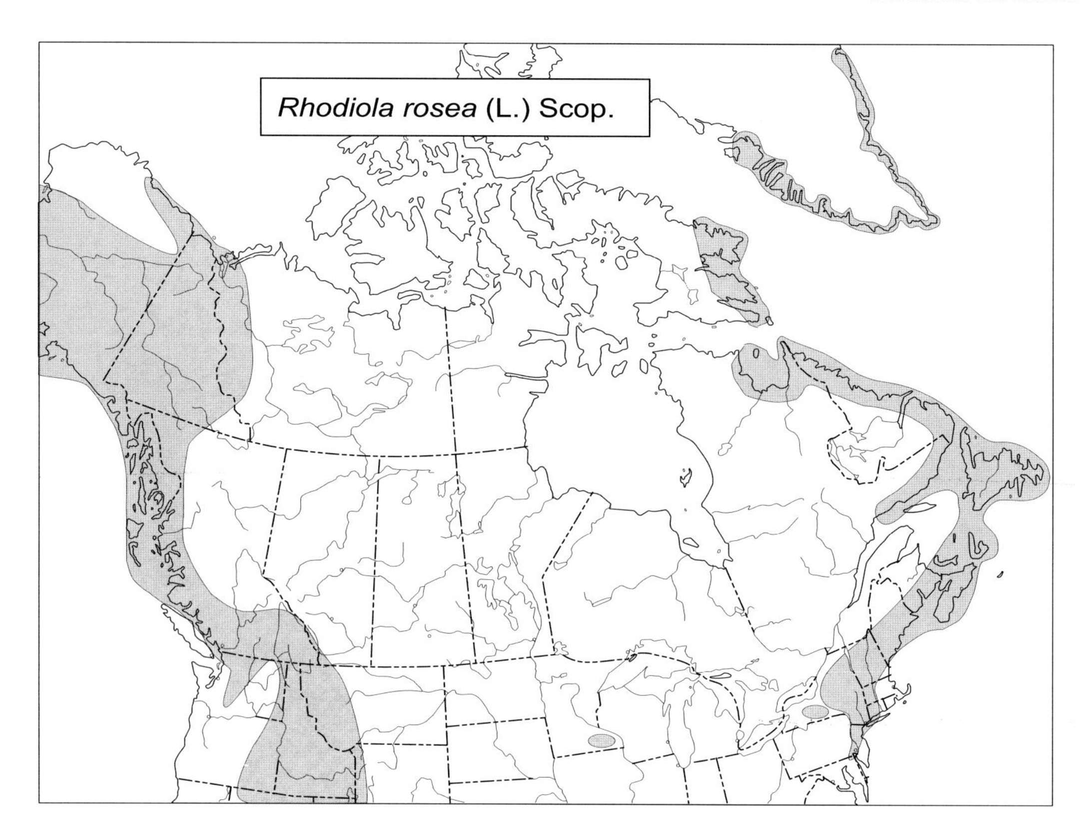

ssp. *leedyi* (Rosendahl & Moore) Clausen. Mostly coastal, eastern plants (isolated in the southern Appalachians of North Carolina), with relatively short yellow or green petals, have been listed as *S. rosea* in the narrow sense. The relationships of North American *S. rosea* in the broad sense with a number of similar Asiatic species is not clear. Some Asian species of *Rhodiola* are also used medicinally much like *R. rosea*, for example *R. quadrifida* (Pall.) Fisch. et Mey., *R. sacra* (Prain ex Hamet) S.H. Fu, *R. kirilowii* (Regel) Regel, and *R. sachalinensis* A. Bor.

Ecology

Rhodiola rosea is found on moist cliffs, ledges, talus, ridges, and dry tundra. Northern plants of North America typically occur in crevices or among mats of moss and other vegetation, often near shores, and sometimes in rather rich substrates; southern plants tend to grow on north-exposed cliffs in alpine regions. The Arctic forms are lower in growth and have fewer flowers.

Medicinal Uses

Roseroot is one of several medicinal plants that have been described as "adaptogens." The world's best known adaptogens are ginseng (certain *Panax* species) and Russian ginseng (*Eleutheroccocus senticosus* (Rupr. & Maxim.) Maxim.); other highly touted adaptogenic plants include roseroot, Schizandra (*Schizandra chinensis* (Turcz.) Baill.), and the mushroom reishi (*Ganoderma lucidum* (Leyss. Fr.) Karst.). The term adaptogen was used by Russian pharmacologists, starting with N.V. Lazarev about 1947, and popularized by his student, I.I. Brekhman. Adaptogenic plant-derived, legal, drug preparations have been credited with improving the performance of elite Russian sports figures and cosmonauts. An adaptogen has been defined as a substance that is "innocuous, causing minimal physiological disorder; non-specific in action, increasing resistance to the adverse influences of a wide range of physical, chemical and biological factors; and capable of a normalizing action irrespective of the direction of the pathological change." The validity of the concept of an adaptogen has been debated, and much of the supportive research to date does not provide unbiased and unequivocal evaluation. Nevertheless, adaptogenic plants are increasingly being used medicinally in response to stress-generated maladies that are becoming more common in western society.

Many articles have been written by Russian scientists on the pharmacological effects of roseroot. Experiments with rats and mice have suggested that the chemically active compounds can improve learning and memory and reduce stress. Experimentation has also suggested that they have anticancer properties, stimulate the central nervous system and protect the liver, for example in carbon tetrachloride intoxication. Methods of determining the authenticity and quality of the rhizome are available, as well as methods for the quantitative analysis of some of the biologically active compounds.

Roseroot is being vigorously marketed today as an almost miraculous tonic, as witnessed by the following advertisement: "Roseroot was used by the Vikings to give them the extra strength they needed for their long, arduous journeys. Today, scientific research confirms what the Vikings experienced. Roseroot helps you adapt and thrive when exposed to stressful conditions, such as overwork, physical exhaustion and mental fatigue. Studies confirm that it helps the body maintain homeostasis by assisting at a cellular level to adapt to environmental stressors, promote mental and physical vitality and increase alertness and physical endurance."

It has been claimed that Roseroot leaves can be used like *Aloe vera* leaves to soothe burns, bites, and other irritations. Consistent with this, a paste or tea from the root has been used to help wounds heal. However, the raw rootstock sometimes causes allergic reactions.

In present-day Ukraine, a medicinal alcoholic drink called "nastojka" is prepared by mixing 40% alcohol (e.g., vodka) and an equal weight of rootstock, and allowing the mixture to stand for a few weeks; only a few teaspoons are consumed daily.

Toxicity

The pharmacological literature we have surveyed does not mention toxic reactions. However, most of the scientific medicinal literature on roseroot comes from Russia, and much of this appears determined to demonstrate the virtues of the plant. Given the lack of familiarity with roseroot medicinal utilization in most of Europe and North America, it seems prudent to assume that while no obvious toxicity has been demonstrated, the safety of using this herb has not been adequately determined.

Chemistry

Some of the purported medical effectiveness of *Rhodiola* extracts have been attributed to a variety of phenolic substances, including rhodioloside, a glycoside.

Flowering branch of *Rhodiola rosea* (roseroot)

Non-medicinal Uses

Roseroot is grown in many countries as an ornamental. The young stems and leaves are also sometimes used as a wild food, either raw or cooked, although this may not be advisable.

Agricultural and Commercial Aspects

Roseroot is cultivated in Russia and some other Eurasian countries. Information on optimal harvesting times with regard to the compounds rosavidine and salidroside have been published. Information is also available on the effects of postharvest treatment on the quality of the rhizome.

In Eurasia where the plant is highly valued, areas for habitat restoration and in situ conservation have been recommended. Roseroot is not presently a medicinal plant of significant value in North America. Its potential importance lies in the fact that it is the only native Canadian plant species besides ginseng that has attracted international attention as an adaptogen. Plants with this reputation have a large potential market, and their economic development deserves periodic evaluation. This Arctic herb could easily be cultivated in cold areas of Canada where few other economic plants can be grown without protection.

Myths, Legends, Tales, Folklore, and Interesting Facts

- The rootstock of *R. rosea* was often used in European folk love potions. The legendary 13th century Ukrainian prince Danila Galitsky, whose reputation rivalled that of Casanova, was believed to have used roseroot as an aphrodisiac.

Selected References

Bocharova, O.A., and Serebryakova, R.V. 1994. Testing plant-based drugs for prevention and nontoxic therapy of cancer diseases on experimental models. Vestn. Rossiiskoi Akad. Med. Nauk **2**: 52–55. [In Russian.]

Clausen, R.T. 1975. *Sedum* of North America north of the Mexican Plateau. Cornell University Press, Ithaca, NY. 742 pp.

Dneprovskii, IU.M, Kim, E.F., and Iumanova, T.P. 1975. Seasonal development and growth of *Rhodiola rosea* L. in relation to introduction [as drug plant]. Biull. Gl. Bot. Sada. **98**: 27–34. [In Russian.]

Evans, R.L. 1983. Handbook of cultivated Sedums. Science Reviews Limited, Northwood, Middlesex, UK. 345 pp.

Hart, H. 1994. The unilacunar two-trace nodal structure of the caudex of *Rhodiola rosea* L. (Crassulaceae). Bot. J. Linnaean Soc. **116**: 235–241.

Kazarinova, N.V. 1977. Ecological and biological characteristics of *Rhodiola rosea* in the Gorno-Altai Autonomous Region. Izv. Sib. Otd. Akad. Nauk. SSSR. Ser. Biol. Nauk. Novosibirsk Dec. **1977**(3): 38–43. [In Russian.]

Kim, E.F. 1976. Experience of cultivation of [the drug plant] *Rhodiola rosea* in the low-mountain area of the Altai. Rastit. Resur. **12**: 583–590. [In Russian.]

Kir'yanov, A.A., Bondarenko, L.T., Kirkin, V.A., and Zapesochnaya, G.G. 1989. Dynamics of the accumulation of rosaviridine and salidroside in the raw material of *Rhodiola rosea* cultivated near Moscow (Russian SFR, USSR). Khim.-farm. Zh. **23**: 449–452. [In Russian.]

Kir'yanov, A.A., Bondarenko, L.T., Kirkin, V.A., Zapesochnaya, G.G., Dubichev, A.A., and Vorontsov, E.D. 1991. Determining biologically active components of *Rhodiola rosea* rhizomes. Khim. Prir. Soedin. **3**: 320–323. [In Russian.]

Kovanda, M. 1976. *Rhodiola rosea* L. Ziva **24**: 12–13. [In Czech.]

Krasnov, E.A., Kuvaev, V.B., and Khoruzhaia,T.G. 1978. Chemosystematic study of *Rhodiola* L. species. Rastit. Resur. **14**: 153–160. [In Russian.]

Krasnov, E.A., Zotova, M.I., Nekhoda, M.F., Aksenova, R.A., and Kolesnikova, N.S. 1978. Stimulating effect of preparations from *Rhodiola* L. species. Rastit. Resur. **14**: 90–92. [In Russian.]

Krylov, G.V., and Kazarinova, N.V. 1973. Productivity of *Rhodiola rosea* and its effective use. In Okhrana Gornykh Landshaftov Sibiri **1973**:162–164. [In Russian.]

Kurkin, V.A., Zapesochnaya, G.C., and Klyaznika, V.G. 1982. Flavonoids of the rhizomes of *Rhodiola rosea*. I. Tricin glucosides. Chem. Nat. Compd. (Consultants Bureau, New York, NY.) **18**: 550–552. [In English and Russian.]

Kurkin, V.A., Zapesochnaya, G.G., and Shchavlinskii, A.N. 1984. Flavonoids of the rhizomes of *Rhodiola rosea*. III. Chem. Nat. Compd. (Consultants Bureau, New York, NY.) **20**: 367–368. [In English and Russian.]

Kurkin, V.A., Zapesochnaya, G.G., and Shchavlinskii, A.N. 1984. Flavonoids of the epigeal part of *Rhodiola rosea*. I. Chem. Nat. Compd. (Consultants Bureau, New York, NY.) **20**: 623–624. [In English and Russian.]

Kurkin, V.A., Zapesochnaya, G.G., Shchavlinskii, A.N., and Nukhimovskii, E.L. 1985. A method for the determination of the authenticity and quality of the rhizomes of *Rhodiola rosea*. Khim.-farm. Zh. **19**: 185–190. [In Russian.]

Kurkin, V.A., Zapesochnaya, G.G., Kir'yanov, A.A., Bondarenko, L.T., Vandyshev, V.V., Mainskov, A.V., Nukhimovskii, E.L., and Klimakhin, G.I. 1989. Quality of *Rhodiola rosea* raw material. Khim.-farm. Zh. **23**: 1364–1367. [In Russian.]

Likharev, V.S. 1980. Simple way of growing the "golden root" *Rhodiola rosea*, drug plant. Stepnye. Prostory (Saratov, Ministerstvo sel'skogo khoziaistvo RSFSR) April (4): 41–42. [In Russian.]

Lovelius, O.L., and Stoiko, S.M. 1990. *Rhodiola rosea* L. In the Ukrainian Carpathians (USSR). Ukrayins"kyi Botanichnyi Zh. **47**(1): 90–93. [In Ukranian.]

Nukhimovskii, E.L. 1974. Ecological morphology of some medicinal plants under the natural conditions of their growth. 2. *Rhodiola rosea* L. Rastit. Resur. **10**: 499–516. [In Russian.]

Nukhimovskii, E.L. 1976. Initial stages of biomorphogenesis of [the drug plant] *Rhodiola rosea* L. cultivated in the Moscow Region. Rastit. Resur. **12**: 348–355. [In Russian.]

Ohba, H. 1977. New or critical species of Asiatic Sedoideae. 1. [*Rhodiola serrata*, *Sedum roseum*, *Rhodiola ludlowii*]. J. Jpn. Bot. **52**: 263–268.

Ohba, H., and Midorikawa, K. 1991. Geographical distribution of *Rhodiola rosea* in Honshu, Japan. Bull. Biogeogr. Soc. Jpn. **46**(1–22): 179–185. [In Japanese.]

Ohba, H. 1981. A revision of the Asiatic species of Sedoideae (Crassulaceae). 2. *Rhodiola* (subgen. *Rhodiola* sect. *Rhodiola*). J. Fac. Sci. Univ. Tokyo Sec. IV Bot. (Tokyo: The University) **13**: 65–119.

Petkov, V.D.,Yonkov, D., Mosharoff, A., Kambourova, T., Alova, L., Petkov, V.V., and Todorov, I. 1986. Effects of alcohol aqueous extract from *Rhodiola rosea* roots on learning and memory. Acta Physiol. Pharmacol. Bulgarica **12**: 3–16. [In Russian.]

Polozhii, A.V., and Reviakina, N.V. 1976. Developmental biology of *Rhodiola rosea* L. in the Katunj range (Altai Mts.) Rastit. Resur. **12**: 53–59. [In Russian.]

Polozhii, A.V., Surov, I.U.P., and Kopaneva, G.A. 1976. Genus *Rhodiola* in Southern Siberia. Arealy. Rast. Flory. SSSR. (Leningrad: Izd-vo Leningradskogo Universiteta.) **1976**(3): 170–173. [In Russian.]

Revina, T.A., Krasnov, E.A., Sviridova, T.P., Stepaniuk, G.I.A., and Surov, I.U.P. 1976. Biological characteristics and chemical composition of [the drug plant] *Rhodiola rosea* L. cultivated in Tomsk. Rastit Resur. **12**: 355–360. [In Russian.]

Saratikov, A.S. 1977. Rose-root stone-crop (*Rhodiola rosea*) [Chemical composition of rhizomes]. Khim. Farm. Zh. **11**(4): 56–59. [In Russian.]

Singh, N.B., and Bhattacharyya, U.C. 1982. Nomenclature notes on *Rhodiola* (Crassulaceae) India. J. Econ. Taxon. Bot. Jodphur. **3**: 631–632.

Tril, V.M. 1988. Wild medicinal plant reserves in Ala Tau near Kuznetsk (Kemerovo Oblast, Russian SFSR, USSR). Rastit. Resur. **24**: 348–352. [In Russian.]

Udintsev, S.N., and Shakhov, V.P. 1991. The role of humoral factors of regenerating liver in the development of experimental tumors and the effect of *Rhodiola rosea* extract on this process. Neoplasma (Bratislava) **38**: 323–332.

Uhl, C.H. 1952. Heteroploidy in *Sedum rosea* (L.) Scop. Evolution **6**: 81-86.

Vasak, V. 1971. *Rhodiola rosea* L. from the mountain. Ziva 19: 45–46. [In Czech.]

Zapesochnaya, G.G., and Kurkin, V.A. 1982. Glycosides of cinnamyl alcohol from the rhizomes of *Rhodiola rosea*. Chem. Nat. Compd. (Consultants Bureau, New York, NY.) **18**: 685–688. [In English and Russian.]

Zapesochnaya,G.G., and Kurkin, V.A. 1983. The flavonoids of the rhizomes of *Rhodiola rosea*. II. A flavonolignan and glycosides of herbacetin. Chem. Nat. Compd. (Consultants Bureau, New York, NY.) **19**: 21–29. [In English and Russian.]

World Wide Web Links

(Warning. The quality of information on the internet varies from excellent to erroneous and highly misleading. The links below were chosen because they were the most informative sites located at the time of our internet search. Since medicinal plants are the subject, information on medicinal usage is often given. Such information may be flawed, and in any event should not be substituted for professional medical guidance.)

Clinical research of *Rhodiola* on healthy people:
http://www.bio-synergy.com/articles/rhodiola.html

Roseroot - *Rhodiola rosea* [just a color photograph]:
http://www.lektor.co.uk/marine/plant6.html

Leedy's roseroot (*Sedum integrifolium* ssp. *leedyi*) [color photograph]:
http://www.fws.gov/r9endspp/i/q8b.html

Herbal adaptogens, Christopher Hobbs [deals with several adaptogens, but not *Rhodiola*]:
http://www.healthy.net/library/articles/hobbs/adaptx.htm

Adaptogens: nature's answer to stress by Morton Walker:
http://www.holoworks.com/Walker.html

Sanguinaria canadensis (bloodroot)

Sanguinaria canadensis L. Bloodroot

The genus name *Sanguinaria*, Latin for bloody, as well as the common names, reflect the orange-red latex which oozes as a juice from damaged areas of all parts of the plant, but especially from the rhizome.

English Common Names

Bloodroot, puccoon (from the name in Algonkian languages, *poughkone*), Indian paint, Indian plant, Indian red paint, coonroot, paucon, pauson, red paint root, red puccoon, red root, sanguinaria, snakebite, sweet slumber, tetterwort (tetter is an Old English word for various skin diseases such as ringworm, impetigo, and eczema).

French Common Names

Sanguinaire, sanguinaire du Canada.

Morphology

Sanguinaria canadensis in the poppy family (Papaveraceae) is the only species of its genus. It is a low-growing (ca. 15 cm high when flowering; to 50 cm at maturity) perennial herb, produced from a branching, horizontal, reddish-brown, succulent rhizome, from which stems arise aerially at various points, while beneath the ground there are many adventitious roots. Each stem generally produces a solitary basal leaf (sometimes two), which in its early development envelops and protects a single long-stalked flower 2.5–5 cm broad (or even larger in cultivated plants), with 4–16 fugaceous white or rarely pink petals (more in a cultigen, as noted below). Flowering occurs in very early spring. As flowering progresses, the leaf continues to grow, developing a striking orbicular or reniform blade 6–30 cm broad, with 6–9 palmately arranged lobes, the lobes themselves sometimes lobed. The petioles arise beneath the ground, and may elongate to as much as 30 cm, so that the mature leaf overtops the fruit. The solitary, narrow capsule is 3–6 cm long, and splits when mature into two persistent valves, releasing the numerous 3 mm long seeds, each with a large crest.

Classification and Geography

Bloodroot is indigenous to eastern and central Canada and the US, occurring from Nova Scotia west to Manitoba and Nebraska, and south to Alabama, Arkansas and Florida. It is more common inland than on the Coastal Plain, and sometimes has become infrequent in areas where it has been gathered for medicine. Plants of the southern US range are sometimes segregated as var. *rotundifolia* (Greene) Fedde.

Ecology

Bloodroot grows best in shaded, cool, moist open hardwood groves and on well-drained woodland slopes, on circumbasic soils. In favorable situations large colonies may be formed, although sometimes only scattered individuals are encountered.

The attractive flowers have no nectar and the plant self-pollinates readily. However, the blossoms can supply some pollen to foraging insects. Several species of ants are attracted by small, oily protuberances on the surface of bloodroot seeds. The ants carry the seeds to their underground nests. It has been suggested that this increases seed survival and seedling establishment, reducing losses from foraging insects, birds and rodents. Dispersal of seeds by ants, i.e., myrmechory, is a widespread symbiotic relationship. The ants benefit by consuming the *elaiosomes* (nutritive tissues of the propagule developed specifically for the ants), while the plants benefit by having their seeds planted, often in areas rich in soil nutrients. Seneca snakeroot, discussed in this work, is also ant-dispersed.

Medicinal Uses

Traditionally, bloodroot was commonly used by North American Indigenous Peoples before the arrival of European settlers, for such ailments as rheumatism, asthma, bronchitis, laryngitis, and fevers. It was frequently employed orally as an emetic and as an expectorant (especially in treating bronchitis), and topically as an ointment for skin ulcers and cancers, both by Indians and settlers, and these usages were continued in modern western medical practice of the early part of this century. Traditional medicinal uses of bloodroot have largely been abandoned because of the toxicity of the plant, although it has been established that bloodroot has antimicrobial and antitumor properties. Despite its toxicity, bloodroot continues to be used in medicine, and is present in more than a dozen commercial preparations marketed in Canada, usually as expectorants, cough syrups and tinctures. Lexat is an Australian product for digestive disorders that contains bloodroot. Bloodroot is no longer recommended as an emetic. The most recent modern

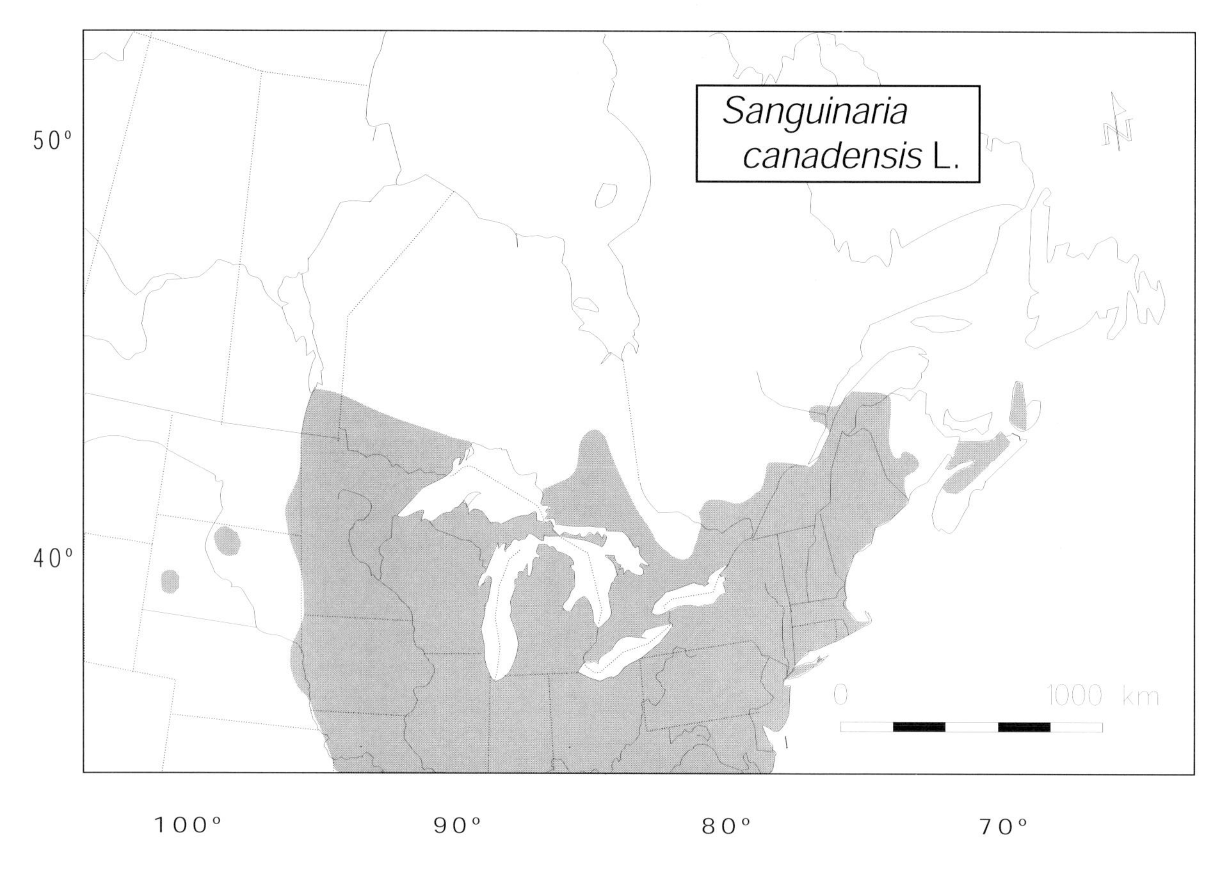

medical use of bloodroot arises from its bactericidal and bacteriostatic properties against oral plaque-forming organisms. Sanguinarine has been the active agent in anti-plaque, anti-gingivitis oral rinses and toothpastes, the most well known of which is the brand Viadent. However, very recently sanguinarine has been removed from Viadent formulations by the manufacturer, Colgate, because of new evidence linking the development of precancerous interoral lesions with its prolonged use (see Damm et al. 1999).

Toxicity

In toxic doses, bloodroot acts as a poison to the voluntary muscles, produces vomiting, sensations of burning in all contacted mucous membranes, tormenting thirst, faintness, vertigo, dimness of vision, cardiac inhibition, and possibly death. Although potentially deadly, fatal consumption of wild plants of bloodroot does not seem to have been recorded in North America either for people or animals, presumably because the bitter, acrid taste discourages consumption of a dangerous amount.

This plant was listed in a 1995 Health Canada document (see "Herbs used as non-medicinal ingredients in nonprescription drugs for human use," http://www.hc-sc.gc.ca/hpb-dgps/therapeut/drhtmeng/policy.html) as a herb that is unacceptable as a nonprescription drug product for oral use.

Chemistry

The physiological activity of bloodroot is due to benzophenanthridine alkaloids, found mainly in the rhizomes, and constituting 3–9% of the rhizomes. Sanguinarine makes up 50% or more of these alkaloids, and is considered the most important. The older rhizomes have more sanguinarine than the younger. Sanguinarine is water-soluble, and is responsible for the orange-red color of the latex of bloodroot. It has been shown that there is a north-south ecocline in the US in sanguinarine content, the southern populations having a higher content. Several of the alkaloids present also occur in opium (not surprising since opium poppy and bloodroot belong to the same family, the Papaveraceae), and have some feeble narcotic potential. Sanguinarine strongly inhibits root rot fungi, and may protect the perennial rhizome from fungal disease.

Non-medicinal Uses

Bloodroot is cultivated in gardens for its showy flower and striking foliage. As an ornamental, it is most useful as an early-flowering border plant in shady locations. Because it is uncommon, it should not be transplanted from the wild, but rather established from seeds or from horticultural suppliers who have propagated their stock, not merely collected wild plants. The petals of wild plants are easily lost to wind and rain, but a popular mutant form with extra petals, called "peony flower" and "double bloodroot," or known by the cultivar names "Multiplex" and "Florepleno," arose in the Midwestern US about 1950. This has extra petals in place of the stamens, but is sterile, and must be cultivated from rhizomes and not from seeds.

The reddish juice of bloodroot is a very effective dye, and was much used by Indigenous Peoples as a body paint, and to decorate baskets, weapons, implements, and clothing (hence the name "Indian paint"). Early colonists also used the plant to dye cloth, and it was exported to Europe for this purpose. Some Indian groups also employed bloodroot in religious services, and as an insect repellent.

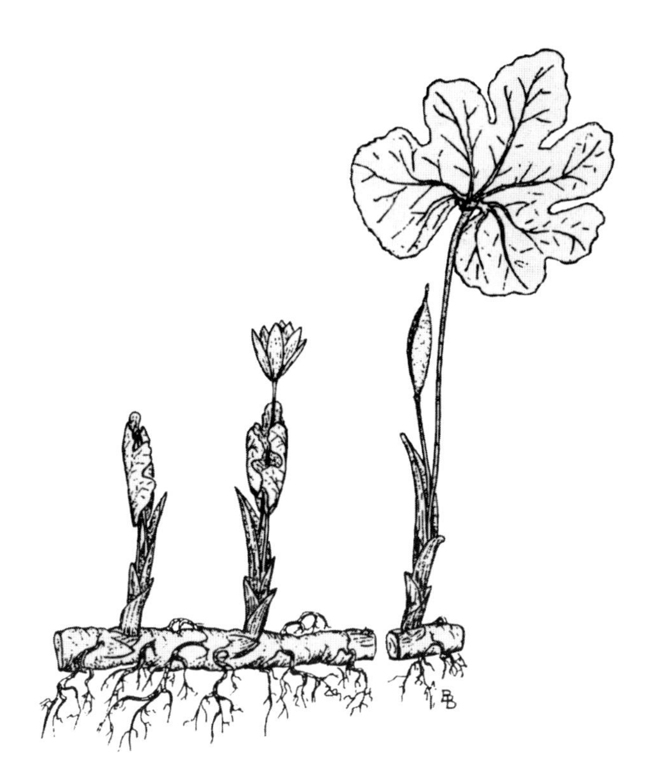

Sanguinaria canadensis (bloodroot)

Agricultural and Commercial Aspects

Most bloodroot rhizomes acquired for commerce come from harvesting wild eastern US plants. Rhizomes of plants that are 2 or more years old are excavated in late summer, and dried after the brittle, wiry roots are removed. In the past it was often recommended that collecting be carried out in the fall, after the foliage dies down, although more recent work suggests that the time of highest concentration of alkaloids is during or immediately after flowering. Collected rhizomes vary from 2 to 7 cm in length, and from 5 to 15 mm in diameter. The orange-red latex oozes from the broken ends, forming a reddish resinous coating upon drying.

In the last decade it has been found that sanguinarine can be obtained from alternative sources, so that cultivating bloodroot for this chemical is doubtfully worthwhile. Sanguinarine occurs in the Asian *Macleaya cordata* (Willd.) R. Br. (*Bocconia cordata*), the plume poppy, and since this species is relatively easily cultivated it is a potential competitive commercial source of the alkaloid. Moreover, technology has advanced to the point that sanguinarine is being produced commercially from tissue culture in large vats, for use in toothpaste and mouthwash. Nevertheless, bloodroot extracts are the subject of considerable ongoing pharmacological research. Although cultivation is of limited agricultural interest in Canada today, the remarkable chemistry of bloodroot is believed to have economic potential.

Myths, Legends, Tales, Folklore, and Interesting Facts

- Bachelors of the Ponca tribe used bloodroot as a love charm. Applying the red juice to the palm and shaking hands with the woman they wanted to marry was believed to induce consent.

Selected References

Becci, P.J., Schwartz, H., Barnes, H.H., and Southard, G.L. 1987. Short-term toxicity studies of sanguinarine and of two alkaloid extracts of *Sanguinaria canadensis*. J. Toxicol. Environ. Health **20**: 199–208.

Bennett, B.C., Bell, C.R., and Boulware, R.T. 1990. Geographic variation in alkaloid content of *Sanguinaria canadensis* (Papaveraceae). Rhodora **92**: 57–69.

Cullinan, M.P., Powell, R.N., Faddy, M.J., and Seymour, G.J. 1997. Efficacy of a dentifrice and oral rinse containing sanguinaria extract in conjunction with initial periodontal therapy. Aust. Dent. J. **42**: 47–51.

Damm, D.D., Curran, A., White, D.K., and Drummond, J.F. 1999. Leukoplakia of the maxillary vestibule - an association with Viadent? Oral Surg. Oral Med. Oral Pathol. Oral Radiol. Endod. **87**: 61-66. [Abstract available at http://www1.mosby.com/mosbyscripts/mosby.dll?action=searchDB&searchDBfor=art&art Type=abs&id=a94688&target=]

Frankos, V.H., Brusick, D.J., Johnson, E.M., Maibach, H.I., Munro, I., Squire, R.A., and Weil, C.S. 1990. Safety of sanguinaria extract as used in commercial toothpaste and oral rinse products. J. Can. Dent. Assoc. **56**(7 Suppl.): 41–47.

Godowski, K.C. 1989. Antimicrobial action of sanguinarine. J. Clin. Dent. **1**(4): 96–101.

Hannah, J.J., Johnson, J.D., and Kuftinec, M.M. 1989. Long-term clinical evaluation of toothpaste and oral rinse containing sanguinaria extract in controlling plaque, gingival inflammation, and sulcular bleeding during orthodontic treatment. Am. J. Orthod. Dentofacial Orthop. **96**: 199–207.

Harkrader, R.J., Reinhart, P.C., Rogers, J.A., Jones, R.R., Wylie, R.E., 2d, Lowe, B.K., and McEvoy, R.M. 1990. The history, chemistry and pharmacokinetics of *Sanguinaria* extract. J. Can. Dent. Assoc. **56**(7 Suppl.): 7–12.

Johnson, K.L. 1983. Rare plants of the eastern deciduous forest. I: Bloodroot (*Sanguinaria canadensis*). Bull. Man. Nat. Soc. **6**(6): 11.

Judd, W.W. 1977. Insects associated with flowering bloodroot, *Sanguinaria canadensis* L., at Fanshawe Lake. Ontario. Entomol. News **88**(1/2): 13–17.

Karlowsky, J.A. 1991. Bloodroot. Can. Pharm. J. **124**: 260, 262-263, 267.

Keller, K.A., and Meyer, D.L 1989. Reproductive and developmental toxicological evaluation of sanguinaria extract. J. Clin. Dent. **1**(3): 59–66.

Kiger, R.W. 1997. *Sanguinaria*. *In* Flora of North America north of Mexico, Vol. 3. *Edited by* Flora of North America Editorial Committee. Oxford University Press, New York, NY. p. 305.

Kopczyk, R.A., Abrams, H., Brown, A.T., Matheny, J.L., and Kaplan, A.L. 1991. Clinical and microbiological effects of a sanguinaria-containing mouthrinse and dentifrice with and without fluoride during 6 months of use. J. Periodontol. **62**: 617–622.

Kuftinec, M.M., Mueller-Joseph, L.J., and Kopczyk, R.A. 1990. Sanguinaria toothpaste and oral rinse regimen clinical efficacy in short- and long-term trials. J. Can. Dent. Assoc. **56**(7 Suppl.): 31–33.

Laster, L.L., and Lobene, R.R. 1990. New perspectives on sanguinaria clinicals: individual toothpaste and oral rinse testing. J. Can. Dent. Assoc. **56**(7 Suppl.): 19–30.

Lehmann, N.L., and Sattler, R. 1993. Homeosis in floral development of *Sanguinaria canadensis* and *S. canadensis* 'Multiplex' (Papaveraceae). Am. J. Bot. **80**: 1323–1335.

Lemire, S.W., and Busch, K.L. 1994. Chromatographic separations of the constituents derived from *Sanguinaria canadensis*: thin-layer chromatography and capillary zone electrophoresis. J. Planar Chromatogr. - Mod. TLC **7**: 221–228.

Lyon, D.L. 1992. Bee pollination of facultatively xenogamous *Sanguinaria canadensis* L. Bull. Torrey Bot. Club **119**: 368–375.

Mallatt, M.E., Beiswanger, B.B., Drook, C.A., Stookey, G.K., Jackson, R.D., and Bricker, S.L. 1989. Clinical effect of a sanguinaria dentifrice on plaque and gingivitis in adults. J. Periodontol. **60**: 91–95.

Marino, P.C., Eisenberg, R.M., and Cornell, H.V. 1997. Influence of sunlight and soil nutrients on clonal growth and sexual reproduction of the understory perennial herb *Sanguinaria canadensis* L. J. Torrey Bot. Soc. **124**: 219–227.

Nikiforuk, G. 1990. The sanguinaria story - an update and new perspectives (overview of the Toronto symposium). J. Can. Dent. Assoc. **56**(7 Suppl.): 5–6.

Rho, D., Chauret, N., Laberge, N., and Archambault, J. 1992. Growth characteristics of *Sanguinaria canadensis* L. cell suspensions and immobilized cultures for production of benzophenanthridine alkaloids. Appl. Microbiol. Biotechnol. **36**: 611–617.

Schwartz, H.G. 1986. Safety profile of sanguinarine and sanguinaria extract. Compend. Contin. Educ. Dent. S212-S217.

Tin-Wa, M., Farnsworth, N.R., Fong, H.H., and Trojanek, J. 1970. Biological and phytochemical evaluation of plants. 8. Isolation of a new alkaloid from *Sanguinaria canadensis.* Lloydia **33**: 267–269.

Walker C. 1990. Effects of sanguinarine and *Sanguinaria* extract on the microbiota associated with the oral cavity. J. Can. Dent. Assoc. **56**(7 Suppl.): 13–30.

World Wide Web Links

(Warning. The quality of information on the internet varies from excellent to erroneous and highly misleading. The links below were chosen because they were the most informative sites located at the time of our internet search. Since medicinal plants are the subject, information on medicinal usage is often given. Such information may be flawed, and in any event should not be substituted for professional medical guidance.)

BloodRoot, *Sanguinaria canadensis*:
http://ncnatural.com/wildflwr/blodroot.html

HealthLink Online Resources, *Sanguinaria*:
http://www.healthlink.com.au/nat_lib/htm-data/htm-herb/bhp728.htm

Blood root, *Sanguinaria canadensis*:
http://www2.best.com/~timj/herbage/A15.htm

Bloodroot, *Sanguinaria canadensis* [color photographs]:
http://www.wildroots.com/bloodroo.htm

Chapter 5, A bloody early bloomer [a well-written, entertaining review]:
http://www.acorn-online.com/hedge/blood.htm

Wildflower notes, Bloodroot, *Sanguinaria canadensis*:
http://www.duc.auburn.edu/~deancar/wfnotes/bludrt.htm

A modern herbal by M. Grieve:
http://www.botanical.com/botanicaL/mgmh/b/bloodr59.html

Sanguinaria canadensis (bloodroot or red puccoon) [color photographs]:
http://www.ansci.cornell.edu/bloodroot.html

Top-rated websites - bloodroot [links to sites dealing with bloodroot]:
http://www.lycos.com/wguide/wire/wire_21329986_71958_3_1.html

Taraxacum (dandelion)

Taraxacum spp. Dandelion

English Common Names

Dandelion, lion's tooth.

The English name dandelion is a corruption of the French *dent de lion*, lion's tooth, generally interpreted as a reference to the coarsely toothed leaves.

French Common Names

Pissenlit, dent de lion, dent-de-lion.

The French *pissenlit*, literally "wet-a-bed," reflects the reputation of dandelion for stimulating the kidneys.

Morphology

Dandelion is a herbaceous perennial with a rosette of jagged, irregularly lobed leaves produced from a long, thick, fleshy taproot that can descend more than 1 m. The leaves may be nearly smooth-margined, saw-toothed, or deeply cut. The single flowering stalk, sometimes over 50 cm tall, is hollow and bears a head of tiny yellow flowers, the whole head referred to as a flower by non-botanists. The flowering stalk is hollow and elongates with age. Fruiting heads produce tiny (3–5 mm) brown "seeds" (achenes), each carried by a "parachute" of white, fluffy hairs on a stalk. White, bitter, milky juice exudes from the plant where it is cut or broken; this stains hands brown and is difficult to remove. Cultivated selections differ in various respects from wild plants, some tending to have broader, more deeply-lobed leaves, others with a very high production of leaves, often semi-erect.

Classification and Geography

Species of *Taraxacum* are perennial herbs, mostly native to north temperate and Arctic regions of the Northern Hemisphere. *Taraxacum* is an extremely complex genus. Many of the species reproduce mostly by apomixis and generate numerous intergrading microspecies. Some taxonomists outside North America have recognized many of the races as different species (over 1500 European species have been described), while others submerge most of the races into only a few species. To avoid the problem of recognizing many trivial apomictic species, rather than referring to species one can cite the sections into which *Taraxacum* is divided. Despite the apparent lack of sexual reproduction in North American weedy dandelions, there is considerable genetic variation present. South of boreal regions native kinds of *Taraxacum* are usually sexual and these (about 100 species) are comparable in distinctiveness to the kinds of species recognized by most botanists. In North America the native species inhabit Arctic and alpine regions. These belong to sections *Taraxacum* or *Borealia* HM. In Canada 3–15 species have been recognized. The introduced *T. palustre*, only recently discovered in Ontario and Quebec, has proven to be widespread. The classification of dandelions in Canada requires much additional work.

Taraxacum officinale Weber, the name that has long been applied to the so-called common dandelion, is not the correct name, but a solution to the problem requires more and possibly extensive study. Almost all North American weedy dandelions belong to one of two sections: *Ruderalia* Kirschner, Ollgaard & Stepanek and *Erythrosperma* (Dahlst.) Lindb. f. Since identifying dandelion species is difficult, and it is unclear which species are used medicinally, we simply refer to the genus *Taraxacum* in this discussion. Most weedy forms of dandelion, and probably most if not all cultivated selections, originate from Europe and Asia. Introduced dandelions occur in all provinces and territories of Canada.

Ecology

The dandelion is now a common plant throughout the world in temperate regions, often in pastures, meadows, gardens, and waste ground, and along roadsides. In North America it is regarded as a serious weed of lawns.

When the sun shines, the flower heads are open, and when the weather turns dull, the flower heads close up. Dandelions are "short-day plants," producing flowers when there are less than 12 hours of light. Flowers are produced mostly in mid-spring, with a much reduced second period of blooming in the fall up until the first frost.

Medicinal Uses

As a medicinal plant, dandelion has been used at least since the time of the Arabian physicians of the 10th and 11th centuries. Root extracts were once used extensively as a diuretic (to promote urination), and are still sometimes so employed. A tradition developed in Europe of taking dandelion as a "cleansing cure" in the spring. Dandelion has also been said to be useful for treating jaundice and other liver ailments. Both of these medicinal properties seem to trace to the Doctrine of

Signatures, whereby aspects of plants are said to signal their medicinal uses. The yellow of dandelion flowers was interpreted as a sign that jaundice (which causes yellow coloration) and other liver diseases could be treated. The juiciness of the dandelion, suggestive of water retention, was interpreted as indicating usefulness as a diuretic. Dandelion juice, expressed from the roots, was once sold by druggists. Dandelion has also been claimed to be useful as an appetite stimulant and as a tonic. Other ailments treated in the past include fever, insomnia, jaundice, rheumatism, eczema and other skin diseases, constipation, warts, cancers, and tumors. Most of the claims for medical effectiveness of dandelion are based on research that is pre-World War II, and there is a need for modern investigation. Nevertheless, medical usage of dandelion remains fairly common.

Toxicity

For most people the only hazard of consuming dandelion is excessive urination. Overuse of diuretics (which stimulate urination) can lower the level of potassium ions and lead to muscular weakness and constipation. Potassium ions are a part of the mechanism for transmitting nervous impulses, and an imbalance of potassium can sensitize the heart muscle to drugs like digitalis, resulting in irregular heartbeat and other symptoms. The German Commission E monograph on dandelion recommends that it be avoided if gallstones are present; it should not be used if bile ducts are obstructed. Dandelion root can cause hyperacidity in some people. Contact dermatitis has been reported from handling dandelions. Dandelions should not be harvested from locations such as lawns and roadsides that may have been sprayed with herbicides, pesticides, or fungicides. Finally, roadside dandelions could accumulate lead or other chemicals from nearby automobile traffic.

Chemistry

The bitter resin in the roots and shoots contains taraxacin, taraxerin, taraxerol, taraxasterol, inulin, gluten, gum, potash, choline, levulin, and putin. Taraxacin and taraxasterol are active ingredients of the roots of dandelion. The dried plant contains 2.8% tannins. The bitter taste of dandelions is due to 11,13-dihydrotaraxine acid-1´O-β-D-glucopyranoside and several other awesomely-named chemicals (Kuusi et al. 1985). Salad cultivars are not as bitter as wild forms, but are likely less suitable medicinally, since the bitter principles are regarded as medicinally effective.

Non-medicinal Uses

For a plant with a reputation as a troublesome weed, dandelion has a remarkable number of virtues. It has been consumed for thousands of years as food. Almost all parts of the dandelion can be eaten. The nutritive value of dandelion greens, particularly for vitamin C, is much higher than that of most other salad plants. In Europe, dandelion was once widely used as a pot herb by the poor, who gathered it from nature. It was also grown in a blanched form and eaten in salads, and the roots were used as a coffee substitute.

Hybrids of T. *kok-saghyz* Rodin, Russian dandelion (from Turkmenistan), have been grown in Russia and elsewhere as a source of rubber, derived from the plant latex. Essentially abandoned as a crop, Russian dandelion has been reconsidered recently in Russia as a source of rubber in order to reduce dependence on imports.

Several species of *Taraxacum* are grown as ornamentals. Dandelion is a valuable bee plant because the flowers bloom early in the spring, and at that time may be the only major source of nourishment for bees. Dandelion is also considered to be an excellent, highly nutritious pasture plant for beef cattle.

As well as having food, medicinal and industrial uses, dandelions have proven to be valuable for research in evolution, ecology and population biology. Among the classical examples was the finding that genotypes from the most disturbed and ephemeral habitats produced the most seeds, adapting them to dispersal and colonization. By contrast, genotypes from stable habitats produced less seeds but were superior competitors in their own habitats.

Agricultural and Commercial Aspects

Despite widespread occurrence, the US has imported over 45 tonnes of dandelion in some years for use in patent medicines, at least as recently as 1957. In Canada, over 50 commercially sold medicinal preparations contain dandelion.

Dandelion is grown commercially as a food plant in Europe and North America. At least a dozen cultivars are available which are much tastier than wild dandelion. Centers of dandelion cultivation in North America include the eastern seaboard states, Florida and Texas. The annual value of dandelion sold in Canadian markets sometimes exceeds a half million dollars. Good yields are 18 800 kg/ha for leaves and 1 100–1 700 kg/ha for roots.

Myths, Legends, Tales, Folklore, and Interesting Facts

- In ancient Greek legend, the Minotaur was a monster shaped half like a man and half like a bull, confined in a labyrinth where he was periodically fed young men and maidens. After killing the Minotaur the Greek hero Theseus ate a dandelion salad.

Taraxacum (dandelion)

- An Algonquin legend relates the story of Shawondesee, the fat lazy south wind. One day he observed a dandelion in the form of a beautiful golden-haired maiden on the meadow near him, but he was too lazy to pursue her. In a few days, he returned, and saw a bent old woman with grizzled white hair. Disappointed, he heaved a tremendous sigh, and observed the old woman's white hair fly away on the breeze. In the spring, the south wind still sighs for the lost beauty who might have been his.
- The Iroquois ascribed sexual symbolism to dandelion roots. Roots with a side root reminiscent of the male sexual organ were tossed backwards while saying the name of one's desired love. Alternatively, roots growing intertwined were boiled, and the water used to wash one's face and fingers in the expectation that this would make one sexually irresistible.
- There are indications that seeds of dandelions were deliberately brought to North America by the *Mayflower* pilgrims, quite possibly to grow as a garden plant.
- The number of inches a child will grow in the coming year is said to be foretold by the tallest dandelion stalk he can find.
- A wish will come true if with one breath you can blow all the white silky fluff off a dandelion in the seed stage.

- Dandelion is so strong as a diuretic that children handling the flowers extensively have absorbed enough of the constituents through their skin to cause bed wetting.
- Dogs are likely to urinate on dandelions in lawns - an interesting irony since the only obvious medicinal use of the plants is to promote urination.
- The dandelion has been declared an endangered wildflower in England. While this may raise eyebrows, we too in Canada doubtless have native species of dandelions meriting protection. Despite their modern reputation as an undesirable weed, dandelions have great value, and it is important to learn more about them.

Selected References

Anon. 1989. Development opportunities in the horticultural sector for selected vegetables products. Agric. Can. Comm. Coord. Dir., Ottawa, ON. 60 pp.

Anon. 1990. Iron-fisted dandelions stand guard over crops. Agri-Features Agric. Can. Comm. Branch, No. **2120**: 11–12.

Bayer, M. 1973. The red dandelion. Nurs. Outlook **21**(1): 32.

Bergen, P., Moyer, J.R., and Kozub, G.C. 1990. Dandelion (*Taraxacum officinale*) use by cattle grazing on irrigated pasture. Weed Technol. **4**: 258–263.

Brunton, D.F. 1989. The marsh dandelion (*Taraxacum* section *Palustria*; Asteraceae) in Canada and the adjacent United States. Rhodora **91**: 213–219.

Davies, M.G., and Kersey, P.J. 1986. Contact allergy to yarrow and dandelion. Contact Dermatitis **14**: 256–257.

Doll, R. 1977. Zur *Taraxacum* - flora Nordamerikas. Feddes Repertorium **88**(1-2): 63-80.

Fernald, M.L. 1933. *Taraxacum* in Eastern America. Rhodora **35**: 369–386.

Fernald, M.L. 1948. The name *Taraxacum officinale*. Rhodora **50**: 216.

Gray, E., McGehee, E.M., and Carlisle, D.F. 1973. Seasonal variation in flowering of common dandelion. Weed Sci. **21**: 230-232.

Haglund, G. 1943. *Taraxacum* in Arctic Canada (East of 100° W). Rhodora **45**: 337–359.

Haglund, G. 1946. Contributions to the knowledge of *Taraxacum* flora of Alaska and Yukon. Sven. Bot. Tidskr. **40**: 325–361.

Haglund, G. 1948. Further contributions to the knowledge of the *Taraxacum* flora of Alaska and Yukon. Sven. Bot. Tidskr. **42**: 297–336.

Haglund, G. 1949. Supplementary notes on the *Taraxacum* flora of Alaska and Yukon. Sven. Bot. Tidskr. **43**: 107–116.

Haglund, G. 1950. *Taraxacum*. Flora. Aka. Yuk. **10**: 1633–1658.

Handel-Mazetti, H. 1907. Monographie der Gattung *Taraxacum*. Deuticke, Leipzig. 175 pp.

Hausen, B.M. 1982. Taraxinic acid 1'-O-beta-D-glucopyranoside, the contact sensitizer of dandelion. Derm. Beruf. Umwelt. **30**(2): 51–53. [In German.]

Kirschner, J., and Stepanek J. 1986. Toward a monograph of *Taraxacum* sect. *Palustia* (Studies in *Taraxacum* 5). Preslia **58**: 97–116.

Kuusi, T., and Autio, K. 1985. The bitterness properties of dandelion (*Taraxacum*): I. Sensory investigations. Lebensm.-Wiss. Technol. **18**: 339–346.

Kuusi, T., Hårdh, K., and Kanon, H. 1982. The nutritive value of dandelion leaves. Altern./Appropr. Technol. Agric. **3**(1): 53–60.

Kuusi, T., Hårdh, K., and Kanon, H. 1984. Experiments on the cultivation of dandelion for salads use. I. Study of cultivation methods and their influence on yield and sensory quality. J. Agric. Sci. Finland **56**(1): 9–22.

Kuusi, T., Hårdh, K., and Kanon, H. 1984. Experiments on the cultivation of dandelion for salads use. II. The nutritive value and intrinsic quality of dandelion leaves. J. Agric. Sci. Finland **56**(1): 23–31.

Kuusi, T., Pyysalo, H., and Autio, K. 1985. The bitterness properties of dandelion. II. Chemical investigations. Lebensm. Wiss. Technol. (Zurich) **18**: 349–349.

Lovell, C.R., and Rowan, M. 1991. Dandelion dermatitis. Contact Dermatitis **25**: 185–188.

Macoun, J.M. 1902. *Taraxacum* in Canada. Ottawa Nat. **15**: 276–7.

Manchev, M. 1981. Use of the leaves of the dandelion (*Taraxacum officinale* L.) in the feeding of the silkworm *Bombyx mori* L. Vet Med. Nauki. **18**(7): 105–110. [In Bulgarian.]

Martinkova, Z., and Honek, A. 1997. Germination and seed viability in a dandelion, *Taraxacum officinale* agg. Ochrana Rostlin **33**: 125–133.

Miller, S.S., and Eldridge, B.J. 1989. Plant growth regulators suppress established orchard sod and dandelion (*Taraxacum officinale*) population. Weed Technol. **3**: 317–321.

Mitich, L.W. 1989. The intriguing world of weeds: common dandelion — the lion's tooth. Weed Technol. **3**: 537–539.

Munro, D.B., and Small, E. 1997. Vegetables of Canada. NRC Research Press, Ottawa, ON. 417 pp. [Chapter on *Taraxacum*: pp. 351–354.]

Nijs, H. den, and Bachmann, K. 1995. Clonal diversity and phylogeography of asexual *Taraxacum* in Europe. Int. Org. Plant Biosyst. Newsl. **25**: 4–7.

Oldham, M.J., Brunton, D.F., Sutherland, D.S., and McLeod, D. 1992. Noteworthy collection: Ontario (*Taraxacum palustre*). Mich. Bot. **31**: 41–42.

Richards, A.J. 1985. Sectional nomenclature in *Taraxacum* (Asteraceae). Taxon **34**: 633–644.

Sackett, C. 1975. Dandelions. Fruit & vegetable facts & pointers. United Fresh Fruit and Vegetable Association, Alexandria, VA. 7 pp.

Schmidt, M. 1979. The delightful dandelion. Org. Gard. **26**(3): 112–117.

Shipley, N. 1956. The Hidden Harvest. Can. Geogr. J. **52**(4): 178–181.

Sherff, E.E. 1920. North American Species of *Taraxacum*. Bot. Gaz. **70**: 329–359.

Small, E. 1997. Culinary herbs. N.R.C. Research Press, Ottawa, ON. 710 pp. [Chapter on *Taraxacum*: pp. 598–605.]

Smith, A. (*Editor*) 1995. A dandy plant. Cornell Plant. **50**(2): 22.

Solbrig, O.T. 1971. The population biology of dandelions. Amer. Sci. **59**: 686-694.

Szabo,T.I. 1984. Nectar secretion in dandelion (*Taraxacum* spp.). J. Apicult. Res. **23**: 204–208.

Szczawinski, A. F., and Turner, N. J. 1978. Edible garden weeds of Canada. National Museums of Canada, Ottawa, ON. 184 pp.

Tanaka, O., Tanaka, Y., and Wada, H. 1988. Photonastic and thermonastic opening of capitulum in dandelion, *Taraxacum officinale* and *Taraxacum japonicum*. Bot. Mag. Tokyo **101**(1062): 103–110.

Tanaka, O., Wada, H., Yokoyama, T., and Murakami, H. 1987. Environmental factors controlling capitulum opening and closing of dandelion, *Taraxacum albidum*. Plant Cell Physiol. **28**: 727–730.

Taylor, R.J. 1987. Populational variation and biosystematic interpretation in weedy dandelions. Bull. Torrey Bot. Club **114**: 109–120.

Williams, C.A., Goldstone, F., and Greenham, J. 1996. Flavonoids, cinnamic acids and coumarins from the different tissues and medicinal preparations of *Taraxacum officinale*. Phytochemistry **42**: 121–127.

World Wide Web Links

(Warning. The quality of information on the internet varies from excellent to erroneous and highly misleading. The links below were chosen because they were the most informative sites located at the time of our internet search. Since medicinal plants are the subject, information on medicinal usage is often given. Such information may be flawed, and in any event should not be substituted for professional medical guidance.)

Fire effects information system:
http://svinet2.fs.fed.us/database/feis/plants/forb/taroff/

Growing and selling the common dandelion for food and medicine in New Zealand [an excellent review article]:
http://www.crop.cri.nz/broadshe/dandel.htm

The health benefits of dandelions:
http://www.herbal-alternatives.com/dandelio.htm

In defense of the dandelion:
http://www.mtnlaurel.com/win96_issue/dandelin.htm

Herbal Materia Medica, Dandelion, David L. Hoffman:
http://www.healthy.net/library/books/hoffman/materiamedica/dandelion.htm

Seed companies selling dandelion seed:
http://www.seedquest.com/seedcobycrop/herb/Dandelion.htm

Saskatchewan Agriculture and Food, Weed Control Notes, Weed Identification Series, Dandelion (*Taraxacum officinale*):
http://www.gov.sk.ca/agfood/weeds/dandelin.htm

A modern herbal by M. Grieve:
http://www.botanical.com/botanicaL/mgmh/d/dandel08.html

The downtrodden dandelion [has several good links]:
http://www.suite101.com/articles/article.cfm/7346

Dandelion (*Taraxacum officinale*):
http://herbsforhealth.miningco.com/library/weekly/aa061298.htm

Taxus brevifolia (Pacific yew)

Taxus brevifolia Nutt. Pacific Yew

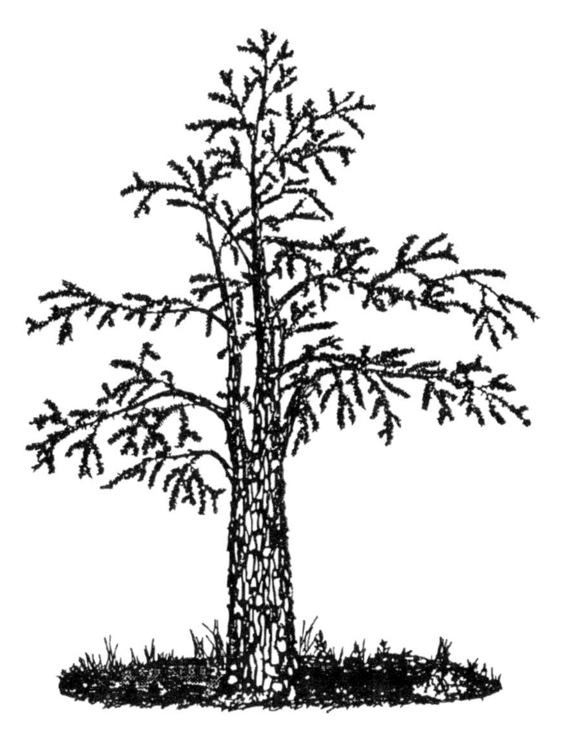

Taxus brevifolia (Pacific yew)

English Common Names

Pacific yew, western yew, American yew, Oregon yew, bowplant, mountain mahogany.

French Common Names

If de l'ouest, if occidental.

Morphology

Pacific yew is an evergreen, spindly tree, usually growing to a height of 6 – 10 m, and a diameter of 15–30 cm. One yew in western Washington had a record diameter of 1.4 m, and some trees are reported to have grown as high as 25 m. In drier, eastern parts of its range Pacific yew is found in open areas as a shrub less than 2 m high. Male trees produce small but abundant yellowish flowers on the underside of the branches. Female trees produce seeds enclosed in pulpy, sweet, red or scarlet arils (fleshy, berry-like structures), not cones as do most other Coniferae. The single seed of the fruit often protrudes beyond its outer cover. The foliage is relatively sparse. The trunk is tapered and usually fluted, and covered by scaly, reddish-brown to purplish-brown bark only 2–6 mm thick. Thin, purple scales of the outer bark are easily removed, exposing a reddish-purple under-bark. Lower branches contacting the soil will root, and cut stumps will sprout, producing clumps of trees. Yew trees are slow-growing and long-lived.

Classification and Geography

Six to 20 species of *Taxus* are recognized, depending on authority, with two species indigenous to Canada. The shrubby Canada yew (*T. canadensis* Marsh., also known as American yew and ground hemlock) occurs from Manitoba eastwards, and southwards into the US. This poisonous plant was used medicinally by North American Indians. Pacific yew is native to the mountains of western North America. It ranges from southeast Alaska to northern California, and from the Pacific coast to interior Idaho and Montana. Occasional trees are found as far south as San Francisco.

Ecology

Pacific yew normally grows inconspicuously and slowly beneath a conifer forest canopy, in dense shade. The species is not abundant, generally occurring in small groups or single trees. It grows best on cool, moist flats along streams, in deep gorges and damp ravines, and where fires are relatively infrequent. The seeds are disseminated by birds. Deer, elk, and moose sometimes browse on the foliage, although *Taxus* species are considered unpalatable to livestock, and poisoning has been reported for some of the species.

Medicinal Uses

Indigenous Peoples of North America used the bark, foliage, and fruits of yew medicinally. Bella Coola Indians used leaf tea for lung ailments; Chehalis Indians employed leaf preparations to induce healthful sweating; Cowlitz used poultices of ground leaves on wounds; and Karok drank twig bark tea to relieve stomach ache.

An anti-cancer compound taxol (paclitaxel) is present in yew trees. Taxol is active against advanced refractory ovarian cancer (for which treatment alternatives are limited), as well as breast cancer, and it is undergoing clinical trials for efficacy against a variety of other cancers. Taxol is therapeutic because it is a mitotic spindle poison which inhibits uncontrolled cancerous growths (i.e. the spindle apparatus which aligns chromosomes during cell division (mitosis) is disrupted, preventing cancer cells from reproducing). Cancer is the second most frequent cause of death in industrialized countries, and improved treatments are urgently needed. It has been estimated that in the future over a quarter million people could be treated yearly with taxol, and that the drug could have a commercial value of the order of $1 billion annually. Taxol treatment is presently expensive, the drug costing between $10 000.00 and $100 000.00 for each patient, depending on number of treatment cycles (one to ten).

Toxicity

Although yew trees are now viewed as a tree of life, ironically they were once known as the tree of death. This is because all parts of all species of *Taxus*, except the fleshy arils, can be quite poisonous to humans and livestock. Most references state that the "fleshy fruits" (i.e., the cup-like arils) are edible, but the seed within each aril can be deadly, and could be ingested unintentionally. Therefore, eating the arils is not recommended. The English yew is the most toxic native plant in Britain. Because yew species are so poisonous, no one should attempt self-medication.

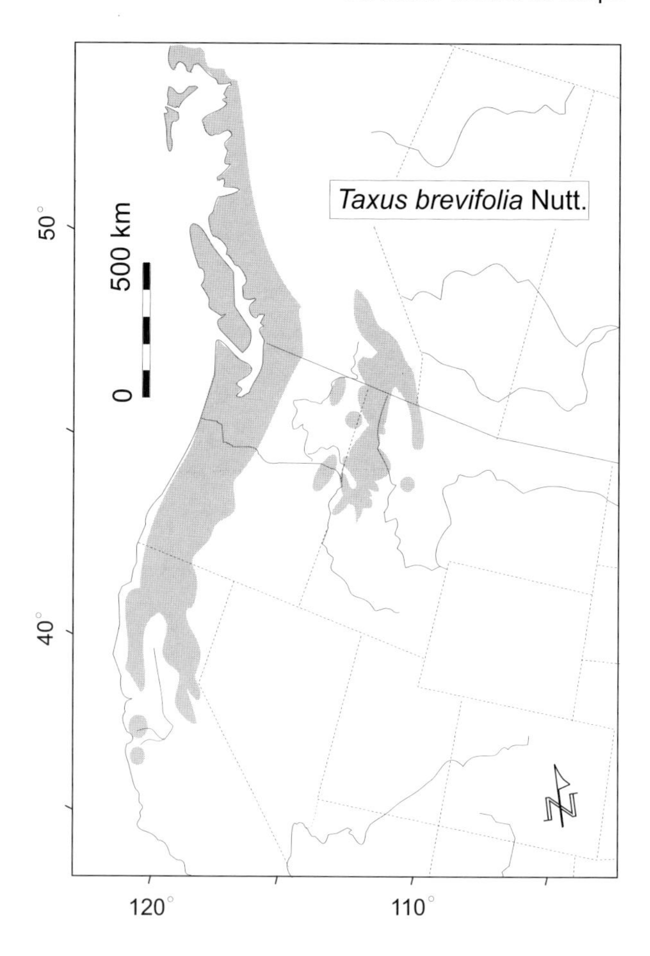

Chemistry

The anti-cancer compound taxol (paclitaxel) is contained in most parts of yew trees, but is especially concentrated in the inner bark (i.e., the cambium). The word Taxol (with a capital T) is a registered trademark name for a drug formulated with paclitaxel (cf. "Coke" and "coke").

Non-medicinal Uses

Traditionally, Native North Americans valued the extremely hard, decay-resistant wood for tools like canoe paddles and fish hooks, weapons like archery bows and spears, and ceremonial and decorative items. Gunstocks, boat decking, veneer, snowshoe frames, furniture, musical instruments, and sculptures have all been made of yew in recent times. Archery bows and canoe paddles continue to be manufactured from yew. The wood is still used occasionally by craftsmen, and for firewood and fence wood, and the plant is sometimes grown as an ornamental. The hybrid yew (*Taxus* × *media* Rehd. = *T. baccata* × *T. cuspidata*) is the most frequently cultivated species in southern Ontario, the largest region of ornamental yew cultivation in Canada (J.B. Phipps, personal communication). In Canada and the northern US, Japanese yew (*Taxus cuspidata* Sieb. & Zucc.) is frequently cultivated, and all parts of this except the fleshy red fruits are very toxic.

Agricultural and Commercial Aspects

Until recently, Pacific yew has been considered a nuisance weed by modern forestry, insufficiently important to harvest for lumber or pulp. It was often burned along with logging slash that remained after timber harvest. In 1962, the National Cancer Institute of the US found that extracts from Pacific yew bark showed in vitro activity against cancer cells. Thus began a period of stardom for this hitherto largely ignored "trash species," which was turned into "Cinderella of the Great Western Woods." Coincidentally, this period of celebrity overlaps that of another rare species of older western forests, the spotted owl. The seriously declining owl population unfortunately led to the polarization of environmentalists and loggers as to whether habitat or jobs is more important. An even more painful ethical dilemma was raised by the yew tree: whether saving trees or victims of cancer is more important. Of course, this is an artificial debate: the maximum yield from natural stands can only be obtained by limiting harvests to sustainable levels, and following habitat-conservation principles.

Yew bark is usually harvested by cutting the tree down, peeling the bark off the trunk and larger limbs using chisels or hatchets, and bagging the bark. Studies of damaged trees have suggested that killing entire trees, in many cases over 100 years old, is unnecessarily wasteful, and that removal of bark from one side allows the tree to survive without seriously affecting growth, and thus provides more bark for future harvests. The bark is chipped, dried, and its taxol extracted. Illegal harvest by poachers on public and private lands has become a major problem in the US and Canada. Unfortunately the yield of taxol from the bark is extremely small: < 0.02% dry weight. A single women suffering from ovarian cancer might need as much as 3 g of taxol, which would require the bark of 7.5 average yew trees. One study indicated an average requirement of six 100-year-old trees per cancer patient. About 7270 kg of bark is required to produce 1 kg of taxol. About 726 000 kg of Pacific yew bark were harvested in 1991. Projected demands over the next 20 years could require sacrificing as many as a million trees a year, a figure which is well beyond the available supply of wild trees. Coupled with the facts that the Pacific yew is rare and very slow growing, the expanding demand has made it clear that wild trees cannot provide adequate supplies. Decimation of trees led to the US government in 1992 passing The Yew Act to ensure sustainable management of the yew harvest.

Several alternatives to harvesting wild Pacific yew to obtain taxol are now being developed. These include: establishing yew plantations, not just of Pacific yew, but of several cultivated species (all other species of *Taxus* of the world produce taxol, although in lower concentrations); breeding high-taxol cultivars of *Taxus*; production using tissue culture; production from *Taxomyces*, a newly-named genus of fungus isolated from *Taxus*, which also produces taxol (an apparent example of natural exchange of the genes responsible for taxol synthesis between higher plants and fungi); full synthesis; and synthesis of chemical analogues. These alternatives are currently being explored by numerous laboratories and intensive cultivation in forest tree nurseries has begun. However, at present the chief means of augmenting the supply of taxol obtained from bark is a partial synthesis of taxol starting with extracted chemicals from the foliage of species of *Taxus* (often collected abroad). This is a welcome development since the foliage can be harvested regularly without sacrificing the plants.

Taxus brevifolia (Pacific yew)

The story of Pacific yew is instructive. It demonstrates how research on biodiversity can uncover invaluable materials and information. It illustrates the need to preserve biodiversity options for future generations. It reminds us of how unchecked human greed can quickly threaten a natural resource. It also shows how wise stewardship of an ecosystem can be achieved to ensure sustainable harvest of a natural resource of great importance to people.

Myths, Legends, Tales, Folklore, and Interesting Facts

- Some Pacific yew trees may live for over 500 years, and it has been suggested that one tree reached the age of 1800 years. It has been claimed that an English yew is the world's second oldest living tree. One English yew tree is believed to be over 4000 years old. [Claims of world's longest living tree have been made for a pair of closely related species in the southern US Rockies, *Pinus aristata* Engelm. (bristlecone pine) and for another pine with which it is often combined, *P. longaeva* Bail. (ancient pine); one tree named "Methuselah" was found to be 4 723 years old and is said to be the world's oldest known living tree. Ironically after studying the world's oldest organism, the discoverer, Edmund Schulman, died of a heart attack at the comparatively young age of 49.]
- Vancouver Island Indians used yew to make paddles for their dugout canoes. Because the paddles were heavy and sharp, they could double as weapons.
- In the Willamette Valley, Oregon, Native Americans were often buried with their yew bows.

- The English yew has the distinctions of being known as the tree from which famous archers such as Robin Hood and William Tell made their bows. Arthur Conan Doyle (1859–1930), creator of Sherlock Holmes, wrote:

  What of the bow?
  The bow was made in England;
  Of true wood, of yew wood,
  The wood of English bows;
  So men who are free
  Love the old yew tree,
  And the land where the yew tree grows.

- In England, yews have a particular association with graveyards. This is because Druids appear to have planted yews at sacred sites, which were taken over by Christians, who built their churches there, establishing nearby graveyards. It has been suggested that the yews are able to live for many centuries because they are nourished by the calcium-rich soil left by the bones of the dead. According to one myth, the roots of yew trees in graveyards enter the mouths of the deceased, grow down their throats, envelope their hearts, and whisper their secrets in the breeze. Alfred Lord Tennyson (1809–1892), Poet Laureate of England, wrote (in his poem *In Memoriam*):

  Old Yew, which graspest at the stones
  That name the underlying dead,
  Thy fibers net the dreamless head,
  Thy roots are wrapt about the bones.

- The International Union for the Conservation of Nature recently issued a report showing that 33 798 of the 270 000 known species of plants (about one in eight) are in danger of extinction. The report stated that 15 of 20 species of yew trees are threatened.

Taxus baccata (English yew)

- Some garden books have recommended burning yew clippings to ward off insects - a dangerous practice since poisonous constituents may be carried in the smoke.
- In 1991 a leathery corpse that has become known as "The Ice Man," deep-frozen for about 5300 years, was found trapped in a glacier on the Austrian/Italian Alps. Beside his body was found a little axe handle and a bow, both made of yew wood.

Selected References

Adams, J.D., Flora, K.P., Goldspiel, B.R., Wilson, J.W., Arbuck, S.G., and Finley, R. 1993. Taxol: a history of pharmaceutical development and current pharmaceutical concerns. J. Natl. Cancer Inst. Monogr. **15**: 141–147.

Anon. 1992. Taxol news. Bioeng. News **13**(11): 5–6.

Anon. 1992. Medicinals from India. Herbalgram **27**: 7, 58–59.

Anon. 1993. Taxol from fungus. Biotech. News **13**(12): 7–8.

Bailey, J.D., and Liegel, L.H. 1997. Response of Pacific yew (*Taxus brevifolia*) to partial removal of the overstory. West. J. Appl. For. **12**(2): 41–43.

Borman, S. 1991. Scientists mobilize to increase supply of anticancer drug taxol. Chem. Eng. News **69**(35): 11–18.

Busing, R.T., Halpern, C.B., and Spies, T.A. 1995. Ecology of Pacific yew (*Taxus brevifolia*) in western Oregon and Washington. Conserv. Biol. **9**: 1199–1207.

Choi, M.S., Kwak, S.S., Liu, J.R., Park, Y.G., Lee, M.K., and An, N.H. 1995. Taxol and related compounds in Korean native yews (*Taxus cuspidata*). Planta Med. **61**: 264–266.

Crawford, R.C., and Johnson, F.D. 1985. Pacific yew (*Taxus brevifolia*) dominance in tall forests, a classification dilemma. Can. J. Bot. **63**: 592–602.

DeFuria, M.D., and Horovitz, Z. 1992. Taxol commercial supply strategy. *In* Proceedings of the second National Institute Workshop on Taxol and *Taxus*, Alexandria, VA. pp. 195–198.

Difazio, S.P., Vance, N.C., and Wilson, M.V. 1996. Variation in sex expression of *Taxus brevifolia* in western Oregon. Can. J. Bot. **74**: 1943–1946.

Difazio, S.P., Vance, N.C., and Wilson, M.V. 1997. Strobilus production and growth of Pacific yew under a range of overstory conditions in western Oregon. Can. J. For. Res. **27**: 986–993.

Donehower, R.C., and Rowinsky, E.K. 1993. An overview of experience with taxol (paclitaxel) in the U.S.A. Cancer Treat. Rev. **19**(Suppl. C): 63–78

Edgington, S.M. 1991. Taxol out of the woods. Bio/Tech. **9**: 933–934.

El-Kassaby, Y.A., and Yanchuk, A.D. 1994. Genetic diversity, differentiation, and inbreeding in Pacific yew from British Columbia. J. Hered. **85**: 112–117.

Elsohly, H.N., Croom, E.M., el-Kashoury, E.S., elSohly, M.A., and McChesney, J.D. 1994. Taxol content of stored fresh and dried *Taxus* clippings. J. Nat. Prod. **57**: 1025–1028.

Georg, G.I, Chen, T.T., Ojima, I., and Vyas, D.M. (*Editors*). 1995. Taxane anticancer agents: basic science and current status. ACS Symposium Series 583, American Chemical Society, Washington, DC. 353 pp.

Göçmen, B., Jermstad, K.D., Neale, D.B., and Kaya, Z. 1996. Development of random amplified polymorphic DNA markers for genetic mapping in Pacific yew (*Taxus brevifolia*). Can. J. For. Res. **26**: 497–503.

Hamon, N.W. 1993. Yew. Can. Pharm. J. **126**: 192, 196, 199-200.

Hartzell, H., Jr. 1991. The yew tree: a thousand whispers: the biography of a species. Hulogosi, Eugene, OR. 318 pp.

Heiken, D.O. 1992. The Pacific yew and taxol: federal management of an emerging resource. J. Environ. Law and Litigation 7: 175–245.

Hils, M.H. 1993. Taxaceae. *In* Flora of North America north of Mexico, Vol. 2. *Edited by* Flora of North America Editorial Committee. Oxford University Press, New York, NY. pp. 423-427.

Hogg, K.E., Mitchel, A.K., and Clayton, M.R. 1996. Confirmation of cosexuality in Pacific yew (*Taxus brevifolia* Nutt.). Great Basin Nat. **56**: 377–378.

Joyce, L. 1991. Scientists take wide variety of approaches to taxol studies. The Scientist **5**(24): 15, 20, 22.

Kingston, D.G.I. 1993. Taxol, an exciting anticancer drug from *Taxus brevifolia*. Am. Chem. Soc. Symp. Ser. **534**: 138–148.

McAllister, D.E., and Haber, E. 1991. Western yew - precious medicine. Can. Biodiversity **2**: 2–4.

Minore, D., and Weatherly, H.G. 1994. Effects of partial bark removal on the growth of Pacific yew. Can. J. For. Res. **24**: 860–862.

Minore, D., and Weatherly, H.G. 1996. Stump sprouting of Pacific yew. U.S. For. Serv. Gen. Tech. Rep. PNW **378**: 1–6.

Minore, D., Weatherly, H.G., and Cartmill, M. 1996. Seeds, seedlings, and growth of Pacific yew (*Taxus brevifolia*). Northwest Sci. **70**: 223–229.

Mitchell, A.K. 1997. Propagation and growth of Pacific yew (*Taxus brevifolia* Nutt.) cuttings. Northwest Sci. **71**(1): 56–63.

Murray, M.D. 1991. The tree that fights cancer. Am. For. **97**(7–8): 52–54.

Nicholson, R. 1992. Death and *Taxus*: the yew is an unlikely candidate for the tree of life. Nat. Hist. **9**: 20.

Nicolaou, K.C., R.K. Guy, and P. Potier. 1996. Taxoids: New weapons against cancer. Sci. Am. **274**(6): 94–98.

Nicolaou, K.C.,Yang, Z., Liu, J.J., Ueno, H., Nantermet, P.G, Guy, R.K, et al. 1994. Total synthesis of taxol. Nature **367**: 630–634.

Piesch, R.F., and Wheeler, N.C. 1993. Intensive cultivation of *Taxus* species for the production of taxol - Integrating research and production in a new crop plant. Acta Hortic. **344**: 219–228.

Rae, C.A., and Binnington, B.D. 1995. Yew poisoning sheep. Can. Vet. J. **36**: 446.

Rao, K.V. 1993. Taxol and related taxanes. I. Taxanes of *Taxus brevifolia* bark. Pharm. Res. **10**: 521–524.

Rao, K.V., Bhakuni, R.S., Hanuman, J.B., Davies, R., and Johnson, J. 1996. Taxanes from the bark of *Taxus brevifolia*. Phytochemistry (Oxford) **41**: 863–866.

Rao, K.V., Hanuman, J.B., Alvarez, C., Stoy, M., Juchum, J., Davies, R.M., and Baxley, R. 1995. A new large-scale process for taxol and related taxanes from *Taxus brevifolia*. Pharm. Res. **2**: 1003–1010.

Scher, S., and Jimerson, T.M. 1989. Does fire regime determine the distribution of Pacific yew in forested watersheds. U.S. Dep. Agric. Pac. Southwest For. Range Exp. Stn. Gen. Tech. Rep. 109. 160 pp.

Stierle, A., Strobel, G., and Stierle, D. 1993. Taxol and taxane production by *Taxomyces andreanae*: an endophytic fungus. Science **260**(5105): 214–216.

Stierle, A., Strobel, G., Stierle, D., Grothaus, P., and Bignami, G. 1995. The search for a taxol-producing microorganism among the endophytic fungi of the Pacific yew, *Taxus brevifolia*. J. Nat. Prod. **58**: 1315–1324.

Strobel, G.A., Stierle, A., and Hess, W.M. 1993. Taxol formation in yew — *Taxus*. Plant Sci. **92**: 1–12.

Suffness, M. (*Editor*). 1995. Taxol science and applications. CRC Press. Boca Raton, FL. 426 pp.

Vance, N.C., Kelsey, R.G., and Sabin, T.E. 1994. Seasonal and tissue variation in taxane concentrations of *Taxus brevifolia*. Phytochemistry **36**: 1241–1244.

Wall, M.E., and Wani, M.C. 1995. Camptothecin and taxol: discovery to clinic - thirteenth Bruce F. Cain Memorial Award Lecture. Cancer Res. **55**: 753–760

Walter-Vertucci, C., Crane, J., and Vance. N.C. 1996. Physiological aspects of *Taxus brevifolia* seeds in relation to seed storage characteristics. Physiol. Plant. **98**: 1–12.

Werth, J.Von Der, and Murphy, J.J. 1994. Cardiovascular toxicity associated with yew leaf ingestion. Br. Heart J. **72**(1): 92–93.

Wheeler, N.C. 1993. Taxology: a study in technology commercialization. J. For. **91**(10): 15–18.

Wheeler, N.C., Jech, K., Masters, S., Brobst, S.W., Alvarado, A.B., Hoover, A.J., and Snader, K.M. 1992. Effects of genetic, epigenetic, and environmental factors on taxol content in *Taxus brevifolia* and related species. J. Nat. Prod. **55**: 432–440.

Wheeler, N.C., Jech, K.S., Masters, S.A., O'Brien, C.J., Timmons, D.W., Stonecypher, R.W., and Lupkes, A. 1995. Genetic variation and parameter estimates in *Taxus brevifolia* (Pacific yew). Can. J. For. Res. **25**: 1913–1927.

Whiterup, K.M., Look, S.A., Stasko, M.W., Ghiorzi, T.J., Muschik, G.M., and Cragg, G.M. 1990. *Taxus* spp. needles contain amounts of taxol comparable to the bark of *Taxus brevifolia*: analysis and isolation. J. Nat. Prod. **53**: 1249–1255.

World Wide Web Links

(Warning. The quality of information on the internet varies from excellent to erroneous and highly misleading. The links below were chosen because they were the most informative sites located at the time of our internet search. Since medicinal plants are the subject, information on medicinal usage is often given. Such information may be flawed, and in any event should not be substituted for professional medical guidance.)

Fire effects information system:
http://svinet2.fs.fed.us/database/feis/plants/shrub/taxbre/

Medical attributes of *Taxus brevifolia* — the Pacific yew, by M. Costello and K. Kellmel:
http://wilkes1.wilkes.edu/~kklemow/Taxus.html

Pacific yew & taxol, by A. Mitchell:
http://www.pfc.forestry.ca/www_users/lgalbraith/mitchelL/yew.html

L'if de l'Ouest et le taxol, par A. Mitchell:
http://www.pfc.cfs.nrcan.gc.ca/www_users/lgalbraith/mitchelL/yewf.html

Taxol (Paclitaxel),The cancerBACUP factsheet:
http://www.cancerbacup.org.uk/info/factsheet/taxol.htm

Taxol: an exciting anticancer compound:
http://c267b.chor.ucl.ac.be/taxol.htm

Taxol, purpose & side effects:
http://www.bmi.net/mcaron/taxol.html

Taxol and the yew tree, by N.J. Lawrence:
http://uchii1.ch.umist.ac.uk/group/subtopics/treeoflife.html

The history of taxol:
http://www.missouri.edu/~chemrg/210w97/taxol_bodypage.htm

Taxol, by N. Edwards:
http://www.bris.ac.uk/Depts/Chemistry/MOTM/taxoL/taxol1.htm

Vaccinium macrocarpon (cranberry)

Vaccinium macrocarpon Ait. Cranberry

Synonym: *Oxycoccus macrocarpos* (Ait.) Pers.

English Common Names

Cranberry, large cranberry, American cranberry.

Red-fruited species of *Vaccinium* are almost universally called cranberries in North America. The taxonomically complex genus *Vaccinium* is often divided into several subgenera or sections, including the subgenus or genus *Oxycoccos*, in which about four species are placed, including *V. macrocarpon* and its close relative *V. oxycoccos* L., sometimes called the "true cranberries." The partridgeberry or mountain cranberry (*V. vitis-idaea* L.) is segregated in a different subgenus of *Vaccinium*, but is somewhat similar. The American cranberrybush, *Viburnum trilobum* Marsh., also called "highbush cranberry," is not related to the true cranberries; this species is a close relative of the elderberry. The Florida cranberry, *Hibiscus sabdariffa* L., is also quite unrelated to *Vaccinium*, but the fleshy calyx which surrounds the mature fruit is cooked and served as "cranberry."

French Common Names

Canneberge, airelle à gros fruits, gros atocas (Îles de la Madeleine: graines, pommes de prée).

Morphology

The cranberry is an evergreen, creeping, mat-forming plant with slender, intricately forking woody horizontal stems 30–150 cm long, and upright flowering stems arising from the leaf axils. Although often referred to as a vine, it is a trailing shrub rather than a true vine. The leaves are 5–18 (generally 7–10) mm long, somewhat leathery, with a very short leaf stalk. One to 10 nodding pink flowers are borne on young upright shoots 4–15 cm high. Each flower is on a pedicel 10–30 mm long, in the axil of a small leaf. Flower buds form in late summer, open in midsummer (late June – early July) of the following year and, after insect pollination, the fruits mature late in the autumn (mid- to late October). The berries are typically bright red at maturity (some plants produce white berries), 1–2.5 cm thick, globose, ellipsoid, or pear-shaped, with hard, shiny skin. Fruits of cranberry cultivars may be bell-shaped, bugle-shaped, or cherry-shaped, and color may vary from a light yellow through to dark red to almost black. The berries are high in acid and pectin, and extremely tart (astringent) due to their low sugar content. The fruits ripen in early to mid-fall, at which time they become loosely attached to the stem.

Classification and Geography

Vaccinium is a large, complex, imperfectly understood genus of perhaps 150 species (estimates range up to 400), mostly in North America and eastern Asia. This genus includes a wide range of edible berries (bearing such colorful names as blueberries, tackleberries, hurtleberries, huckleberries, farkleberries, sparkleberries, deerberries, and southern gooseberries). Perhaps the best known medicinal species is *V. myrtillus* L., the bilberry, treated elsewhere in this work.

Vaccinium vitis-idaea L., variously known as the lingonberry (lingenberry), partridgeberry, foxberry, mountain cranberry, and rock cranberry, grows well in very cold climates, and indeed is circumboreal. It is a minor wild-collected crop of Newfoundland (over 100 000 kg/year) and Nova Scotia (about 5000 kg/year). The species is grown in the Scandinavian countries and is currently being developed as a crop in Poland and the Soviet Union. In Europe, lingonberry sauce is often labelled for export as cranberry. The lingonberry has been occasionally used medicinally as a urinary antiseptic, as described in some detail here for the large cranberry.

The large cranberry, *V. macrocarpon*, is native to North America. It is distributed from Newfoundland to central Minnesota, south to Nova Scotia, New England, Long Island (New York), West Virginia, northern Ohio, central Indiana, northern Illinois, and rarely Arkansas, and the Appalachian Mountains of Tennessee and North Carolina. The large cranberry is cultivated within its native range, but also in parts of British Columbia, Washington and Oregon, and in these areas it has escaped and established in the wild. Plants are also cultivated in parts of northern and central Europe, and the species has also escaped and become naturalized in Britain, Germany, the Netherlands and Switzerland. Numerous cultivars were selected in the 19th century. Most cranberry cultivars originated as single vine selections from native populations, but breeding has produced some improved cultivars. Cultivars include 'Early Black,' 'Howes,' 'Stevens,' 'Searles,' 'McFarlin,' 'Bergman,' 'Crowley,' and 'Ben Lear.'

Vaccinium oxycoccus, the small cranberry or mossberry, is closely related and very similar to

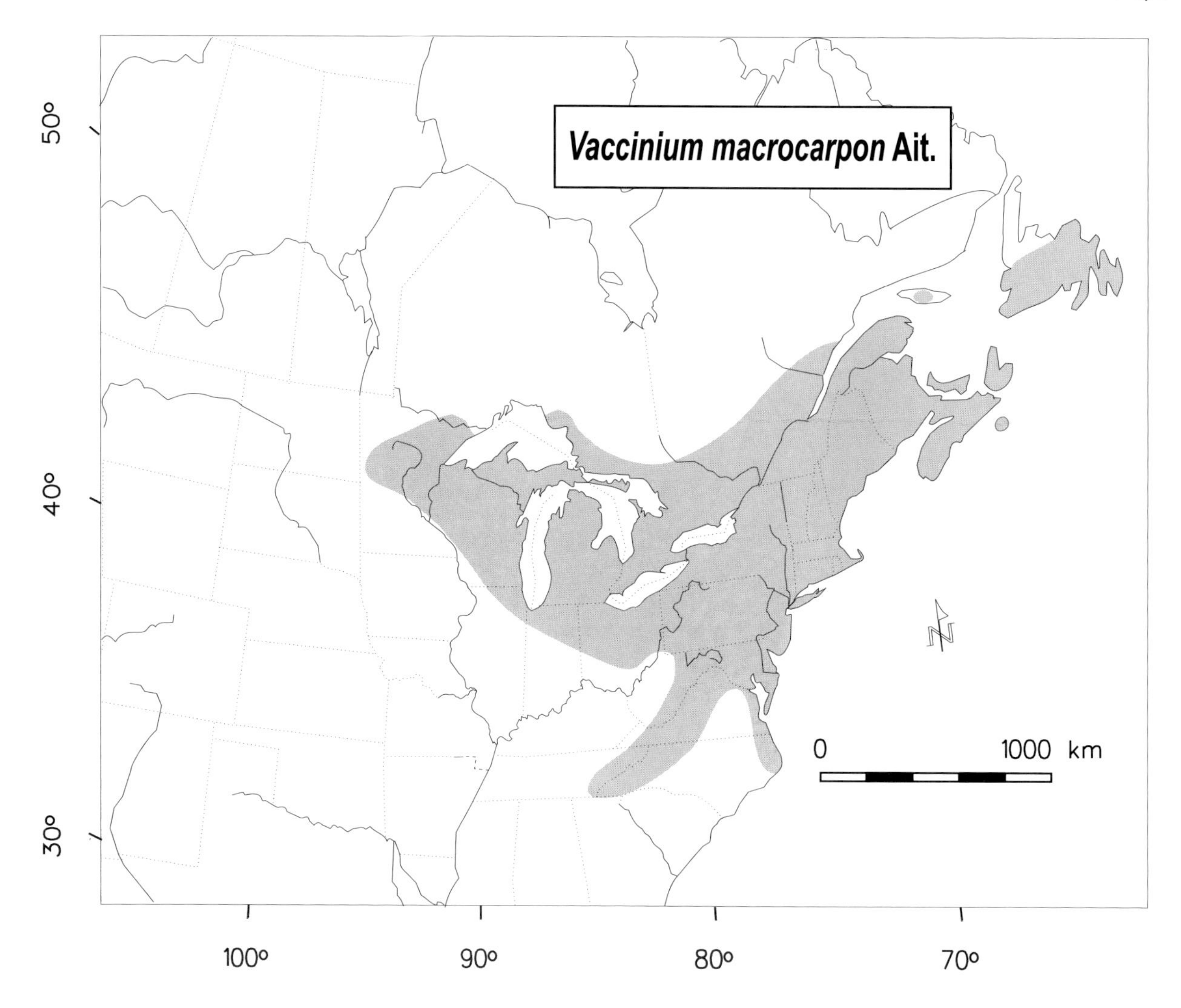

the large cranberry. The small cranberry is a very widespread species of boggy or peaty soil, occurring from Greenland and Labrador to Alaska, south to New Jersey, Pennsylvania, Ohio, Indiana and Minnesota, in the west to California, and also in Eurasia. The species is variable and is sometimes treated as a complex of several species. It has been interpreted as a tetraploid hybrid (with 48 chromosomes) of the diploid *V. macrocarpon* (with 24 chromosomes) and the diploid *V. microcarpon* (Turcz.) Hook. of boreal America and Eurasia. The small cranberry may be distinguished from the large cranberry by the pedicels bearing a pair of smaller red bractlets less than 1 mm broad near or below their middle (whereas the bractlets are closer to the flowers in *V. macrocarpon*, green and mostly 1–3 mm broad) and the smaller fruit (5–13 mm in diameter vs. 10–25 mm for *V. macrocarpon*). The leaves of the small cranberry are pointed whereas those of the large cranberry are blunt or rounded at the tip. Berries are gathered from the wild, but the smaller fruits and relatively limited productivity of the small cranberry have discouraged interest in its domestication.

The genetic diversity of the large cranberry is threatened as a consequence of the general decline of natural wetland habitats. Genetic diversity appears to be low in the northern part of the range but relict populations in the southern Appalachians south of land glaciated during the Pleistocene period may be more variable. A recent study suggested that DNA analysis (ssRAPDs) can be used effectively in assessment of genetic diversity and varietal identification.

Ecology

Cranberry occurs in open areas and wet shores, preferring open acid bogs, swamps and damp heaths. It is adapted to highly acidic soils (thriving at pH 3.2–4.5). The cranberry can self-pollinate, although bees increase fruit yield, and for this purpose bee colonies are frequently brought into cranberry bogs during flowering. A variety of natural pollinators has been reported, most important of which are bumble bees. The berries of *V. macrocarpon* are dispersed by water and shore birds. Muskrat and deer are also fond of the berries.

Medicinal Uses

European colonists appreciated the value of cranberry in warding off scurvy (a legitimate medicinal use, since the berries are high in vitamin C). Cranberry also acquired a folk reputation as a treatment for urinary tract problems, and this application has also been validated. Cranberry is a strong diuretic (promoting urination), and the juice is often prescribed as dietary treatment for urinary tract infections, kidney disorders and other conditions where the passing of fluids is desirable. Many women suffer from cystitis, an inflammation and (or) infection of the bladder, at some time in their lives, with as many as one in five women estimated to have urinary tract discomfort at least once a year. Cystitis is most frequently caused by bacteria that travel from the urethra, and because women have a shorter urethra which is more easily contaminated by organisms from the vagina and surrounding areas, urinary tract infections are about 50 times more common in females than in males. Antibiotics are an effective therapy, but antibiotic use sometimes results in adverse reactions, can be costly, and can lead to the development of resistant organisms. Cranberry can be used both as a preventive and as an adjunct treatment for urinary tract infections. Cranberry juice therapy may require drinking a liter (or pint) of straight cranberry juice daily, which few are willing to do. Fortunately cranberry capsules are easily swallowed. Consumers should be aware that some of the "cranberry juice" on the market may have much sugar and water added, but the pure form or a concentrate is likely to be more beneficial. In addition to preventing urinary tract infections, cranberry is reported to have the added benefit of deodorizing the urinary tract.

There are several theories to explain why cranberry is effective for maintaining urinary tract health: cranberry juice may make the urine more acidic, and the bacteria that cause infection are not likely to flourish in an acid environment (there is relatively little support today for this explanation); cranberries may cause relatively large amounts of

Vaccinium macrocarpon (cranberry)

the bacteriostatic hippuric acid to be excreted in the urine; components of cranberry juice may interfere with the ability of bacteria to stick to the lining of the bladder and urethra, and so they are washed away in the urinary stream. The last interpretation is generally favored today. It appears that the most common bacterium responsible for urinary tract infection, *Escherichia coli*, produces constituents known as adhesins to anchor to tissues, and that there are anti-adhesin factors in cranberry juice.

Preliminary tests conducted by University of Wisconsin researchers have suggested that cranberry juice may have an anti-oxidant effect on clogged heart arteries, thus reducing cardiovascular disease.

Toxicity

Cranberries are not considered toxic. Drinking more than 4 liters a day can result in diarrhea and other gastrointestinal problems.

Chemistry

Cranberries are extremely high in vitamin C, moderately high in vitamin A, and quite high in fiber and anthocyanins, all components that have health-giving qualities. Hippuric acid is an important medicinal constituent of the fruit, as the metabolism of this compound produces low pH urine, unlike most other fruits, which as noted above may explain why cranberry juice is a useful urinary antiseptic. (Compare our treatment of bearberry, which is effective as a urinary antiseptic only if the urine is alkaline.)

Non-medicinal Uses

Cranberries have long been in demand as a food plant. The fruit was prized by Native Americans, who used it in many ways, including in pemmican, a dried mixture of animal fat and fruit, which was the precursor of the dehydrated foods used by present-day hikers during camping trips. Benzoic acid in the berries likely aided in preservation. Today, cranberries provide a tasty sauce to accompany meat and poultry, as well as being incorporated in salads and numerous products, including juice, cocktails, pies, tarts, jellies and preserves. The cranberry is a beautiful fruit, and this has led to its use especially during the Christmas and Thanksgiving festive seasons. Most cranberries are frozen for storage or marketing.

Agricultural and Commercial Aspects

Cranberry culture was developed in the New World. Henry Hall of Dennis, Massachusetts, on Cape Cod, was the first to cultivate the crop, at the beginning of the 19th century. Commercial (engineered) bogs became common soon after that. By the late 19th century, commercial cultivation had spread north to Nova Scotia.

The center of cranberry cultivation and production is Massachusetts, but considerable quantities are also raised in the peatlands of British Columbia, New Jersey, Washington and Oregon. Considerable cranberry culture also occurs in Wisconsin, and in limited degree in Ontario, Quebec, and the maritime provinces. In Canada native people have harvested cranberries on a small scale for many years, for example in the Parry Sound region of Ontario. For a long period the major area of cranberry production in Canada was Lulu Island at the mouth of the Fraser River. Currently in Canada cranberry growing is carried out primarily in the lower portions of the Fraser River Valley of British Columbia, an area near Drummondville Quebec, and in several parts of Nova Scotia. There are over 1000 cranberry growers in the United States, utilizing almost 14 000 ha and producing over $1 billion in retail sales annually. Ocean Spray Inc., a growers' cooperative comprising 950 cranberry and grapefruit growers in the US and Canada, markets about 90 percent of the cranberries grown in North America.

Cranberry bogs

Cranberries are confined to cool, moist regions, and culture of them is largely restricted to acid soils along the edges of streams and ponds, and in bogs of temperate North America. Cranberry culture is a highly specialized form of small fruit production. Although cranberries are adapted to moist habitats, too high a water level encourages rushes, sedges and other competing plants, which then crowd the crop and interfere with pest control and picking. *Sphagnum* bogs are unsuitable until drained and the substrate allowed to decay to muck. Muck soils are particularly suitable because of their moisture-holding capacity. Cranberries have also been successfully planted on relatively dry soils near the sea, where relatively low summer temperatures retard evaporation. Bogs are occasionally built on sand or clay with little or no muck, but these usually require considerable fertilization, normally supplied by the decaying organic matter.

The development of a cranberry bog is a complex operation requiring considerable capital outlay. Typically a series of planting beds about a ha in size, serviced by a single reservoir of water, is laid out. A cranberry bog is usually established by pressing stem cuttings into the surface of a freshly sanded bed. A coating of sand is provided annually or every 2 years. Sand is a favorable rooting medium, suppressing weeds, and protecting the plants from mild frosts by holding absorbed heat from the sun. Cranberry bears some fruit in the second and third season, but 3–5 years are necessary for full production. Because of high labor costs and a short picking season, mechanical harvesters are widely used. Bogs are typically flooded at harvest time to a depth of 15–20 cm. The berries are shaken off the bushes and as they float on the surface they are easily raked or vacuumed into containers for cleaning and sorting. All of these operations can be carried out by some mechanical harvesters. Flooding during the winter season is to protect the plants from freezing damage (encasing the vines in ice insulates the buds against winter injury). Properly maintained bogs are almost permanent, some plantings remaining productive for over 75 years. Yield usually ranges from 8 000 to 10 000 kg/ha.

Cultivation

Pests of cranberry crops may account for almost 35% of field costs. The pests include at least 26 insects, 35 fungi, 6 nematodes, several viruses, and many weeds (mostly grasses and sedges), as well as two parasitic dodder plants (genus *Cuscuta*). Larvae of the cranberry fruit worm moth consume the berries. These and some of the fireworm moth pests can be effectively controlled by appropriate cycles of flooding. Frost damage has been a major cause of periodic crop losses. Yield of cranberries has increased by five times between 1909 and the present due to gradual and continuous improvements in management, including frost injury prevention, improved harvest technology and modern control of pests. The substantial increases in productivity have been stimulated by a combination of scientific research and producer innovations.

The future

In 1955 20% of US households used cranberry juice products. By 1985 this had increased to 70%, and the popularity of cranberry continues to expand. There has been concern that environmentally sensitive wetlands might be eliminated or damaged by development of new cranberry bogs, but technology is available to control damage to wetland sites, and even to develop bogs on dryland sites. Demand for the fruit has almost always exceeded the supply, keeping prices high. However, the potential for increasing productivity is good. There are a number of locally adapted cultivars, and newer hybrid cultivars are expected to become increasingly available as a result of accelerated breeding programs. Experts have suggested that considerable improvements could be realized with a sustained breeding program for insect resistance, as well as other traits. A wealth of information on cultivation procedures exists in books and regional agriculture department pamphlets. Current markets are reported to require several thousand hectares of new plantings in the near future, and demand is expected to continue to increase. Although the cost of initial establishment is relatively high and costs are not recovered quickly, cranberry appears to be a good investment. With a large potential area of cultivation in the eastern provinces and British Columbia, cranberry could become a much more important Canadian crop.

Myths, Legends, Tales, Folklore, and Interesting Facts

- "Cranberry" is a corruption of "crane-berry," an early American name for the plant. The flower bud was said to resemble a crane (the slender curving pedicel, calyx and corolla of the flower bud, before it opens, respectively simulate the neck, head and beak of a crane).
- The Pilgrims on arriving in the New World observed cranberries growing profusely in the area about Cape Cod, and noted that the Indians used the fruit as a source of a brilliant red dye for their clothes.
- Damaged cranberries often sink in water, so that simply washing berries in water can reveal the superior berries. Another method that has been recommended to judge the quality of cranberries is to bounce them: the more times they bounce, the better the berry (http://www.kiwiseed.com/library/112097/112097.htm).
- The "Craisin™," a dried, sweetened cranberry, is currently being marketed in breakfast cereals and fruit mixes by Ocean Spray Inc.
- At least 700 consumer products contain cranberries in one form or another.
- Although most cultivated cranberry is used to produce juice, about 50 million kg of cranberry sauce is consumed in the US each year.
- A sauce was made with white cranberries as a prelude to a new marketing strategy, but it was of an amber color and had an insipid taste. Because the dark pigments of the fruit contribute substantially to the taste and medicinal quality of cranberries, the value of white cranberries is debatable.
- In 1959 many cranberry growers in the United States applied the weed killer aminotriazole prematurely, before the cranberries were harvested, rather than following the usual practice of waiting until the crop was removed. As a result the berries were contaminated. The incident led to sensationalistic publicity after it was disclosed that the chemical was capable of producing cancer in mice, and the resulting suspension of cranberry sales led to millions of dollars of losses for the cranberry industry.
- In 1970, cranberry juice was named the state beverage of Massachusetts. A fifth-grade class adopted the cause of making the cranberry the official berry of the state. Their 2 years of lobbying, petitions, and hearings were finally rewarded in 1994.

Selected References

Ahuja, S., Kaack, B., and Roberts, J. 1998. Loss of fimbrial adhesion with the addition of *Vaccinium macrocarpon* to the growth medium of P-fimbriated *Escherichia coli*. J. Urol. **159**: 559–562.

Avorn, J., Monane, M., Gurwitz, J.H., Glynn, R.J., Choodnovsky, I., and Lipsitz, L.A. 1994. Reduction of bacteriuria and pyuria after ingestion of cranberry juice. JAMA **271**: 751–754.

Bomser, J., Madhavi, D.L., Singletary, K., and Smith, M.A. 1996. In vitro anticancer activity of fruit extracts from *Vaccinium* species. Planta Med. **62**: 212–216. [Evidence that components of lowbush blueberry, cranberry, and lingonberry exhibit potential anticarcinogenic activity as evaluated by in vitro screening tests.]

Bruederle, L.P., Hugan, M.S., Dignan, J.M., and Vorsa, N. 1996. Genetic variation in natural populations of the large cranberry, *Vaccinium macrocarpon* Ait. (Ericaceae). Bull. Torrey Bot. Club **123**: 41–47.

Bureau, L. 1970. Un exemple d'adaptation de l'agriculture à des conditions écologiques en apparence hostiles: l'atocatière de Lemieux. Cahiers de Géographie de Québec **33**: 383-394.

Buszek, B.R. 1978. The cranberry connection: cranberry cookery with flavour, fact, and folklore, from memories, libraries, and kitchens of old and new friends and strangers. 2nd ed. Greene Press, Brattleboro, VT. 208 pp.

Calvin, L. 1997. Cranberry supply expands in response to higher demand. Agric. Outlook (Herndon, VA) **246**: 8–10.

Cross, E.C. 1973. Cranberries - the last one hundred years. Arnoldia **33**: 284–291.

Dana, M.N. 1983. Cranberry cultivar list. Fruit Var. J. **37**: 88–95.

Dana, M.N. 1989. The American cranberry industry. Acta Hortic. **241**: 287–294.

Eaton, G.W., Shawa, A.Y., and Bowen, P.A. 1983. Productivity of individual cranberry uprights in Washington and British Columbia, Canada, *Vaccinium macrocarpon*, yield. Sci. Hortic. (Amsterdam) **20**: 178–184.

Eck, P. 1990. The American cranberry. Rutgers University Press, New Brunswick, NJ. 420 pp.

Elle, E. 1996. Reproductive trade-offs in genetically distinct clones of *Vaccinium macrocarpon*, the American cranberry. Oecologia **107**: 61–70.

Fleet, J.C. 1994. New support for a folk remedy: cranberry juice reduces bacteriuria and pyuria in elderly women. Nutr. Rev. **52**(5): 168–178.

Fiander-Good Associates Ltd. 1993. The feasibility of establishing a cranberry industry in New Brunswick. Fiander-Good Associates Ltd., Fredericton. [Canada/New Brunswick Cooperation Agreement on Planning, 1993.] 126 + 23 pp.

Fitzpatrick, S.M., and Troubridge, J.T. 1993. Fecundity, number of diapausing eggs, and egg size of successive generations of the blackheaded fireworm (Lepidoptera: Tortricidae) on cranberries. Environ. Entomol. **22**: 818–823.

Galetta, G.J. 1975. Blueberries and cranberries. *In* Advances in fruit breeding. *Edited by* J. Janick and J.N. Moore. Purdue University Press, West Lafayette, IA. pp. 154–196.

Hall, I.V. 1971. Cranberry growth as related to water levels in the soil. Can. J. Plant Sci. **51**: 237–238.

Hall, I.V., and Nickerson, N.L. 1986. The biological flora of Canada. 7. *Oxycoccus macrocarpus* (Ait.) Pers., large cranberry. Can. Field-Nat. **100**: 89–104.

Hall, I.V., Blatt, C.R., Lockhart, C.L., and Wood, G.W. 1982. Growing cranberries. Agriculture Canada Publication 1282/E. Ottawa, ON. 28 pp.

Jackson, B., and Hicks, L.E. 1997. Effect of cranberry juice on urinary pH in older adults. Home Healthc. Nurse **15**(3):198–202.

Jamieson, A.R., Murray, R.A., Hall, I.V., and Brydon, A.C. 1990. Performance of cranberry cultivars at Aylesford, Nova Scotia. Fruit Var. J. **44**: 155–157.

Jorgensen, E.E., and Nauman, L.E. 1994. Disturbance in wetlands associated with commercial cranberry (*Vaccinium macrocarpon*) production. Am. Midl. Nat. **132**: 152–158.

Jukes, T.H. 1970. The global "cranberry incident." Clin. Toxicol. **3**: 147–149.

Kevan, P.G., Gadawski, R.M., Kevan, S.D., and Gadawski, S.E. 1984. Pollination of cranberries, *Vaccinium macrocarpon*, on cultivated marshes in Ontario. Proc. Entomol. Soc. Ont. **114**: 45–53.

Kuzminski, L.N. 1996. Cranberry juice and urinary tract infections: is there a beneficial relationship? Nutr. Rev. **54**(11 Pt 2): S87-S90.

Liebster, G., and Delor, H.W. 1974. Bibliography of the international literature on the cranberry, *Vaccinium macrocarpon* Ait. Technishe Universität Berlin, Bibliographische Reiche. Band 3. 52 pp.

Luby, J.J., Ballington, J.R. Draper, A.D., Kazimierz, P, and Austin, M.E. 1991. Blueberries and cranberries. *In* Genetic resources of temperate fruit and nut crops. *Edited by* J.N. Moore and J.R. Ballington. International Society for Horticultural Science, Wageningen, The Netherlands. pp. 393–456.

MacKenzie, K.E. 1994. Pollination requirements of the American cranberry. J. Small Fruit Vitic. **2**: 33–44.

MacKenzie, K.E., and Averill, A.L. 1995. Bee pollinators (Hymenoptera: Apoidea) of cranberry (*Vaccinium macrocarpon* Aiton) in southeast Massachusetts. Ann. Entomol. Soc. Am. **88**: 334–341.

Mahr, D.L., Jeffers, S.N., Stang, E.J., and Dana, M.N. 1988. Cranberry pest control in Wisconsin. Univ. Wisconsin Cooperative Extension Service Publication. 15 pp.

Nazarko, L. 1995. Infection control. The therapeutic uses of cranberry juice. Nurs. Stand. **9**(34): 33–35.

Novy, R.G., and Vorsa, N. 1995. Identification of intracultivar genetic heterogeneity in cranberry using silver-stained RAPDs. HortScience **30**: 600–604.

Novy, R.G., and Vorsa, N. 1996. Evidence for RAPD heteroduplex formation in cranberry: implications for pedigree and genetic-relatedness studies and a source of co-dominant RAPD markers. Theor. Appl. Genet. **92**: 840–849.

Novy, R.G., Kokak, C., Goffreda, J., and Vorsa, N. 1994. RAPDs identify varietal misclassification and regional divergence in cranberry (*Vaccinium macrocarpon* (Ait.) Pursh). Theor. Appl. Genet. **88**: 1004–1010.

Ofek, I., Goldhar, J., and Sharon, N. 1996. Anti-*Escherichia coli* adhesin activity of cranberry and blueberry juices. Adv. Exp. Med. Biol. **408**: 179–183.

Oertel, B. 1996. A conservation plan for every cranberry grower. Land and Water: The magazine of natural resource management and restoration **40**(4): 41–43.

Porsild, A.E. 1938. The cranberry in Canada. Can. Field-Nat. **52**: 116–117.

Rodale, J.I. 1960. Cranberries. Org. Gard. **7**(3): 21–25. [Reports 1959 contamination of cranberries by aminotriazole.]

Sarracino, J.M., and Vorsa, N. 1991. Self and cross fertility in cranberry. Euphytica **58**: 129–136.

Sapers, G.M., Philips, J.G., Rudolf, H.M., and DiVito, A.M. 1983. Cranberry quality: selection procedures for breeding programs. J. Am. Soc. Hortic. Sci. **108**: 241–246.

Sapers, G.M., Graff, G.R., Phillips, J.G., and Deubert, K.H. 1986. Factors affecting the anthocyanin content of cranberry (*Vaccinium macrocarpon*). J. Am. Soc. Hortic. Sci. **111**: 612–617.

Schmidt, D.R., and Sobota, A.E. 1988. An examination of the anti-adherence activity of cranberry juice on urinary and nonurinary bacterial isolates. Microbios **55**(224/225): 173–181.

Serres, R, and McCown, E.L. 1994. Rapid flowering of microcultured cranberry plants. HortScience **29**: 159–161.

Serres, R.A., McCown, B.H., Zeldin, E.L., Stang, E.J., and McCabe, D.E. 1993. Applications of biotechnology to cranberry: a model for fruit crop improvement. Acta Hortic. **345**: 149–156.

Shawa, A.Y. 1980. Control of weeds in cranberries (*Vaccinium macrocarpon*) with glyphosate and terbacil phytotoxicity. Weed Sci. **28**: 565–568.

Sibert, I. 1996. Cranberry pollination. Am. Bee J. **136**: 363–364.

Siciliano, A.A. 1996. Cranberry. HerbalGram. **38**: 51–54.

Stang, E.J. 1993. The North American cranberry industry. Acta Hortic. **346**: 284–298.

Stark, R, Hall, I.V., and Murray, R.A. 1974.The cranberry industry in Nova Scotia. Can. Agric. **19**(3): 30–31.

Strik, B.C., and Poole, A. 1992. Alternate-year pruning recommended for cranberry. Hortic. Sci. **27**(12):1327.

Swartz, J.H., and Medrek, T.F. 1968. Antifungal properties of cranberry juice. Appl. Microbiol. **16**: 1524–1527.

Thomas, J.D. 1990. Cranberry harvest: a history of cranberry growing in Massachusetts. Spinner Publications, New Bedford, MA. 224 pp.

Vander Kloet, S.P. 1983. The taxonomy of *Vaccinium* and *Oxycoccus*. Rhodora **85**: 1–44.

Vander Kloet, S.P. 1988.The genus *Vaccinium* in North America. Research Branch, Agriculture Canada, Ottawa, ON. 201 pp.

Walker, E.B., Barney, D.P., Mickelsen, J.N., Walton, R.J., and Mickelsen, R.A., Jr. 1997. Cranberry concentrate: UTI prophylaxis. J. Fam. Pract. **45**(2): 167–168.

White, J.J. 1916. Cranberry culture. Orange Judd Co., New York, NY. 131 pp.

Wilson T., Porcari, J.P., and Harbin, D. 1998. Cranberry extract inhibits low density lipoprotein oxidation. Life. Sci. **62**(24): PL381-PL386. [A pioneering study of the effect of cranberry juice on heart disease.]

Zafriri, D., Ofek, I., Adar, R., Pocino, M., and Sharon N. 1989. Inhibitory activity of cranberry juice on adherence of type 1 and type P fimbriated *Escherichia coli* to eucaryotic cells. Antimicrob. Agents Chemother. **33**: 92–98.

World Wide Web Links

(Warning. The quality of information on the internet varies from excellent to erroneous and highly misleading. The links below were chosen because they were the most informative sites located at the time of our internet search. Since medicinal plants are the subject, information on medicinal usage is often given. Such information may be flawed, and in any event should not be substituted for professional medical guidance.)

Cranberry home page [excellent general source of information; many links]:
http://www.scs.carleton.ca/~palepu/cranberry.html

Efficacy of cranberry in prevention of urinary tract infection in a susceptible pediatric population [a research report which did not find evidence of the medical usefulness of consuming cranberry juice]:
http://www.duj.com/Article/Foda.html

Healthline: cranberry juice and urinary tract infections [reports on a study that did find evidence of the medical usefulness of consuming cranberry juice]:
http://www.health-line.com/articles/hl940803.htm

Ocean Spray news room [Ocean Spray Inc. home page]:
http://www.oceanspray.com/home.htm

Diseases of cranberry:
http://www.scs.carleton.ca/~palepu/CRANBERRY/disease.html

Cranberry bibliography [a long list of mostly agriculturally-oriented publications]:
http://www.nemaine.com/rc&d/bib.htm

Cranberry agriculture in Maine [home page, has many links]:
http://www.nemaine.com/rc&d/cranberry.htm

Cranberry tablets for offensive urine odour:
http://www.pharmacyweb.com.au/rxconsult/98/988.html

Cranberry production in Wisconsin:
http://www.wiscran.org/productn.html

Vaccinium core catalog [information on germplasm holdings of The National Clonal Germplasm Repository at Corvallis, Oregon; has good list of cultivars]:
http://www.ars-grin.gov/ars/PacWest/Corvallis/ncgr/catalogs/vaccat.html

Cranberries and cranberry growers [a page of links]:
http://www.scls.lib.wi.us/mcm/subjects/cranberries.html

1995 harvested commercial cranberry acreage in United States [a map showing amounts harvested in various states]:
http://www.mda.state.mi.us/hot/cranberry/map.html

Tests show cranberry juice beneficial [1997 report of a preliminary study suggesting cranberry juice may be useful against heart disease]:
http://www.thonline.com/News/100797/Features/78253.htm

Vaccinium myrtillus (bilberry)

Vaccinium myrtillus L. Bilberry

Synonym: *V. oreophilum* Rydb.

The genus name (*Vaccinium*) is of disputed origin; it may have arisen from the Latin *vacca*, cow, allegedly because cows like eating the plant, or from *bacca*, berry, in allusion to the numerous berries. The specific term (*myrtillus*) is a reference to the myrtle-like leaves.

Vaccinium myrtillus should not be confused with *V. myrtilloides* Michx., the velvet-leaf or sour-top blueberry, which is Canada's most widespread wild species of blueberry, occurring from coast to coast.

English Common Names

Bilberry, dwarf bilberry, blaeberry (a Scottish name meaning blue berry), mountain bilberry, whinberry, whortleberry, whortles, myrtle whortleberry, tracleberry, huckleberry.

Most of these common names have also been applied to other species of *Vaccinium*, and so are frequently misinterpreted. However, in the context of medicinal usage, "bilberry" normally refers to *V. myrtillus*.

French Common Names

Airelle, brimbelle, myrtille, myrtille commune, myrtillier commun.

The above names are used in European French. The species is absent from Quebec, and an established French-Canadian name is unavailable, although the name "airelle" is generally used when bilberry extract is sold in stores in Canada. The French-Canadian botanist Bernard Boivin provided the name "myrtille" in his Flora of the Prairie Provinces.

Morphology

Bilberry is a small (5–30, rarely 60 cm tall), somewhat spreading, perennial, deciduous, shrub with slender angular branches arising from a creeping rhizome. Plants are usually shorter at higher elevations. The roots are slim (1.5–2 mm in diameter), much-branched, and often form an interconnected mat in the top 5 cm of substrate. The leaves are 10–30 mm long and bright green. The flowers are globular and waxy, with pale-green or pinkish petals 5–6 mm long. The bluish-black (occasionally reddish, bluish, or blackish), globose, flat-topped berries are 5–10 mm in diameter, sometimes covered when ripe with a delicate grey bloom; they have a slightly acid, sweet flavor. The berries may contain up to 40 seeds, although generally only half this number are viable.

Classification and Geography

For information on the genus *Vaccinium*, see the chapter on *V. macrocarpon* (cranberry).

Vaccinium myrtillus is classified in *Vaccinium* section *Myrtillus*, characterized by a 5-celled berry and stamens enclosed within the petals when the pollen is released. In Europe, hybrids (known as *V.* × *intermedium* Ruthe) with *V. vitis-idaea* L. have been reported. Bilberry has at least seven North American relatives in section *Myrtillus* centered in the Pacific Northwest. Of these, the mountain bilberry (*V. membranaceum* Douglas ex Hooker) has the largest fruit, ranging to 20 mm in diameter, and thus could be employed to increase yield. The Cascade bilberry (*V. deliciosum* Piper) is of potential interest in breeding and crop development because of its excellent fruit flavor, cold-hardiness and blossom frost resistance.

Bilberry is found in most of Europe, but only on mountains in the south. There are some populations in southwestern Greenland that are thought to be ancient European introductions. The species also occurs in Asia. In North America it is found in two areas of the Rocky Mountains (one in southeastern British Columbia and southwestern Alberta through Washington to central Oregon, the other in central Colorado, adjacent Utah, north-central New Mexico, and southern Arizona).

Ecology

Bilberry occurs in various climates in damp woodlands, heaths, and moors. It prefers filtered shade and moist, acidic, fertile soil. In North America it is found in open, moist coniferous woods, hillsides, hummocky seepage slopes, and moraines above 1600 m. The plants are cold-tolerant at least to –20°C, and much lower when protected by snow cover.

Clear-cutting of the trees may be harmful to the plants as a result of mechanical injury, as well as increased frost and drought injury. Although bilberry is relatively fire-tolerant and resprouts after fires, it is most frequent where burn intervals are relatively long (up to 100 years). Fire suppression may lead to dense conifer forests providing insufficient light for berry production. Similarly, areas

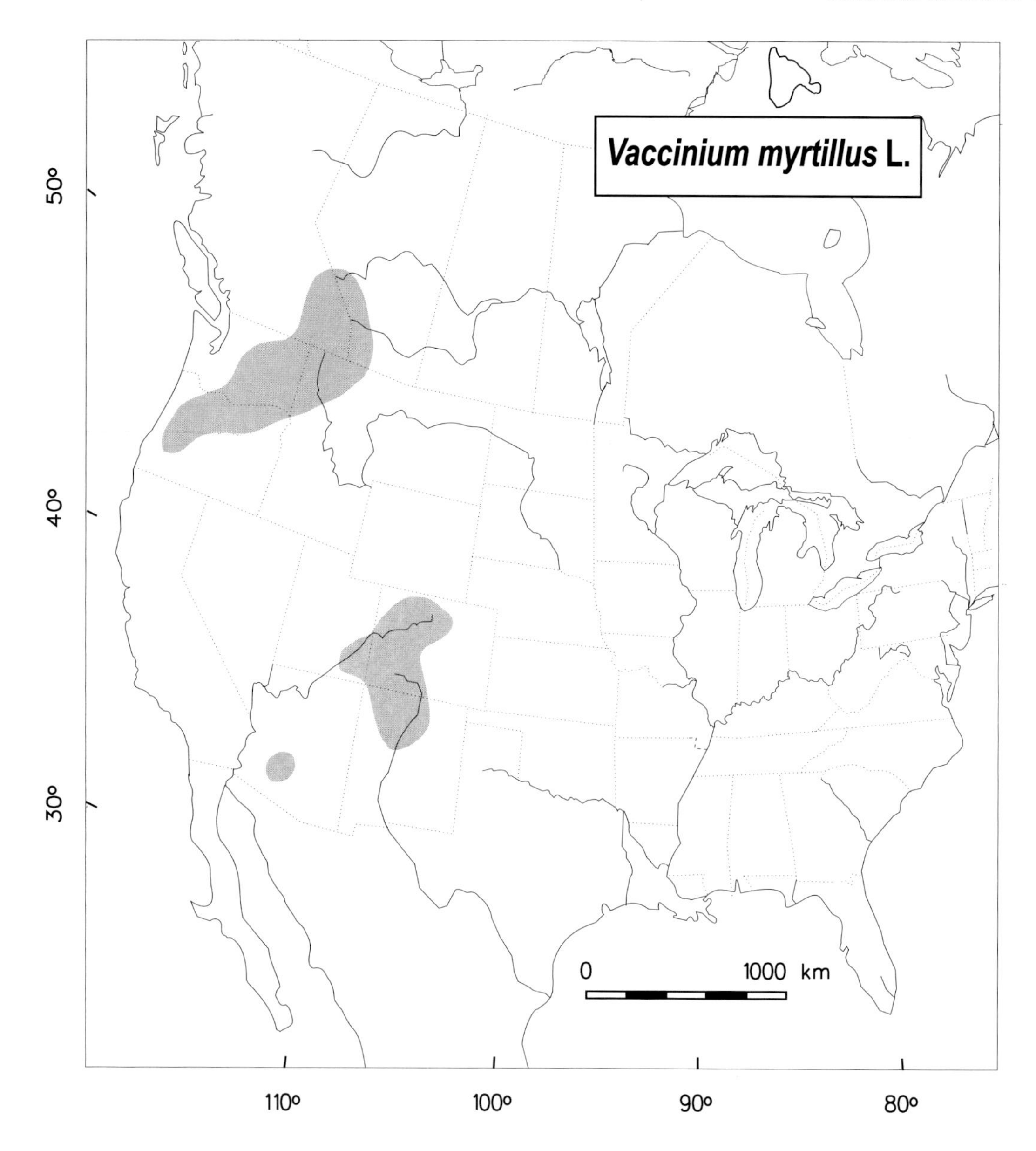

replanted with conifers may not supply sufficient light for good growth.

Pollination is by bees. The berries are eaten by many birds and mammals, including bears, and the foliage and twigs are browsed by some mammals. As with many other species of *Vaccinium*, birds and mammals are known to disperse the seeds. Reproduction is both by seeds and by vegetative spread.

Medicinal Uses

The bilberry has been used in folk medicine for centuries, perhaps even for millennia (although it is difficult to establish the precise identity of some *Vaccinium* species used in past times). Bilberry extracts are popular and are commonly marketed in Europe, as well as by numerous North American supermarkets, drug stores and health stores. Diverse medicinal claims are made for extracts of bilberry; the chief applications are noted below. Numerous pharmacological studies (more than 70) have been conducted that provide at least some support to the various claims.

Strengthening of cardiovascular system

Bilberry is believed to strengthen capillaries, allowing them to stretch without breaking or leaking, so that red blood cells can squeeze through tighter vessels, thereby benefitting blood flow. This is said to reduce bruising, and also varicose veins and "spider" veins which result from leakage of blood from capillaries. In past times in Europe, the herb was used prior to surgery to lessen bleeding, and even today some European doctors prescribe bilberry extracts before operations to reduce post-operative bleeding. Claims have been made that bilberry may lower blood pressure.

Vision

Bilberry has been shown to improve vision, especially night vision. During World War II, pilots and navigators of the English Royal Air Force consumed large quantities of bilberry preserves, particularly before night missions, in the belief that this improved vision. The proposed explanation is improved blood flow to the capillary-rich retina of the eye. The resulting improved eye function has been said to reduce eye fatigue and lessen the incidence of cataracts and common myopia (near-sightedness).

Diabetes

The berries and especially the leaves of bilberry have been used in folk-medicine treatment of diabetes. Bilberry leaf has a reputation as a "blood sugar-reducing" drug, useful in "antidiabetic" (against diabetes) teas. Although it is not widely used to treat diabetes today (insulin therapy has become standard treatment), it is interesting that very high levels of chromium are found in bilberry leaves, and that there has been suspicion that chromium may have a role to play in the treatment of diabetes. Also, myrtillin (methoxylated glucoside of gallic acid) in bilberry leaf has been shown to reduce hyperglycemia and normalize blood sugar level. An allied question that has not yet been sufficiently clarified is whether or not the improvements in blood circulation associated with the use of bilberry (discussed above) are useful for the treatment of circulatory disorders associated with diabetes. As noted below, because of toxicity the consumption of bilberry leaf is generally discouraged today. Insulin therapy can debilitate some people, and herbal treatments capable of safely reducing the need for insulin could be advantageous.

Digestive disorders and mucous membrane inflammation

Bilberry (both leaf and berry) tea has been used as a treatment for diarrhea and as a relief for nausea and indigestion. Bilberry has considerable tannin content, and the astringent action of the tannins is believed to be responsible for its effectiveness in treating digestive disorders, as well as for topical

Vaccinium myrtillus (bilberry)

(surface) treatment of mild inflammation of the mucous membranes of the mouth and throat. Numerous herbs with tannins are also known to have the same useful properties, and bilberry is not especially useful for these purposes, although some over-the-counter herbal preparations with bilberry are available in Europe specifically for digestive disorders. As noted below, the leaf is toxic and prolonged consumption is hazardous. Fresh berries seem less effective than dried, older berries for treating digestive disorders, perhaps because the tannins are not in a suitable chemical form until after drying has taken place.

Urinary tract health

Bilberry has at least some of the beneficial properties of cranberry for maintaining urinary tract function and curing infections (see chapter on cranberry for details).

Antioxidant

Flavonoids such as found in bilberry fruits are natural antioxidants, which are medicinally effective because they disarm damaging free radicals. Free radicals are highly reactive chemical fragments produced during metabolism in the body, that can impair cell function.

Toxicity

Fresh berries may cause diarrhea in some individuals, but are not considered toxic. However, the leaves must not be consumed (for example in tea) over a long period of time, because they are poisonous. Although bilberry leaf tea is sometimes encountered in herbal therapy, its use is hazardous and not recommended.

Chemistry

Anthocyanidin flavonoids responsible for the deep bluish color (often in the form of colorless precursors, proanthocyanidins, in the fresh plant) are believed to be the key medicinal compounds of bilberry. The most significant bioflavonoid in bilberry appears to be myrtocyan. Dried berries of bilberry contain about 0.7% anthocyanins. The dried berries of several other wild species of *Vaccinium* have been shown to have much higher content of anthocyanins, so that it is possible that future research may show that these other species are even more desirable as medicinal plants than bilberry. Other medicinally significant constituents (tannins, myrtillin, chromium) are discussed above.

Non-medicinal Uses

The berries were extensively eaten by Native Americans, including the Kootenay, Carrier, and Shuswap tribes. As with other *Vaccinium* species, bilberry fruit has long been popular in both Europe and North America for jams, jellies, preserves, pie fillings, liqueurs, and wines. Bilberry is of value as a nurse crop for Douglas fir seedlings. This attractive, hardy, shade-tolerant, low-growing shrub makes an interesting edible landscape, which is very ornamental in the autumn when the leaves turn red. Root cuttings can be transplanted to disturbed sites to reduce erosion, at the same time providing a natural food and cover for wildlife. In some parts of the Rocky Mountains problems with black and grizzly bears could be reduced through effective management of natural stands of bilberry. A larger supply of the berries would reduce bear foraging in populated areas and campsites.

Agricultural and Commercial Aspects

Bilberry is harvested mostly from native stands in North America and Europe and has consequently received much less attention from plant breeders than the cultivated species of *Vaccinium*, including the cranberries and blueberries. Most cultivated blueberries are stimulated to produce many young flowering and fruiting branches by heavy pruning of the shrubs. However, this practice is unsuitable for bilberry, which produces most of its fruit at the base of older branches.

Bilberry has a well-deserved reputation as a useful herbal medicine and therefore promises a growing market. North American wild bilberry is not readily accessible (the plant grows in mountainous areas) and so cultivation is desirable. Although not extensively cultivated in North America, natural stands are known to produce up to 100 kg (fresh weight) of berries/ha; a much higher production is likely possible under cultivation. There is considerable information in the European (foreign language) literature on cultivation and management techniques. While there is a large available supply from Europe, the development of a North American supply is a worthwhile enterprise.

Myths, Legends, Tales, Folklore, and Interesting Facts

- The name bilberry is derived from the Danish *bollebar*, meaning dark berry. However, not all berries are dark, and a white-fruited ornamental form is known.
- North American Indians preserved berries of *Vaccinium* species in various ways. In the North, berries were placed in seal oil, or stored in leather bags deposited in the permafrost. Indians commonly dried berries in the sunshine or by a fire.
- During the latter part of the First World War, England experienced a shortage of aniline dye, formerly imported from Germany, and the pigments from blue-black berries were substituted. So much of the bilberry crop was purchased by dye manufacturers that there was little available for making jam.

Selected References

Amouretti, M. 1972. Therapeutic value of *Vaccinium myrtillus* anthocyanosides in an internal medicine department. Therapeutique **48**: 579–581. [In French.]

Azar, M., Verette, E., and Brun, S. 1987. Identification of some phenolic compounds in bilberry (*Vaccinium myrtillus*) juice. J. Food Sci. **52**: 1255–1257.

Badescu, G, and Badescu, L. 1977. Basic elements in the cultivation technology of bilberries (*Vaccinium myrtillus*). Prod. Veg. Hortic. **26**(2): 54–60. [In Romanian.]

Bertuglia, S., Malandrino, S., and Colantuoni, A. 1995. Effect of *Vaccinium myrtillus* anthocyanosides on ischaemia reperfusion injury in hamster cheek pouch microcirculation. Pharmacol. Res. **31**: 183–187.

Bonati, A, and Crippa, F. 1978. Stability of anthocyanosides from *Vaccinium myrtillus* L. in pharmaceutical formulations. Fitoterapia **49**: 10–15.

Cignarella, A., Nastasi, M., Cavalli, E., and Puglisi, L. 1996. Novel lipid-lowering properties of *Vaccinium myrtillus* L. leaves, a traditional antidiabetic treatment, in several models of rat dyslipidaemia: a comparison with ciprofibrate. Thromb. Res. **84**: 311–322.

Colatuoni, A., Betuglia, S., Magistretti, M.J., and Donato, L. 1991. Effects of *Vaccinium myrtillus* anthocyanosides on arterial vasomotion. Arzneimittel-Forschung **41**: 905–909.

Contestabile, M.T., Appolloni, R., Suppressa, F., D'alba, E., and Pecorelli, B. 1991. Prolonged treatment with high dosage of *Vaccinium myrtillus* anthocyanosides: electrophysiological response in myopic patients. Boll. Oculist. **70**: 1157–1169. [In Italian.]

Cristoni, A., and Magistretti, M.J. 1987. Antiulcer and healing activity of *Vaccinium myrtillus* anthocyanosides. Farm. Ed. Pratica **42**(2): 29–44.

Colantuoni, A., Bertuglia, S., Magistretti, M.J., and Donato, L. 1991. Effects of *Vaccinium myrtillus* anthocyanosides on arterial vasomotion. Arzneimittelforschung **41**: 905–909.

Dierking, W., and Dierking, S. 1993. European *Vaccinium* species. Acta. Hortic. **346**: 299–304.

Dombrowicz, E., Zadernowski, R., and Swiatek, L. 1991. Phenolic acids in leaves of *Arctostaphylos uva-ursi* L., *Vaccinium vitis-idaea* L., and *Vaccinium myrtillus* L. Pharmazie **46**: 680–68.

Flower-Ellis, J.G.K. 1971. Age structure and dynamics in stands of bilberry (*Vaccinium myrtillus* L.). Institutionen for skogsekologi, Skogshogskolan, Stockholm, Sweden. Rapporter och uppsatser No. 9. 57 pp.

Fraisse, D., Carnat, A., and Lamaison, J.L. 1996. Polyphenolic composition of the leaf of bilberry. Ann. Pharm. Fr. **54**: 280–283. [In French.]

Friedrich, H., and Schonert, J. 1973. Tannin-producing substances in the leaves and fruits of the bilberry. Arch. Pharm. (Weinheim) **306**: 611–618. [In German.]

Friedrich, H., and Schonert, J. 1973. Phytochemical investigation of leaves and fruits of *Vaccinium myrtillus*. Plant Med. **24**(1): 90–110. [In German, English summary.]

Giba, Z., Grubisic, D., and Konjevic, R. 1995. The involvement of phytochrome in light-induced germination of blueberry (*Vaccinium myrtillus* L.) seeds. Seed Sci. Technol. **23**(1): 11–19.

Jacquemart, A.L., Mahy, G., Raspe, O., and DeSloover, J.R. 1994. An isozyme study in bilberry (*Vaccinium myrtillus*) 2. Mating system and genetic structure. Belg. J. Bot. **127**(2): 105–114.

Jayle, G.E., Aubry, M., Gavini, H., Braccini, G., and De la Baume, C. 1965. Study concerning the action of anthocyanoside extracts of *Vaccinium myrtillus* on night vision. Ann. Ocul. (Paris) **198**: 556–562. [In French.]

Laplaud, P.M., Lelubre, A., and Chapman, M.J. 1997. Antioxidant action of *Vaccinium myrtillus* extract on human low density lipoproteins, in vitro initial observations. Fundam. Clin. Pharmacol. **11**(1): 35–40.

Lietti, A., and Forni, G. 1976. Studies on *Vaccinium myrtillus* anthocyanosides. II. Aspects of anthocyanins pharmacokinetics in the rat. Arzneimittelforschung **26**: 832–835.

Lietti, A., Cristoni, A., and Picci, M. 1976. Studies on *Vaccinium myrtillus* anthocyanosides. I. Vasoprotective and antiinflammatory activity. Arzneimittelforschung **26**: 829–832.

Lorek, E. 1978. Contents of manganese and vitamin C in the fruits of bilberries (*Vaccinium myrtillus* L.) and red berries (*Vaccinium vitis-idaea* L.) growing in highly industrialized areas. Rocz. Panstw. Zakl. Hig. **29**: 381–387. [In Polish.]

Luby, J.J., Ballington, J.R. Draper, A.D., Kazimierz, P, and Austin, M.E. 1991. Blueberries and cranberries. *In* Genetic resources of temperate fruit and nut crops. *Edited by* J.N. Moore and J.R. Ballington. International Society for Horticultural Science, Wageningen, The Netherlands. pp. 393–456.

Magistretti, M.J., Conti, M., and Cristoni, A. 1988. Antiulcer activity of an anthocyanidin from *Vaccinium myrtillus*. Arzneimittelforschung **38**: 686–690.

Morazzoni, P., and Bombardelli, E. 1996. *Vaccinium myrtillus* L. Fitoterapia **67**: 3–29.

Morazzoni, P., and Magistretti, M.J. 1990. Activity of myrtocyan, an anthocyanoside complex from *Vaccinium myrtillus* (VMA), on platelet aggregation and adhesiveness. Fitoterapia **61**: 13–22.

Ritchie, J.C. 1956. Biological flora of the British Isles: *Vaccinium myrtillus* L. J. Ecology **44**: 290–298.

Rogers, L. 1976. Effects of mast and berry crop failures on survival, growth, and reproductive success of black bears. Trans. N. Am. Wildl. Conf. **41**: 431–438.

Sjors, H. 1989. *Vaccinium myrtillus*, a plant portrait. Svensk Bot. Tidskr. **83**: 411–428. [In Swedish.]

Slosse, P., and Hootele, C. 1978. Structure and absolute configuration of myrtine, a new quinolizidine alkaloid from *Vaccinium myrtillus*. Tetrahedron Lett. **4**: 397–398.

Tolan, L., Barna, V., Szigeti, I., Tecsa, D., Gavris, C., Csernatony, O., and Buchwald, I. 1969. The use of bilberry powder in dyspepsia in infants. Pediatria (Bucur.) **18**: 375–379. [In Romanian.]

Vander Kloet, S.P. 1983. Seed and seedling morphology of *Vaccinium myrtillus*. Nat. Can. **110**: 285–292.

Vander Kloet, S.P. 1988.The genus *Vaccinium* in North America. Research Branch, Agriculture Canada, Ottawa, ON. 201 pp.

Woodward, F.I. 1986. Ecophysiological studies on the shrub *Vaccinium myrtillus* L. taken from a wide altitudinal range. Oecologia **70**: 580–586.

Zaparaniuk, A.E. 1984. Effect of mineral fertilizers on the yields of *Vaccinium myrtillus* L. fruit. Rastit. Resur. **20**: 358–362. [In Russian.]

World Wide Web Links

(Warning. The quality of information on the internet varies from excellent to erroneous and highly misleading. The links below were chosen because they were the most informative sites located at the time of our internet search. Since medicinal plants are the subject, information on medicinal usage is often given. Such information may be flawed, and in any event should not be substituted for professional medical guidance.)

Fire effects information system [general botanical information; excellent!]:
http://svinet2.fs.fed.us/database/feis/plants/shrub/vacmyr/

A modern herbal by M. Grieve:
http://www.botanical.com/botanicaL/mgmh/b/bilber37.html

Mining Company guide to herbs for health - bilberry [a bilberry feature, with lots of links (mostly non-functional during 1998)]:
http://herbsforhealth.miningco.com/library/weekly/aa081598.htm

Bilberry's many healing powers [has many references to scientific studies]:
http://www.herbsociety.ca/times3.html

Vaccinium myrtillus [comments on medicinal aspects]:
http://www.herb.com/Files/bberry.html

MEDICINAL CAUTIONS

The traditional medical and folk uses and modern medicinal values of the plants described in this publication are given for information only. Medicinal use of the plants mentioned should be carried out only under the care of a well-informed, qualified physician. Please note that some plants included in this book are poisonous and others may cause toxic reactions in susceptible individuals.

The unusual chemistry of pharmacologically useful plants is associated with unusual abilities to alter human metabolism. The famous Swiss-born alchemist and physician Paracelsus (1493–1541) taught that the only difference between a medicine and a poison was the dose. All substances that are medicinal can also be poisonous, and so the questions of safety and medicinal use are closely related. In North America and elsewhere, there has been an increasing medicinal use of herbs without the advice of conventional doctors. Numerous "health food" or "natural food" stores and popular self-help books are fuelling this movement. As pointed out by Tyler (1984), much of popular herbal literature can only be classified as "advocacy literature," with almost every herb "good" for almost everything. The problems of self-medication are manifold. First is the question of medical competence: maintaining health and curing diseases of the human body are so obviously important that one wants the help of the best medical professionals, guided by thorough scientific knowledge. "He who self-medicates has a fool for a doctor" (Duke 1985). Second, what works for the majority may not work for you, as humans vary greatly in their responses to plant constituents, and what may be harmless or even beneficial to most will harm others. Third is the question of dosage. Natural (unrefined) plant materials and crude plant extracts and mixtures can vary widely in the concentration of active principles, and may be rather unpredictable in terms of interactions of the principles. In pointing out the potential danger of personal use of plants for health purposes, we do not by any means suggest that the growing health food industry based on medicinal plants is an unwelcome development. We simply wish to emphasize that consumers and retailers need to be informed and cautious. This chapter provides guidelines for employing and marketing medicinal herbs (also see WHO 1996 for additional advice on development and regulation, although not necessarily for Canada).

National Council Against Health Fraud Position Paper on Over-the Counter Herbal Remedies

The National Council Against Health Fraud (NCAHF) is an American non-profit health agency that focuses on health fraud, misinformation and quackery as public health problems. The NCAHF website is at: http://www.hcrc.org/ncahf/ncahf.html. The following paragraphs are reproduced from NCAHF Newsletter Volume 18, Number 4 (July-August 1995), and provide an excellent set of guidelines for everyone concerned with herbal remedies (but keep in mind that this is written for the US). The article may be reprinted provided appropriate credit is given to NCAHF. For a copy of the complete NCAHF Position Paper on OTC Herbal Remedies send $2 (American) and a stamped, self-addressed envelope to P.O. Box 1276, Loma Linda, CA, 92354–1276, USA.

The over-the-counter (OTC) herbal remedies business is reported to be well over $1.5 billion in current sales with an estimated annual growth rate of 15%. In 1994, $813.8 million of the health food stores' $4.815 billion in sales (17%) were from herbal remedies. Herbal product vendors benefit from society's romanticized view that equates "natural" with "safe." Unfortunately, the assumption that natural products are safe is false. It is precisely because herbs are a source of potent drugs that responsible people are concerned about the manner in which herbal remedies are being marketed. Consumers are being denied the most fundamental information and assurances of quality. By law, drug labels must provide essential information, but herbal remedies are being marketed as "dietary supplements" with little of the type of information needed to enable people to use these remedies properly. The herbal industry blames current regulatory policies for some of these problems. They say that FDA (Food and Drug Administration) regulations prevent them from supplying drug information because their products are regulated as dietary supplements. Herbal remedies cannot be profitably marketed if they have to meet the full requirements of drug approval. Reformers argue for herbal remedies to be given special regulatory consideration. The FDA is bound by the law to regulate products that make medical claims as drugs. NCAHF finds the present situation untenable, but believes that there is room for regulatory adaptation without sacrificing consumer protection principles. Recommendations are directed at legislators and regulators, manufacturers and marketers, physicians, and consumers.

Recommendations to Legislators and Regulators

Establish a special category of OTC medicines called "Traditional Herbal Remedies" (THRs) regulated as follows:

1. Labels must alert consumers to the fact that herbal remedies are held to a lower standard than that applied to standard medicines. Suggested wording: This product is regulated as a Traditional Herbal Remedy, a special category of medicines not required to meet the full stipulations of the US Food, Drug & Cosmetic Act applied to standard medicines.
2. Limit THR products to those with properties sufficiently documented in the pharmacognosy literature to assure an acceptable measure of safety and efficacy.
3. Limit herbal remedy products to those known not to have lethal or damaging side-effects when taken in overdose, or over an extended time period.
4. Limit THRs to the treatment of nonserious, self-limiting ailments.
5. Require THRs remedies to meet the same labeling standards for all drug products.
6. Require plant sources to be identified by their scientific names.
7. Require that all active ingredients (items that cause an effect) be quantitatively and qualitatively identified on the label.
8. Require herbal remedy products to contain sufficient amounts of pharmacologically active substances for the product to perform as expected. Only those expectations that can be supported by science should be permitted on labels. The FDA should develop a set of acceptable claims just as it has done with health claims for foods.
9. Require labels to inform consumers about what effects they should expect. Suggested wording: (e.g., valerian) The active ingredient in this product is valerian. Traditional folk medicinal uses for this substance include: as a sleep aid, and a relaxant. Valerian has been shown to depress the central nervous system at the doses indicated.
10. Require a highly visible, easily accessible post-marketing surveillance system for tracking unanticipated adverse reactions. The system must enable consumers as well as health professionals to report, and regulators to gather and disseminate information on adverse effects. A good candidate for the agency to receive reports is the US Pharmacopeia Practitioner Reporting System which passes reports on to the FDA and Poison Control Centers. Suggested wording: Adverse reactions associated with the use of this products should be reported to 1–800–638–6725.
11. Require manufacturers to mark product batches for identification, testing, and tracking.
12. Require warnings about dangers of self-treatment on labels and (or) package inserts. Suggested wording: Caution: Self-treatment may delay proper health care. See a medical doctor if health problems persist.
13. Require substantial representation from outside of the herbal industry to assure sufficient skepticism in herbal regulation.
14. Impose strong penalties for adulterating herbal products with potentially dangerous substances.

Recommendations to Herbal Remedy Product Manufacturers and Marketers

NCAHF believes that herbal remedy producers would be wise to take a long range approach to marketing. Just as compliance with food and drug regulations has enabled the American pharmaceutical industry to establish itself as the standard of quality, by adopting high standards, herbal marketers will assure themselves a superior competitive position and longevity.

1. Adopt the most scientific standards as are practical in your manufacturing and promotions, including the rejection of pseudoscientific herbal products such as homeopathic concoctions.
2. Conduct ongoing consumer education warning consumers that: Information found in traditional herbal books is not necessarily reliable as a guide to self-treatment.

The designations "herbologist," "herbalist," "Master Herbalist," and similar terms, have no legal or scientific meaning, and may be indicative of quackery. Likewise, sources issuing credentials in "herbology," "herbalism," and the like, are usually "diploma mills."

Recommendations to Physicians

1. Routinely ask patients about their use of herbal products.
2. Report adverse effects utilizing either the US Pharmacopeia Practitioner Reporting System or FDA's MedWatch Form 3500, or fax to 1–800-FDA-0178.
3. Write up and submit for publication case reports in which patients have experienced adverse effects from herbal product use. Include information on how the patient learned of the product and how its use was discovered.

Physicians may obtain information on herbal drugs and drug interactions from Registered Pharmacists at hospital pharmacies or from Poison Control Centers. Information on some herbal drugs, interactions, or synergistic effects may be obscure or unavailable. These are among the additional risks that those who choose to use medicinal herbal products currently face.

Recommendations to Consumers

1. Do not assume that herbal remedies are safe simply because they are natural. Herbal remedies contain substances that can have powerful effects upon the mind and (or) body. Use even greater cautions than when taking standard medications.
2. Be cautious about taking medicinal herbs if you are pregnant or attempting to become pregnant.
3. Be cautious about taking medicinal herbs if you are breast feeding a baby; herbal drugs in the body may be transferred to breast milk.
4. Do not give herbs to infants or children.
5. Do not take large quantities of any herbal preparation.
6. Do not take any medicinal herb on a prolonged daily basis.
7. Buy only preparations that identify plants on the label and state contraindications for use.
8. Become familiar with the names of potentially dangerous herbs and be cautious about their use.
9. If you are taking medications, do not use medicinal herbs without checking with your doctor.
10. Do not trust your health to unqualified practitioners who use unregulated titles such as "herbalist," "herb doctor," "Master Herbalist," "herbologist," "Natural Health Counselor," or the like.
11. Beware of exaggerated claims for the benefits of herbal remedies.
12. Insist that herbal marketers meet basic consumer protection standards of labeling, safety and efficacy.

Adverse reactions

For an extensive World wide Web site dealing with the toxic potential of medicinal plants, see "Adverse effects of herbs" - http://medherb.com/ADVERSE.HTM.
An extensive and detailed analysis of potentially harmful interactions of medicinal plants and pharmaceuticals is "Herbal-medical contraindications" - http://www.herb.com/contra.htm. For perspective, it may be noted that adverse drug reactions from pharmaceutical drugs alone represent a very significant health hazard, far exceeding the danger posed by medicinal plants (although pharmaceuticals are generally employed to treat more serious conditions than are herbals). In the US, some estimates hold that over 1% of patients suffer adverse drug reactions from pharmaceuticals, costing thousands of dollars on average, and perhaps associated with over 100,000 fatalities annually (for references to key literature, see:
http://www.ama-assn.org/sci-pubs/sci-news/1997/snr0122.htm#oc6106; also see the Journal of the American Medical Association (JAMA), which regularly has articles and letters treating this important subject; an archive of JAMA articles is at:
http://www.ama-assn.org/public/journals/jama/past_iss.htm). The American Herbal Products Association's botanical safety handbook (McGuffin et al. 1997) provides safety data and labeling recommendations for over 600 commonly sold herbal products (for the US). Authoritative guidelines for the safe use of herbs as medicines are provided by the German Federal Health Agency's "Commission E" which has produced more than 300 detailed herbal monographs (for information on English translations see Blumenthal et al. 1996;
http://www.herbalgram.org/commission _e.html).

Herbal remedies and Canadian law

Anyone contemplating growing, manufacturing, or selling medicinal herbs in Canada must be acquainted with Health Canada's regulations. These regulations have changed recently, and in response to considerable public and private sector concerns, may undergo additional changes or refinement. The Canadian Food and Drug Act provides for the sanitary manufacture of food and drugs, the prohibition of sale of harmful foods and drugs, and the accurate labeling and advertising of foods and drugs. Products can be classified as foods or drugs depending on their purpose. Garlic, for example, is mostly sold as a condiment (i.e., as food), but is also commonly promoted as a drug capable of lowering cholesterol and having other properties beneficial to health. Herbal products can therefore, at least in theory, be marketed either as food or as drugs. Herbal drugs, however, may also be considered as a special category, as noted below.

"Traditional herbal remedies" (THMs) are defined as "finished drug products intended for self-medication that contain, as the active principles, herbal ingredients that have received relatively little attention in world scientific literature, but for which traditional or folkloric use is well-documented in herbal references." Commercial drugs may not be present in THMs. THMs can be marketed in Canada as over-the-counter drugs intended to treat given ailments, if certain conditions are met, such as scientific proof of efficacy, standardization of dosage, a statement of contraindications, and proof of adherence to acceptable manufacturing and quality standards. Examples of herbal remedies discussed in this work include: dandelion for use as a diuretic, cascara as a laxative, and hop as a sedative. To qualify as a THM, the herbal remedy must be intended to treat only a minor ailment (one that if untreated would not usually have serious consequences), and the symptoms and the success of the treatment should be easily diagnosable by the user. A number of ailments are listed in the Food and Drugs Act that are considered too serious to be treated by THMs (such as alcoholism, appendicitis, arthritis, asthma, cancer, diabetes, epilepsy, glaucoma, menstrual flow disorder, and prostate disease). Some THMs are approved with restrictions; for example, arnica may only be used externally. There are various general restrictions, for example limitation of THM products to those over 11 years of age, and on the circumstances when specific dosages can be recommended to pregnant or breastfeeding women. Homeopathic use of herbs is governed by separate regulations (homeopathy in its narrow sense treats illness by applying highly diluted doses of alleged remedies that in higher concentrations produce symptoms of the disease treated; see *Homeopathy* in the glossary).

Herbal drugs whose safety and efficacy are well accepted scientifically are not considered to be THMs. Some of these may already be accepted and registered as drugs (such as May-apple), but if not, they can be registered under the New Drug Submission procedures. Registration consists of issuing a Drug Identification Number (DIN) or General Product (GP) Number upon approval, for either a traditional herbal medicine or a New Drug Submission; approvals for THMs cost less than a thousand dollars, while approvals for new drugs cost more than a 100 thousand dollars (the fee for evaluation of supporting data). Information to produce and register a THM is available from: Health Canada Drug Directorate, Tunney's Pasture, Address Locator #0702A, Ottawa, ON, K1A OL2. Also see Health Canada's world wide web site at
http://www.hc-sc.gc.ca/hpb/drugs/drhtmeng/index.html (for specific information on the topics discussed above, see:
http://www.hc-sc.gc.ca/hpbdgps/therapeut/drhtmeng/guidmain.html#direct). An excellent discussion of the above information is Marles (1997), which is available on the internet at:
http://www.agric.gov.ab.ca/cropsspecial/medconf/#top. Another useful discussion (although not sympathetic to the use of herbs as medicines) is Kozyrskyj (1997). Appendix H in Holm and MacGregor (1998) has reprints of the following very useful documents (obtainable from Health Canada's web site mentioned above): 1) "Draft regulatory framework for natural health products (Interim report of the Advisory Panel on Natural Health Products, Presentation to the House of Commons Standing Comittee on Health, Feb. 3, 1998)"; 2) "Good manufacturing practices (Supplementary guidelines for the manufacture of herbal medicinal products, final version, October 1996, Health Canada)"; 3) "Good manufacturing practices (Supplementary guidelines for homeopathic preparations, final version, October 1996, Health Canada"; 4) "Drugs Directorate Guideline, traditional herbal medicines, revised October 1995)."

Many natural health products are regulated as foods, because they do not make medical claims, although they may in fact either be sold or bought for therapeutic purposes. Regulation of these is a matter of current dispute, with some in the health product industry concerned about the possibility of excessively restrictive regulations limiting consumer choice; on the other hand, there are those who are concerned about the consumer being misled or exposed to harmful substances as a result of insufficiently restrictive regulations. Most herbal products sold as foods are found in health food stores and pharmacies, and are often listed as "dietary aids" or with labeling indicating in a general way purported benefits. Unqualified personnel (i.e., without formal academic and licensing requirements) sometimes promote the use of given herbal products to treat disease, although such promotion is illegal in Canada. Moreover, it is clear that many consumers are attempting self-medication for conditions that require professional care. The debate extends to some herbal products that are currently restricted, some believing that the comparatively much less restrictive environment in European nations is desirable. Health Canada's "Advisory Panel on Natural Health Products" is currently evaluating these issues. An excellent examination of the situation is: Hearings on the regulation of natural products in Canada: determining the appropriate regulatory framework for herbal remedies. A brief presented to the Standing Committee on Health, March 12, 1998, by A.R. McCutcheon and D. Awang, representing the Canadian Herb Society. Herbal Times (Canadian Herb Society Journal) - http://www.Herbsociety.ca/times.html.

Some herbal medicines are being sold without the approval of Health Canada, and could be hazardous.

Serious liver and kidney disease, and some types of cancer are known to have been caused by certain herbs. Some illegally imported medications from Asia have been found to contain dangerous levels of heavy metals such as arsenic and mercury. Several illegal herbal preparations have been found to be adulterated with prescription drugs such as benzodiazepines, steroids, hormones, diuretics and anti-inflammatories. An eight digit Drug Identification Number or General Product Number signifies that the product has been approved for sale in Canada. Adverse reactions should be reported (to your doctor, pharmacist, and (or) directly to Health Canada), in order to promote the safer use of herbals.

Herbal remedies and American law

From a Canadian perspective, it is important to appreciate the regulatory and legal framework of medicinal herbs in the US because: a) this is Canada's largest export market, and b) the US has a major influence on public opinion in Canada, hence on Canadian regulation of herbals. The Dietary Supplement Health and Education Act passed in 1994 is the essential legislation governing herbal products (for detailed information, see: http://web.health.gov/dietsupp/ch1.htm; an advertisement for a book analysing the act is at: http://www.fdli.org/pubs/dshea.htm; updated information from the Food and Drug Administration (FDA) on dietary supplements is found at: http://vm.cfsan.fda.gov/~dms/supplmnt.html). This act allows for the marketing of many herbals as dietary supplement ingredients without FDA approval, provided that the manufacturer states that the ingredients have a history of dietary use and have not been sold in the US as drugs. The category "dietary supplements" was created to distinguish such substances from either "foods" (consumed to support the body's structure and function) or "drugs" (designed to treat disease). Dietary supplements include vitamins, minerals, and amino acids, in addition to herbs. Labels for dietary supplements can list the part or function of the body affected, but therapeutic claims are not permitted (unless the statements are backed by clinical research, or "if such claims are based on current, published, authoritative statements from certain federal scientific bodies, as well as from the National Academy of Sciences"; see http://vm.cfsan.fda.gov/~dms/hclmguid.html).

The internet as a source of medical misinformation

"Science and snake oil may not always look all that different on the Net."
Silberg et al. (cited below)

The "medical writer's guide to online resources" [http://www.journalism.iupui.edu/caj/jrodden/HealthOnline.html#anchor1717333] contains several useful warnings about the merit of internet information, for example: "Despite the abundance of online sites, the quality of information continues to pose problems, especially in the science and medical fields, where legitimate information is normally scrutinized during peer review before publication. On the Web, however, it's difficult to distinguish the caliber of one site from another. Human error, conflict of interest, inaccurate data - none of these are readily evident, but all can add to the problem of medical misinformation."

The following article is particularly important as a warning to those who use the internet to obtain medicinal information, and hence to those concerned with herbal information currently on the web: [Juhling McClung, H., Murray, R.D., and Heitlinger, L.A. 1998. The internet as a source for current patient information. Pediatrics 101(6): p. e2; electronic article: http://www.pediatrics.org/cgi/content/full/101/6/e2]. Based on the treatment of childhood diarrhea, these authors surveyed internet articles, and found a low concurrence with recommended treatment. They concluded that: "Patients must be warned about the voluminous misinformation available on medical subjects on the Net," and "Practitioners need to warn their patients about the need for a very critical review of all medical information obtained from the Web, even when it seems to be from a 'reliable' source."

Another excellent review (with links to related articles) is: Silberg, W.M., Lundberg, G.D., and Musacchio, R.A. 1997. Assessing, controlling, and assuring the quality of medical information on the internet. Journal of the American Medical Association 277:1244–1245 [http://www.ama-assn.org/sci-pubs/journals/archive/jama/vol_277/no_15/ed7016x.htm].

For additional links to the issue of internet medical misinformation, see "University of Pennsylvania Cancer Centre's 'Source Reliability Issues':" - http://www.oncolink.upenn.edu/resources/reliabilty.html

"The greatest service which can be rendered any country is to add a useful plant to its culture."
Thomas Jefferson (1743–1826, 3rd president of the US)

The business of medicinal plants

This chapter is particularly oriented to growers of medicinal plants, and associated agri-businesses. The American oil tycoon J. Paul Getty (1892 – 1976), one of the richest men the world has known, had the following formula for success: "Rise early, work hard, strike oil." Always on the search for new ways of making money, Mr. Getty even had pay telephones installed in his English country mansion. The equivalent of striking oil in agriculture is finding extremely profitable (so-called "green gold") crops. Choosing new profitable crops is a difficult exercise, and advice is given here. However, good business practices are also indispensable, and many common sense business tips are provided in this chapter. Finally, a guide to resources for locating helpful information is furnished.

Marketplace timing: the most important key to profitability

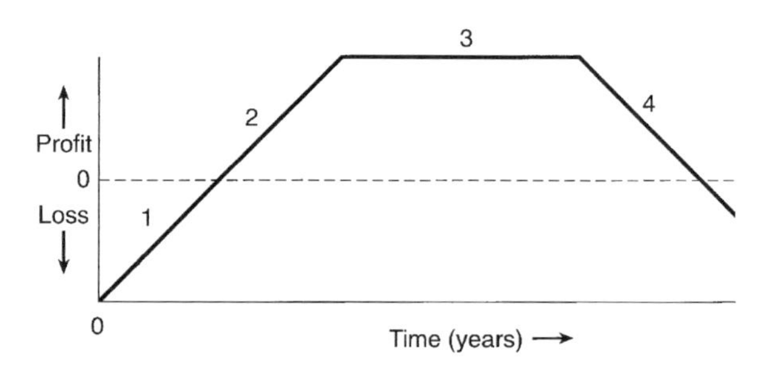

The above figure illustrates in a hypothetical fashion the relationship of profitability and time for many new patent-free products in a free market economy (the situation that prevails for most medicinal crops). Phase 1 is the period of investment in research and development, necessary to bring the new product to the point of profitability (most new products in fact do not survive this foundational period). During phase 2 the market expands, along with profitability. Phase 3 is a stable time of profitability, at the end of which decline in profitability occurs (phase 4). The most common cause of the decline in profitability is copy-cat competition from those who have observed the profitability of the item. The more profitable and popular a given product becomes, the faster competition develops. A frequent consequence of such competition is oversaturation of the market, the generation of surpluses, and business failures of those who remained dependent on the sale of the once-profitable item. It is extremely important that those who contemplate getting into the business of producing medicinal crops understand this sequence of events, because it is particularly pertinent to such crops, which often have a limited life-span of profitable popularity. Developing an entirely new crop (phase 1) by oneself, when others are not making the same investment in that crop, is extremely risky, but offers the greatest potential reward, because the inertia from having a large head start may allow one to rapidly capture a large proportion of the potential market, and to hold that market for a long period. Copying the example of those who started to cultivate the new crop, during phase 2 when the crop is expanding in profitability, is the safest strategy, for although it is somewhat late to get into the game, one has avoided the most dangerous period when most new items simply fail to ever establish profitability. The most common mistake that is made with attempts to cultivate new medicinal crops is to get into the game late, when sustained profitability has been demonstrated, but producers are beginning to flood the market with the product. Frequently when a medicinal plant is very well known, it is too late to get into high-profitability production of that crop.

Some principal determinants of the commercial importance of medicinal plants

As with all crops grown for the marketplace, a variety of considerations bear on commercial value. Market demand in relation to supply is the most obvious factor, but relative stability of demand and supply are also important, since strong fluctuations can be costly to both producers and marketers. Of course, crops differ in their suitability for given regions and climates, and indeed for markets. Ginseng species are especially illustrative of the fact that crops may differ greatly in terms of domestic production, domestic use, exports, and imports. American ginseng is Canada's most important medicinal crop, but most of it is exported. Very little of the Asian species is grown in Canada, but a substantial amount is imported. American ginseng also provides an example of how crops change, since domestic consumption has been increasing in recent years (although the current market is stagnant).

Another factor is the relationship of the quantity of material that needs to be grown in order to extract commercial amounts. For example, as noted in the chapter on Pacific yew, a huge quantity of material must be harvested in order to extract a very small amount of the medicinal constituent, taxol. Ginkgo (from *Ginkgo biloba*), one of the most popular of medicinal plants, is a very large tree, easily and widely grown as an ornamental, and it might seem that adequate material could easily be obtained from a small number of plants. However, large commercial plantations have already been established (e.g., 400 ha in South Carolina) and others are being planted in order to obtain the large amount of foliage necessary.

Many medicinal plants have limited crop value specifically for medicinal products, but are nevertheless valuable as crops because of food or industrial value that exceeds their medicinal worth. For example, alfalfa is considered to be one of the most important of medicinal plants, but because huge quantities are produced as forage, and indeed for seeds for production of alfalfa (*Medicago sativa* L.) sprouts, growing alfalfa solely as a medicinal plant is not practical. Another example is red clover (*Trifolium pratense* L.), which like alfalfa is widely grown as forage, but happens to be considered a strongly medicinal plant. Extracted flavorings, oils, dyes, and industrial chemicals are the primary economic reason for growing many herbs which happen to have some or even considerable medicinal value, and so growing the crops strictly as medicinal plants is generally not profitable. Sometimes, however, different cultivars may be more appropriate for medicinal use than the type of plant that is widely raised as a crop. For example, large quantities of catnip (from *Nepeta cataria* L.) are grown for cats, but lemon-flavored catnip is popular in some medicinal teas. The general point to be emphasized is that wide availability of plant species (or at least varieties of those species) strongly tends to lower their profit potential specifically for medicinal purposes, and vice-versa. The key considerations are price and volume - i.e., whether a good-sized crop can be grown profitably specifically for medicinal purposes. Nutraceuticals that are extracted from crops may have most of their value determined by the process of extraction and market preparation, and those who engage in these aspects of the medicinal industry are not limited by the fact that the source crops may be widely and commonly available at low prices.

Medicinal plants - just one category of chemical crop that can be grown profitably

Motives for growing different categories of plants vary. From an agri-business viewpoint, those who contemplate raising medicinal crops should be aware that this type of crop is very similar in numerous respects to other types of crop (which also tend to be potentially high-value, labor- and (or) knowledge-intensive, niche-market crops harvested because of valuable chemical constituents, often grown on small acreages and marketed in low volumes). Such crops, therefore, may be equally or even more suitable for particular growers, and before choosing medicinal crops, the other types should also be considered. A comparison of merits based on personal or corporate preferences, as well as economic and business considerations, should determine whether or not to pursue medicinal crops preferentially. To help choose the type of new crop to grow, we suggest a perusal of the general literature and Web sites provided. The choice includes "aromatic plants" (which supply extracts to the flavoring and perfume industries), culinary herbs and spices (i.e., plants consumed in small quantities for flavor), "tea crops" (such as catnip, rose hip, lemon balm, and chamomile, which are equally both culinary herb and medicinal plants), "bioactive" crops (e.g., producing pesticides or hormones), and nutraceutical or functional food crops which, as explained in the Introduction, includes an extremely wide variety of preparations from plants that have health benefits. "Industrial crops" is a vague category that could include the above-mentioned plants, but typically refers to high-acreage crops producing oils and other chemicals of industrial value. Below, we list some possible alternatives to growing the medicinal crops we discuss in detail in this work.

Some potentially high-value niche-market crops that can be grown in Canada for chemical constituents.

Aromatic (essential oil) crops

(All of these are used medicinally, although the crops are primarily used for food, flavoring, or ornament.)

Bergamot (*Monarda didyma* L.)
Caraway (*Carum carvi* L.)
Catnip (*Nepeta cataria* L.)
Coriander (*Coriandrum sativum* L.)
Dill (*Anethum graveolens* L.)
Hyssop, anise (*Agastache foeniculum* (Pursh) O. Kuntze)
Peppermint (*Mentha* × *piperita* L.)
Spearmint, Scotch (*Mentha* ×*gracilis* Sole)

Insecticidal crops

Pyrethrum (*Tanacetum cineriifolium* (Trev.) Schultz-Bip.)

"Nutraceutical" or "functional food" crops

Common name	Scientific name	Purposes for which the crop is widely grown	Medicinal crop purpose
Wheat	*Trititicum aestivum* L.	cereal	bran extract
Oats	*Avena sativa* L.	cereal, fodder	bran extract
Sunflower	*Helianthus annuus* L.	edible oil, birdfeed	high oleic edible oil (a high monosaturated oil with low levels of both saturated and polyunsaturated fats)
Rapeseed (canola)	*Brassica napus* L.	edible oil	extracted phytosterol (skin moisturizer and trans-dermal barrier transport agent)
Fenugreek	*Trigonella foenum-graecum* L.	Fodder, spice	extracted sex hormone precursors (for making birth control pills)
Flax	*Linum usitatissimum* L.	linseed oil, flax fiber	seed oil is a source of omega-3 fatty acids, especially the essential alpha-linolenic acid; rich sources of lignans, a type of phytoestrogen that some suspect of anti-cancer activity
Hemp	*Cannabis sativa* L.	fiber	high-GLA (essential fatty acid) edible oil from seeds
Borage	*Borago officinalis* L.	ornamental	high-GLA (essential fatty acid) edible oil from seeds
Garlic	*Allium sativum* L.	culinary herb	garlic extract (used in tablets consumed for health benefits)
Horseradish	*Armoracia rusticana* P. Gaertn., B. Mey. & Scherb.	culinary herb	horseradish is grown in Canada to extract the enzyme peroxidase, used for the diagnosis of the AIDS virus
Psyllium	*Plantago psyllium* L.; *P. ovata* Forsk. (blond psyllium)	no other uses	widely used laxative from seeds

Some non-native medicinal plants that might be grown profitably in Canada

(Also see the above list of "nutraceutical" crops, all of which are of foreign origin; most can be grown profitably in Canada.)

Common name	Scientific name	Predominant medicinal use
Burdock	*Arctium lappa* L	Used as over-the-counter traditional herbal treatment for various disorders
Calendula (pot marigold)	*Calendula officinalis* L.	Used to relieve minor skin conditions
Chamomile, English	*Chamaemelum nobile* (L.) All.	A general anti-inflammatory, anti-infection herbal remedy
Chamomile, German	*Matricaria recutita* L.	A general anti-inflammatory, anti-infection herbal remedy
Feverfew	*Tanacetum parthenium* (L.) Schultz-Bip.	Used to treat fever
Milk thistle	*Silybum marianum* (L.) Gaertn.	Used to treat liver conditions
Nettle	*Urtica dioica* L	Used to treat a variety of maladies, including prostate problems and inflammation of mucous membranes of nose
Peony	*Paeonia* species	Used in Chinese herbal medicine, and seems to be arousing interest in Western medicine
st. John's-wort	*Hypericum perforatum* L.	An anti-depressant
Valerian	*Valeriana officinalis* L.	A sedative

Why are indigenous Canadian plants economically important?

Our focus on indigenous medicinal plants reflects our belief that they offer Canada the most obvious crops deserving attention and development. True, there are also numerous foreign medicinal plants that can be grown as crops in Canada, and these will be the subject eventually of a separate guide. As noted by the asterisks in the following list, three of the top-selling medicinal herbs in the US are native plants of Canada (dealt with in this work).

Cultivating medicinal herbs as a small enterprise

Most major medicinal herbs are marketed by "big business," but much of the cultivation is by small enterprises. Large processing companies often contract the cultivation and harvest of medicinal herbs to private growers or simply bid on the open market. Popular medicinal crops (such as echinacea, valerian and St. John's wort) may be grown on large (e.g., 10 ha or more) acreages. Plants grown for gamma linolenic acid (an essential fatty acid lacking in perhaps 15% of humans), which includes borage, evening primrose, and hemp, can also fall into this category of large-acreage crop. Aromatherapy (the external application of essential oils for healing, pleasure, and stress reduction) employs plant essential oils, most of which are grown as large crops, especially in hot climates. However, many medicinal plants are grown in quite small acreages. Notwithstanding the economic importance of corporations, small entrepreneurs are of special importance to the future development of the medicinal herb industry in North America. There are opportunities for small private enterprises, including the cultivation of medicinal herbs as part of the spectrum of crops raised by farmers. Many medicinal herbs are well suited to small scale production because of the relatively limited volumes of produce and the specialized growing conditions required. The relatively high crop values also contribute to the fact that they can frequently be grown economically on small acreages without sophisticated machinery or at least with machinery that can be adapted. Some medicinal herbs are rather weedy, and can be grown on marginal land, although generally medicinal herbs require good land like other crops.

Top six medicinal herbs in US (sales for 1995)

(Cited by P. Brevoort, 1996, HerbalGram 36 (Spring): 54. Also cited in (http://www.attra.org/attra-pub/herb.html)

Echinacea*
Garlic
Goldenseal*
Ginseng*
Ginkgo
Saw palmetto[1]

[1] The supply of saw palmetto berries (*Serenoa serrulata* (Michx.) Hook. f., a palm), popular as an over-the-counter herbal remedy for prostate problems, is largely obtained by harvesting wild stands in Florida, although there is some cultivation. Goldenseal also is mostly obtained from wild supplies, although increasingly it is cultivated. Medicinal supplies of the remaining species are derived mostly or entirely from cultivation.

Preparation of extracts and dried products is specialized, competitive, and often requires large investments. Dehydrating, processing, and extracting generally demand considerable knowledge and experience, machinery and equipment. Dehydration or extraction normally is carried out immediately after harvesting, and in close proximity to the area of cultivation. Large firms have established high quality and processing standards, and tend to dominate the market. However, there are companies specializing in transforming a medicinal crop into saleable products (e.g., tinctures, elixirs and capsules) on behalf of the grower, who then markets the product, either at the wholesale or retail level. This can be very advantageous to the grower, since farm gate prices for medicinal crops can be quite low, while the profit on products can be quite high. Moreover, liability insurance for marketing medicinal preparations may be obtainable only within such an arrangement where there is demonstrable pharmacological expertise. For some medicinal crops, growers can retain profits, that would otherwise go to market intermediaries, by acquiring necessary extraction, processing, and marketing skills. Several body care products such as soaps and salves can be produced relatively easily by small producers, who can directly market by means such as mail order and craft fairs.

Risks and problems for the grower

Risk is particularly high for medicinal crops. The degree of risk associated with a particular medicinal crop decreases as the associated industry and market infrastructure develop. At the same time, competition often increases significantly and profit margins sometimes decline as the crop becomes less risky.

Commercial-scale medicinal plant production for certain herbs (e.g., ginseng) may involve relatively high capital investment (e.g., for planting stock, machinery, and drying equipment). For many species, only limited information may be available on appropriate production practices and quality standards. Markets may be uncertain, seed supply sources may provide stocks that are unsuitable for one reason or another, and regulations concerning pesticides, herbicides, and fungicides may not have been established. Overproduction or shortages can dramatically affect prices, which can drop or escalate overnight. Markets for a particular medicinal crop can disappear almost immediately should it be deemed unsafe for consumption. Still other undesirable possibilities are the chemical synthesis of a chemical produced by a given medicinal plant, or discovery of another species that more productively produces that compound. This could make growing the medicinal species on which one has specialized suddenly obsolete (for example, see chapter on bloodroot).

Many plant species can only be cultivated practically in tropical areas. However, hundreds of medicinal plants can or could be grown in Canada as crops. Perhaps for the majority of these, traditional expertise for their production has been developed in foreign countries. Europe has a long history of supplying world medicinal plant markets, and it is difficult to compete with such entrenched competition. Of all factors restricting profitability of medicinal crops, overseas competition is the most significant, and this can make it extremely difficult for the enterprising small entrepreneur to break into the market. Countries around the Mediterranean Sea, eastern Europe, Latin America and India are the world's major suppliers of medicinal herbs. Very low labor costs often mean that medicinal herbs are placed on the market at low prices. However, lack of purity of much of the foreign supply and lack of certification that the material is "organic" or pesticide-free provides a very large advantage for domestic growers. Pharmaceutical companies often look to domestic contract growers to provide a certified organic crop.

Still another problem particularly associated with the medicinal plant industry is the irregularity of market demand. This makes it essential for a small private entrepreneur to establish a close working relationship with a buyer or company purchasing the product.

The production of seeds of medicinal plants is another very problematical subject. Imported seed and seed of major herbs produced by bulk North American suppliers once again tend to make this area of the herb business difficult for the small entrepreneur. Germination rates of many common herbs are rather poor and, coupled with poor seed production by many species, the profitability of seed production can be marginal. Unfortunately, for many medicinal plants there is an urgent need for production of high-quality, reliably identified seed, that is often not met by suppliers. Imported seeds may be mixtures of cultivars or contaminated by weed seeds. Growers need to be especially sensitive to the issues of quality, purity, and identity, since the success of their crops depends on these factors. It may be necessary to produce one's own seed to ensure the crop is successful.

Research and information needs

Research of various types is needed for most medicinal herbs:

- Collecting germplasm, characterizing its variation, and protecting it in nature and in storage facilities
- Evaluation of phytochemistry
- Pharmacological testing to establish effectiveness and safety
- Screening and breeding of more productive varieties adapted to local conditions
- Investigations of cultural and post-harvest requirements, including weed and pest control, soil nutrition, planting densities, packaging, and critical storage temperatures
- Greenhouse cultivation.

Greenhouse production deserves special consideration. For some high-value medicinal herbs,. This is a very important subject in Canada, where low temperatures and short seasons limit productivity. Obviously it is expensive to maintain a greenhouse, but under some circumstances this may be justified by the savings realized in avoiding importation. Greenhouses allow excellent regulation of temperature, moisture, humidity, diseases and insects, and indeed serve to avoid most insects and weeds, as well as birds, dust, and atmospheric pollution. A major issue is whether or not particular herbs can be grown in such artificial conditions and still produce the desired chemicals. It is well known that in many plants environmental and cultural conditions can alter the concentrations of economic constituents. Agriculture and Agri-Food Canada's Greenhouse & Processing Crops Research Centre at Harrow, Ontario, is the largest centre of greenhouse crop research in Canada, and medicinal plant cultivation is a current concern (for information, see: http://res.agr.ca/harrow/#top).

Advice regarding medicinal crops that was circulated in Canada more than half a century ago

The following set of recommendations (last revised in 1947 by W.J. Cody and H.A. Senn) was distributed as a mimeographed sheet for many years, in reply to inquiries about medicinal plants, by the Botany and Plant Pathology unit of the Division of the Central Experimental Farm at Ottawa (a precursor of our organization, the Eastern Cereal and Oilseed Research Centre). Although more than half a century old, the advice is sufficiently relevant to today that it is worth noting here. This also provides an interesting opportunity to analyse recent changes.

1. The successful cultivation of medicinal plants requires considerable care. Equipment for harvesting and preparing the material for market is highly specialized. Consequently, the prospective grower should, if possible, have some experience in the production of drug plants before he embarks on a venture of his own.
2. Some species and varieties occupy the land for two or more years before a marketable crop is available. Often they require a careful attention during this time.
3. The idea that high prices are usually paid for medicinal plants seems to prevail in some districts. Although prices for some types did rise considerably during the war, it is seldom that more than a reasonable profit is realized from these crops. Very often the profit is small or a loss is experienced.
4. The market for medicinal plant products has been very unstable. A grower would be well advised to have a contract with a prospective buyer to insure against having no market for his crop. The wholesale drug companies are not eager to buy from the small collector or producer since they cannot be certain of the continuity of supply or the identity and uniformity of the product.
5. As a rule the total demand for any of these plant products is very small. Thus a small acreage usually will suffice to supply the market. If a high price were obtainable for any particular plant, the probability is that many people would grow it. Thus the market would be flooded, a consequent fall in price would result, and only the most favorably situated producers would be successful in making a profit.
6. The present high cost of labor in Canada makes profitable gardening of any sort more expensive than formerly. Almost everywhere labor is very difficult to obtain. Moreover, now that the war is over, Canadians will not likely be able to compete with the much cheaper labor and expert culture of countries where plants have been normally produced.

The above recommendations, while pertinent to current circumstances, need to be viewed in the light of recent developments. Herbal remedies have become extremely popular in health food stores, drugstores, and supermarkets, and this is quite unlike past times. Consequently, much of points 3 and 5, regarding market size and profitability, is less of a concern. Although farm gate prices for many botanicals remain low, growers now have the option of personally adding value to their crop to take advantage of the much higher profit margins available to sellers. Point 6, cheap foreign labor, is also not as important as formerly, because quality control and organic cultivation give domestic producers a strong advantage. Current demographic trends also suggest a promising future for medicinal crops: the population is aging, becoming much more concerned about health issues, and turning to over the counter medicinal products as a supplement to conventional medical treatment.

Advice regarding growing medicinal crops today in Canada

Probably the best advice that can be given to those contemplating getting into the business of medicinal herbs is to spend about a year doing research. Libraries can be good sources of information. Although travel is expensive, attendance at market-oriented conferences can provide key ideas and contacts. Marketing publications, agricultural departments of government and universities, and established growers can provide invaluable information and can save you from mistakes. A personal business plan should be created (for advice on business plans, see for example Dove 1990, Dove 1994, Hankins 1994), and as much expert help as possible sought in preparing this. Keep in mind the grower's motto "find your market before you sow your seed." The following points may help those considering getting into the production of medicinal crops.

Tips and questions for the potential grower of medicinal crops

- A definite plus is high enthusiasm, indeed a love for herbs and plant cultivation in general and medicinal plants in particular. Do you have this kind of motivation?
- While medicinal plant production can be a satisfying hobby or sideline, generally this requires a substantial commitment in time, energy, and resources, acceptance of risk, and business talent. All of these requirements also apply to farming enterprises in general, but are more demanding with medicinal plant crops because of higher levels of risk. Does this fit in with your personal situation and abilities?
- In practice, farmers who already own a farm and have equipment and experience growing crops are in the best position to undertake the growing of medicinal crops. Well-recommended established farmers, preferably organic, are most likely to obtain contracts to grow particular crops. Does this describe you?
- Consider your own grower experiences and preferences, and size up the competition. Would you benefit from partnership with another, more experienced grower?
- Carefully consider the benefits of hiring a professional processor/manufacturer to convert your crop into marketable commodities that you take responsibility for selling. Farm gate profitability of most medicinal crops can be very low, while retail profits can be very high.
- Consider the advantages of organic production, and learn what is necessary for certification (for information on organic agriculture in Canada, see "Canadian Organic Growers" - http://www.gks.com/cog/). Consider weed and pest control strategies consistent with organic production. It is important to avoid areas where chemical residues are present, where spray drift occurs, and where hard to control weeds are established.
- Consider the maxim "Don't put all your eggs in one basket!" New crop ventures are always risky, and the market for medicinal plants is known to be very variable; therefore there is wisdom in trying several promising possible medicinal crops rather than just one.
- Acquire as much information as you are able on crops you decide to grow. Consult other growers, and provincial and federal government and university advisory personnel. If unknown pests appear or diseases develop, who will be able to identify these and provide advice?
- Government (both federal and provincial) and university crop research specialists in your area regularly conduct studies on such aspects of crops as physiology, soils, genetics, and development of mechanical harvesters, sorting machines, dryers, etc. Find out what crop research is going on, and whether aspects of this research can benefit your planned enterprise.
- Elected local politicians (municipal, provincial, federal) can be particularly valuable allies in starting up new ventures. To be taken seriously, do lots of research before contacting them, and try to be specific in your requests for assistance (e.g., for names and addresses of contacts, for letters of introduction, even for assistance in developing your business). Be persistent - politicians are there to serve you.
- Successful producers of new crops often spend more time marketing than growing their crops. Because of the necessity of devoting time to production and processing during the summer, market development may best be conducted during the off-season. Make market contacts, including alternative health providers who use herbs,

local processing companies, brokers, manufacturing companies, and health food stores. Remember, targeted personal contacts are more effective than mailed advertisements.

- Consider the suitability of your location (including distance to processing facilities and market), climate, soil (including drainage), rotation of the medicinal crops with complementary crops, topography (suitability for machinery), windbreaks (for tall herbaceous crops that could be damaged by wind), winter snowfall accumulation (which may be essential for survival of perennials), availability of water for irrigation, and available acreage.
- Carry out small-scale pilot studies of cultivation and harvesting. Consider testing several seed sources for suitability. If necessary, experimental plantings can be used to determine optimal conditions. Fertilization is known to decrease concentration of desired chemicals in many medicinal plants. Have chemical analysis performed to test market suitability of samples from your pilot study.
- Take a course in how to run a small business.
- A business plan is wise, and may be indispensable if financing is sought. Conventional lenders such as banks will recognize that new crop enterprises are high-risk ventures. Therefore, before attempting to borrow money, it is a good idea to have achieved initial productivity and market success. A recent market study may also be desirable. Seek professional financial advice if necessary.
- Consider the cost and (or) adaptability of required harvesting equipment, a drier and drying shed, sorting or processing equipment, and storage facilities.
- Plan out a schedule of activities, including seedbed preparation, seeding, fertilization, weed control, harvesting, storage, and processing.
- Consider market demands, timing of supply, volumes required, prices, and labor costs and availability. Trade groups and market newsletters can be invaluable for this kind of information (check the reference material cited at the end of this chapter).
- Determine product specifications of the market. If chemical or other types of professional analysis are required, who will do this?
- Production budgets are unavailable for most minor medicinal crops, and an attempt should be made to estimate probable costs. Factors to be considered include input costs, labor requirements, expected yields, number of years to harvest, and probable prices. By comparing opportunity costs for the money you would invest in the medicinal plant enterprise, and for use of the land, as well as your own labor, you can get a rough idea of how worthwhile the project is.
- Remember, the suitability of planting stock is critical to success. Many medicinal plants are completely wild (undomesticated) and special seed treatment is necessary for commercial planting. If seeds are unreliable or in very short supply, vegetative propagation may be necessary.
- Focus on long-term goals. Several years may be required to establish consistent and reliable crops and develop profitable markets.
- Consider concentrating on crops with similar specialized growth or preparation requirements. Many native Canadian medicinal plants are shade-loving and slow-growing, and can be grown in the same type of artificial shade. Culture under the natural canopy of woodland, the traditional method of growing many shade-loving species, is an alternative where considerable forested land is available. Crops that can be grown in greenhouses is one possible venture (although presently requiring research). Crops marketed as teas (and therefore requiring drying facilities) is another possibility.
- Invest in a computer and a software programs appropriate to your needs. Basic accounting programs will allow you to keep track of expenses and sales. Other software will enable you to economically produce a newsletter and advertising circulars, and labels for products. Color sells, and a color printer may be a smart investment.
- Access the World Wide Web. "There's a new Web site every 4 seconds" (Steven Calcote). The amount of information on all aspects of medicinal crops and indeed on all kinds of agricultural and business endeavours is staggering, and you will surely find much that is useful.
- Join organizations that represent your interests. There is strength in number, especially in attempting to influence governmental policies with respect to restrictive regulations concerning the manufacture and sale of herbal medicines.
- Treat employees honestly and fairly; they will return the favor.
- Strive to build a favorable reputation. Buyers are loyal to suppliers of a consistent, clean product that is available at the right time, place, and price, and meets industry standards and specifications of quality.
- Try to develop new markets. Political instability, crop failures, and contamination of products that

happen to competitors are opportunities to acquire new clients.

Literature useful for business aspects of medicinal plants

Berzins, R., and Richter, C. (*Editors*). 1997. Richters first commercial herb growing conference — transcripts. Richters, Goodwood, ON. 93 pp.

Berzins, R., Snell, H., and Richter, C. (*Editors*). 1998. Richters second commercial herb growing conference — transcripts. Richters, Goodwood, ON. 189 pp.

Davis, D. (Editor) 1995–96. Drug and cosmetic industry directory issue. Advanstar Communications, New York, NY. 300 pp. [Biennial directory issue of this periodical for the drug and cosmetic trade; this lists buyers/sellers of botanicals for industrial uses; American.]

Dove, J.S. 1990. Formulating a business plan. *In* Herbs '90. Proceedings fifth annual conference, International Herb Growers & Marketers Association. *Edited by* J.E. Simon, A. Kestner and M.R. Buehrle. International Herb Growers & Marketers Association, Silver Springs, PA. pp. 123–126.

Dove, J.S. 1994. Business plan outline. *In* Herbs '94. Proceedings ninth national herb growing and marketing conference. *Edited by* A. Kestner, J.E. Simon and A.O. Tucker. International Herb Growers & Marketers Association, Mundelein, IL. pp. 125–130.

Ference, D., and Associates Ltd. 1989. Economic Opportunities for Canada in Essential Oils and Medicinal Crops. Policy Branch, Agriculture and Agri-Food Canada Working Paper 10/89. 287 pp.

Gibson, E. 1994. Sell What You Sow! The grower's guide to successful produce marketing. New World Publishing, Carmichael, CA. 302 pp. [Does not deal with medicinal plants, but an excellent guide to many aspects of business, oriented to the novice agricultural entrepreneur.]

Hankins, A.G. 1994. Market options for herbs and spices. *In* Herbs '94. Proceedings ninth national herb growing and marketing conference. *Edited by* A. Kestner, J.E. Simon and A.O. Tucker. International Herb Growers & Marketers Association, Mundelein, IL. pp. 118–130.

Holm, W.R., and MacGregor, D. 1998. Processing guide for specialty crops. Science council of British Columbia – Okanagan. Irregularly paginated (ca. 150 pp.)

Miller, R.A. 1985. The potential of herbs as a cash crop. Acres U.S.A., Kansas City, MO. 230 pp.

Reppert, B. 1994. Growing Your Herb Business. Storey Communications, Inc., Burlington, VT. 186 pp.

Sturdivant, L. 1988. Profits from your backyard herb garden. San Juan Naturals, Friday Harbor, WA. 118 pp. [4th edition.]

Sturdivant, L. 1994. Herbs for sale. Growing and marketing herbs, herbal products, and herbal know-how. San Juan Naturals, Friday Harbor, WA. 246 pp.

Sturdivant, L., and Blakely, T. 1999. Medicinal herbs in the garden, field & marketplace. San Juan Naturals, Friday Harbor, WA. 323 pp.

Wallin, C. 1989. Backyard Cash Crops. Homestead Design, Inc., Friday Harbor, WA. 232 pp.

World Wide Web Resources and Sites of Interest with special reference to business development (also see General guide to WWW)

ICAR - Inventory of Canadian agri-food research [a wonderful tool for finding (by keyword search) what researchers are investigating particular crops; however, coverage is by no means complete or up to date]:
http://www.agr.ca/icar/E/main.html

AgroPharm Technologies [The company states: "We are committed to producing world-class medicinal plant products with the enabling agricultural and technological resources available in Canada. Founded in 1994. Currently operating from Waterdown and Simcoe, Ontario. Ongoing development of health promoting botanicals to diversify from tobacco production focused on small acreage high intensity cropping. Development of standardized procedures to produce drug grade quality, development of practices to meet government compliance for Traditional Herbal Medicines, development of ecologically minded 'Organic' cultivation practices. 100% Canadian grown and manufactured products."]:
http://www.agropharm.com/contents.html

Richters Herb Specialists - home page [Richters is Canada's largest supplier of seeds and plants of medicinal herbs; this site offers a free catalogue, a newsletter, and various information features]:
http://www.richters.com/

Natural Factors Nutritional Products Ltd. [Canadian branch of this company has large medicinal plant farms in western Canada]:
http://www.naturalfactors.com/

Growing herbs for essential oils, cooking and medicine. Summary of the potential for commercial herb production in New Zealand including an assessment of agronomic and economic factors and market prospects [excellent general information for anyone considering going into the medicinal herb business]:
http://www.crop.cri.nz/broadshe/herbs.htm

Nova Scotian wildcrafted medicinal herbs for sale [a small private Canadian business, exemplifying the combination of wild-collecting local medicinal plants and selling by internet]:
http://www.agro-lanka.org/messages/33.htm

Taiga Herbs, Darwell, Alberta, the Labrador tea [a Canadian outlet marketing dried wild-collected leaves of the widespread Canadian native shrub *Ledum groenlandicum* as a vitamin-C-rich herbal tea]:
http://www.farm-online.com/taiga/lab-tea.htm

News article featured in Ontario Farmer Production Section / Tuesday, June 11, 1996 [news reports about an Ontario farming enterprise that has undertaken the cultivation of medicinal plants]:
http://www.agropharm.com/ontfarmer.html

Farmer invests in plant said to relieve migraines [experiences of an Ontario farmer who recently started to cultivate and market feverfew]:
http://www.agropharm.com/expositor.html

Herbs & flowers. This page contains books on echinacea, dried flowers, growing and selling flowers, growing and marketing herbs, starting an herb business. [a good selection of practical books on the commercial growing of medicinal herbs and other possibilities]:
http://www.vfr.net/~newworld/herbsflow.html

Openair-market net: the world wide guide to farmers' markets, street markets, flea markets and street vendors [many links, some of which are useful when searching for ideas and materials related to growing and marketing crops]:
http://www.openair.org/

The Business of herbs. The international news and resource service for herb businesses. The Business of Herbs is an international publication for the herb business [advertisement for this herb business journal]:
http://gardennet.com/BOH/default.htm

The Herb, Spice, and Medicinal Plant Digest [advertisement for a journal with important professional information on medicinal plants]:
http://www-unix.oit.umass.edu/~herbdig/

Essential oils industry, Alberta Agriculture, Food and Rural Development [considerable practical information on essential oil plants, some of which are medicinal, with specific reference to Alberta]:
http://www.agric.gov.ab.ca/agdex/100/8883001.html

Special crops conference, Alberta Agriculture, Food and Rural Development [considerable practical information on a variety of crops, some of which are medicinal]:
http://www.agric.gov.ab.ca/crops/speciaL/conf/

Saskatchewan herb and spice industry, introductory information guide - part 1 [general information with particular reference to Saskatchewan. A more extensive document is listed below]:
http://www.gov.sk.ca/agfood/farmfact/herbspc1.htm

Farmfacts, Saskatchewan herb and spice industry, introductory information guide [an excellent overview with particular reference to Saskatchewan; amongst other information, this document shows how to calculate whether or not replacing an old crop with a new one will be profitable (assuming sufficient information is available)]:
http://www.agr.gov.sk.ca/saf/farmfact/herbspc1.htm

Farmfacts, marketing non-traditional (niche) crops [written for Saskatchewan, but contains excellent general business advice on new crop development]:
http://www.agr.gov.sk.ca/saf/farmfact/fmm0196.htm

United States Department of Agriculture, Cooperative State Research Education and Extension Service, Office for Small-Scale Agriculture - herbs [provides general American sources of information on herbs; some of this is relevant to medicinal plants]:
http://www.sfc.ucdavis.edu/pubs/brochures/Herbs.html

Appropriate technology transfer for rural areas [a long, very well prepared document, loaded with useful information and references on herbs in general]:
http://www.attra.org/attra-pub/herb.html

Canada Business Service Centres [a huge guide to business-related programs, services and regulations in Canada; some of the links identify numerous support programs]:
http://www.cbsc.org/fedbis/index.html

Medicinal garlic, an industry overview [written for Saskatchewan]:
http://www.agr.gov.sk.ca/saf/crops/garlic.htm

Information sources

Canadian expert contacts

These have been arranged by province of residence, from west to east, but as documented for each expert listed, many will provide information to all Canadians, and indeed to non-Canadians. A list of this kind is never complete, and the authors would appreciate suggestions of additional experts to be included in future revisions.

BRITISH COLUMBIA

Dr. W.G. Bailey

Department of Geography
Simon Fraser University
Burnaby BC
V5A 1S6
Tel. 604–291–4425
Fax 604–291–5841
E-mail: bailey@.ca
Crop specialties: North American ginseng (*Panax quinquefolius*), Asian ginseng (*Panax ginseng*)
Professional specialties: all aspects (production, harvest and post-harvest, drying, marketing, international trade)
Information will be provided to: everyone

Dr. Pat Bowen

Research Scientist
Pacific Agri-Food Research Centre
Agriculture and Agri-Food Canada
PO Box 1000
BC
V0M 1A0
Tel. 604–796–2221 Ext. 225
Fax 604–796–0359
E-mail: bowenp@em.agr.ca
Crop specialties: *Echinacea spp.*, *Tanacetum parthenium* (feverfew), *Glycyrrhiza spp.* (licorice), new crops
Professional specialties: production (plant physiology, crop quality), cultural system design
Information will be provided to: herb researchers, herb crop extension agents, Canadian herb growers

Ms. Debra Lees

Marketing Specialist
728 Cambridge Cres.
Kamloops BC
V2B 5B6
Tel. 250–554–1040
Fax 250–554–1080
E-mail: dalees@mail.ocis.net
Crop specialties: ginseng, medicinal plants
Professional specialties: marketing
Information or advice will be provided to: everyone connected with development and marketing

Dr. Thomas S.C. Li

Research Scientist
Pacific Agri-Food Research Centre
Agriculture and Agri-Food Canada
Hwy. 97S
Summerland BC
V0H 1Z0
Tel. 250–494–6375
Fax: 250–494–0755
E-mail: lit@em.agr.ca
Crop specialties: ginseng, echinacea, medicinal plants, alternative crops, nutraceutical crops
Professional specialties: production, control crop pathology, general information, greenhouse culture, breeding, agronomy, organic cultivation, chemistry
Information will be provided to: Canadians

Dr. Allison R. McCutcheon

Research Associate
Department of Botany
University of British Columbia
#3529 - 6270 University Blvd.
Vancouver BC
V6T 1Z4
Tel. 604–222–3488
Fax 604–222–9613
E-mail: allison@imag.net
Crop specialties: small scale propagatation, production and processing of medicinal plants, potted plant production
Professional specialties: North American native medicinal plants, medicinal plant taxonomy and pharmacognosy (scientific evaluation of herbs in terms of their identity, purity, potency, and pharmacological activity); lecturer in pharmacognosy and alternative medicines
Information will be provided to: Canadians; consulting services for herbal medicines, quality assessments & certifications, botanical identity and chemical analyses
[Also see (below) "The Canadian Herb Society" under: Organizations, Canadian]

Mr. Al Oliver

Special Crops Horticulturist, Provincial Ginseng Specialist
British Columbia Ministry of Agriculture, Fisheries and Food
162 Oriole Road
Kamloops BC
V2C 4N7
Tel. 250–371–6050
Fax: 250–828–4631
E-mail: Al.Oliver@gems3.gov.bc.ca
Crop specialties: special crops, ginseng
Professional specialties: production
Information will be provided to: everyone

Dr. Zamir K. Punja

Director, Pest Management
Department of Biological Sciences
Simon Fraser University
8888 University Drive
Burnaby, BC
V5A 1S6
Tel. 604–291–4471
Fax: 604–291–3496
E-mail: punja@sfu.ca
Crop specialties: ginseng
Professional specialties: tissue culture and control of ginseng diseases
Information will be provided to: everyone

Mr. Max Xiao

Marketing & Trade Officer
Market and Industry Services Branch
Agriculture and Agri-Food Canada
620 Royal Avenue, Suite 204
New Westminster BC
V3M 1J2
Tel. 604–666–9353
Fax: 604–666–7235
E-mail: xiaom@em.agr.ca
Crop specialties: ginseng
Professional specialties: marketing
Information will be provided to: Canadians, medicinal herb growers

ALBERTA

Dr. Stan Blade

New Crops Research Scientist
Crop Diversification Centre North
Alberta Agriculture, Food and Rural Development
17507 Fort Road
Edmonton
AB T5B 4K3
Tel. 780–415–2311
Fax: 780–422–6096
E-mail: blade@agric.gov.ab.ca
Crop specialties: pulses, alternate field crops, spices, medicinal plants, hemp
Professional specialties: plant breeding and agronomy
Information will be provided to: everyone, with priority to Alberta residents

Dr. Kan-Fa Chang

Research Scientist - Plant Pathologist
Crop Diversification Centre - South
Alberta Agriculture, Food and Rural Development
S.S. #4
Brooks AB
T1R 1E6
Tel. 403–362–1334
Fax 403–362–1326 or 403–362–1306
E-mail: changk@agric.gov.ab.ca
Crop specialties: all medicinal plants
Professional specialties: disease control
Information will be provided to: everyone

Mr. Dean Dyck

Farm Management Specialist
Alberta Agriculture, Food and Rural Development
#301, 4920 - 51 Street
Red Deer AB
T4N 6K8
Tel. 403–340–7007
Fax 403–340–4896
E-mail: dyck@agric.gov.ab.ca
Crop specialties: all new crops for Alberta
Professional specialties: economics, cost of production
Information will be provided to: everyone

Dr. Ronald J. Howard

Plant Pathologist and
Leader, New Crop Development Unit
Crop Diversification Centre - South
Alberta Agriculture, Food and Rural Development
S.S. #4
Brooks AB
T1R 1E6
Tel. 403–362–1328 or 403–362–1300
Fax 403–362–1326 or 403–362–1306
E-mail: ron.howard@agric.gov.ab.ca
Crop specialties: all medicinal plants

Professional specialties: disease control
Information will be provided to: commercial growers, researchers, extension specialists, educators

Dr. Sheau-Fang Hwang

Senior Research Scientist, Plant Pathology
Crop and Plant Management
Alberta Research Council
Hwy. 16A, 75 Street
P.O. Bag 4000
Vegreville AB
T9C 1T4
Tel. 403–632–8228
Fax 403–632–8612
E-mail: hwang@arc.ab.ca
Crop specialties: all medicinal plants
Professional specialties: disease control
Information will be provided to: everyone

Dr. Mohyuddin Mirza

Greenhouse Crops Specialist
Alberta Agriculture, Food and Rural Development
Crop Diversification Centre North
R.R. 6
Edmonton AB
T5B 4K3
Tel. 403–415–2303
Fax 403–422–6096
E-mail: mirza@agric.gov.ab.ca
Crop specialties: echinacea, calendula, St. John's wort, gotukola
Professional specialties: greenhouse culture
Information will be provided to: anyone interested in the production of medicinal plants

SASKATCHEWAN

Dr. Branka Barl

Professional Research Associate and
Leader, Herb Research Program
Department of Plant Sciences
University of Saskatchewan
51 Campus Drive
Saskatoon SK
S7N 5A8
Tel. 306–966–5868
Fax 306–966–5015
E-mail: barlb@duke.usask.ca
Crop specialties: medicinal plants in general, particularly echinacea and other species suitable for the prairies
Professional specialties: general information, quality assessment, production
Information will be provided to: primarily residents of Saskatchewan, but also other Canadians

Mr. Harvey Clark

Market Analyst
Saskatchewan Irrigation Diversification Centre
901 McKenzie Street South
P.O. Box 700
Outlook SK
S0L 2N0
Tel. 306–867–5402
Fax 306–867–9656
E-mail: pf22407@em.agr.ca
Crop specialties: ginseng, echinacea, all medicinal crops, fruit, vegetables, potatoes, spice crops, all new crops
Professional specialties: marketing, economics, finance, international trade
Information will be provided to: everyone (but mandate concentrates on areas with irrigation)

Mr. John Kort

Shelterbelt Biologist/Agroforester
Agriculture and Agri-Food Canada
Prairie Farm Rehabilitation Administration
PFRA Shelterbelt Centre
Box 940
Indian Head SK
S0G 2K0
Tel. 306–695–2284
Fax: 306–695–2568
E-mail: pf21800@em.agr.ca
Crop specialties: ginseng
Professional specialties: production
Information will be provided to: Canadians

Dr. H.R. Kutcher

Plant Pathologist
Melfort Research Farm
Research Branch
Agriculture and Agri-Food Canada
Box 1240
Melfort SK
SOE 1AO
Tel. 306–752–2776 Ext. 232
Fax 306–752–4911
E-mail: kutcherr@em.agr.ca
Crop specialties: canola pulse crops, new crops for Parkland area (beans, some spice crops)
Professional specialties: plant pathology
Information will be provided to: everyone

Dr. Alister Muir

Research Scientist (Crop Utilization)

Saskatoon Research Centre
Research Branch
Agriculture and Agri-Food Canada
107 Science Place
Saskatoon SK
S7N 0X2
Tel. 306–956–7265
Fax: 306–956–7247
E-mail: muira@em.agr.ca

Crop specialties: nutraceutical, medicinal and industrial crops

Professional specialties: chemistry, extraction and processing, identification of active principles

Information will be provided to: companies interested in processing and (or) quality control

Mr. Brian Porter

Horticulture Development Specialist
Sustainable Production Branch
Saskatchewan Agriculture and Food
125 Walter Scott Building
3085 Albert Street
Regina SK
S4S 0B1
Tel. 306–787–4670
Fax 306–787–0428

Crop specialties: all herbs and medicinal plants

Professional specialties: production, general information

Information will be provided to: residents of Saskatchewan

Dr. Ken Richards

Research Manager
Plant Gene Resources of Canada
Saskatoon Research Centre
Research Branch
Agriculture and Agri-Food Canda
107 Science Place
Saskatoon SK
S7N OX2
Tel. 306–956–7641
Fax 306–956–7246
E-mail: richardsk@em.agr.ca

Crop specialties: crop germplasm; all major Canadian agricultural crops, some minor crops; some native Canadian plant species including medicinal plants; oil and nutraceutical crops

Professional specialties: crop germplasm conservation

Information will be provided to: all Canadians; germplasm to researchers (breeders, plant scientists) nationally and internationally for legitimate reasons

Dr. A.E. Slinkard

Crop Development Centre
University of Saskatchewan
51 Campus Drive
Saskatoon SK
S7N 5A8
Tel. 306–966–4978
Fax: 306–966–5015
E-mail: sharon.stevens@usask.ca

Crop specialties: anise, borage, caraway, coriander, dill, fenugreek, pulse crops

Professional specialties: breeding and management

Information will be provided to: everyone

Dr. Jazeem Wahab

Horticultural and Specialty Crops Agronomist
Saskatchewan Irrigation Diversification Centre
901 McKenzie Street South
P.O. Box 700
Outlook SK
S0L 2N0
Tel. 306–867–5406
Fax 306–867–9656
E-mail: pf22406@em.agr.ca

Crop specialties: medicinal plants, culinary herbs, essential oil crops

Professional specialties: agronomic research for both irrigation and dryland production

Information will be provided to: everyone

Dr. Neil D. Westcott

Research Scientist
Crop Utilization
Saskatoon Research Centre
Research Branch
Agriculture and Agri-Food Canda
107 Science Place
Saskatoon SK
S7N OX2
Tel. 306–956–7266
Fax 306–956–7247
E-mail: westcottn@em.agr.ca

Crop specialties: alternate field crops, medicinal and industrial crops

Professional specialties: chemistry, processing

Information will be provided to: Canadians

MANITOBA

Dr. Campbell G. Davidson

Manager, Morden Research Centre
Research Branch
Agriculture and Agri-Food Canada

Unit 100 - 101 Route 100
Morden MB
R6M 1Y5
Tel. 204–822–7201
Fax: 204–822–7209
E-mail: cdavidson@em.agr.ca
Crop specialties: woody and herbaceous landscape plants
Professional specialties: breeding
Information will be provided to: those wishing information relating to research and plant breeding

Dr. Jack Moes

Principal Consultant
The Great AgVenture
20 Marquis Crescent
Brandon MB
R7B 3R8
Tel. 204–571–1631
Fax 204–727–3964
E-mail: agventure@mb.sympatico.ca
Crop specialties: hemp, alternative field crops, nutraceutical and medicinal crops, essential oil crops
Professional specialties: production/agronomy, primary and secondary processing, marketing, end-use quality
Information will be provided to: everyone (fee basis)

ONTARIO

Professor John Thor Arnason

Department of Biology
University of Ottawa
Ottawa ON
K1N 6N5
Tel. 613–562–5262
Fax 613–562–5765
E-mail: jarnason@science.uottawa.ca
Crop specialties: echinacea, ginseng, feverfew, St. John's wort, eleuthero, other domestic and tropical medicinal plants
Professional specialties: phytochemistry, quality assurance, biological activity, germplasm

Dr. Dennis V.C. Awang

President, MediPlant Consulting Services
P.O. Box 8693, Station T
Ottawa ON
K1G3J1
Tel. 613–741–6606
Fax 613–741–6732
E-mail: awangd@netcom.ca
Crop specialties: ginseng, echinacea, feverfew
Professional specialties: chemistry, phytochemical analysis, chemical/bioactivity/clinical effect
Information will be provided to: everyone (fee basis)

Dr. J.E. Brandle

Plant Geneticist
Southern Crop Protection and Food Research Centre
Research Branch
Agriculture and Agri-Food Canada
1391 Sandford Street
London ON
N5V 4T3
Tel. 519–663–3326
Fax: 519–663–3454
E-mail: brandleje@em.agr.ca
Crop specialties: evening primrose, ginseng, St. John's wort, stevia
Professional specialties: breeding and molecular genetics

Dr. Daniel C.W. Brown

Research Scientist, New Crops Program
Southern Crop Protection and Food Research Centre
Research Branch
Agriculture and Agri-Food Canada
1391 Sandford Street
London ON
N5V 4T3
Tel. 519–457–1470 Ext. 228
Fax: 519–457–3997
E-mail: browndc@em.agr.ca
Crop specialties: feverfew, ginseng, stevia
Professional specialties: plant biotechnology, micropropagation
Information will be provided to: everyone

Mr. Michael J. Columbus

Alternative Crops Specialist
Ontario Ministry of Agriculture, Food and Rural Affairs
Horticultural Experiment Station
P.O. Box 587
Simcoe ON
N3Y 4N5
Tel. 519–426–7120
Fax: 519–428–1142
E-mail: mcolumbu@omafra.gov.on.ca
Crop specialties: new crops for Ontario, including edible, medicinal, industrial
Professional specialties: production, marketing, value-added processing

Information will be provided to: residents of Ontario

Dr. Richard D. Reeleder

Research Scientist - Plant Pathologist
Southern Crop Protection and Food Research Centre
Research Branch
Agriculture and Agri-Food Canada
P.O. Box 186
Delhi ON
N4B 2W9
Tel. 519–582–1950
Fax: 519–582–4223
E-mail: reelederr@em.agr.ca
Crop specialties: ginseng, goldenseal, shade medicinals, essential oil crops, new crops
Professional specialties: crop pathology
Information will be provided to: priority to Canadians

Mr. L. Bruce Reynolds

Research Biologist, Agronomy
Research Branch
Agriculture and Agri-Food Canada
Southern Crop Protection and Food Research Center
Box 186
Delhi ON
N4B 2W9
Tel. 519–582 -1950
Fax 519–582 - 4223
E-mail: reynoldsb@em.agr.ca
Crop specialties: flue-cured tobacco agronomy, ginseng post-harvest conditioning & drying, sweet potato crop production, paste tomato crop production
Professional specialties: transplant production (bed, plug & float systems), irrigation, tobacco & ginseng curing and drying, nematode biocontrol via crop rotations
Information will be provided to: research professionals, growers, industry, press

Ms. Jan Schooley

Ginseng and Medicinal Herb Specialist, Ontario Ministry of Agriculture, Food and Rural Affairs
Horticultural Experiment Station
Box 587
Simcoe ON
N3Y 4N5
Tel. 519- 426–7120
Fax: 519–428–1142
E-mail: jschoole@omafra.gov.on.ca
Crop specialties: medicinals
Professional specialties: field/forest production/protection
Information will be provided to: growers, anyone interested in the production of medicinal plants

Mr. Hilton Virtue

Marketing and Trade Officer
Market and Industry Services Branch
Agriculture and Agri-Food Canada
174 Stone Road West
Guelph ON
N1G 4S9
Tel. 519–837–5860
Fax 519–837–9782
E-mail: virtueh@em.agr.ca
Crop specialties: ginseng
Professional specialties: general information, marketing, trade
Information will be provided to: ginseng industry in Ontario

QUEBEC

Dr. Edward R. Farnworth

Research Scientist (flavors and neutraceutical chemistry)
Food Research and Development Centre
Research Branch
Agriculture and Agri-Food Canada
3600 Casavant Blvd. West
Saint-Hyacinthe PQ
J2S 8E3
Tel. 450–773–1105
Fax: 450–773–8461
E-mail: farnworthed@em.agr.ca
Crop specialties: Jerusalem artichoke
Professional specialties: composition, chemistry, physiological effects
Information will be provided to: general public, industrial collaborators

NOVA SCOTIA

Dr. Willy Kalt

Research Scientist - Food Chemistry
Atlantic Food and Horticulture Research Centre
Research Branch
Agriculture and Agri-Food Canada
32 Main St.
Kentville NS
B4N 1J5
Tel. 902–679–5757
Fax 902–679–2311
E-mail: kaltw@em.agr.ca

Crop specialties: blueberries, cranberries, lingonberries, other *Vaccinium* species, strawberries, raspberries, exotic northern fruit

Professional specialties: chemistry, physiology, postharvest storage, processing

Information will be provided to: research and extension workers, interested growers (primarily Eastern Canadian), the press, industry members, marketing groups

Professor Lloyd Mapplebeck

Department of Plant Science
Nova Scotia Agricultural College
P.O. Box 550
Truro NS
B2N 5E3
Tel. 902–893–6683
Fax 902–897–9762
E-mail: lmapplebeck@cox.nsac.ns.ca

Crop specialties: chamomile, lemon balm, feverfew, milk thistle, roseroot, valerian, etc. (all medicinal plants except ginseng), native plants used in traditional mi'kmaq medicine

Professional specialties: horticultural production: greenhouse, nursery and specialty crops

Information will be provided to: growers, potential growers, general public

The following applies to the three specialists listed below:

Ms. Beverley MacPhail

(Greenhouse Specialist)

Mr. Charles N. Thompson

(Vegetable Specialist)

Mr. John Wilson

(Horticulturist - Organic Specialist)
Nova Scotia Department of Agriculture and Marketing
Kentville Agricultural Centre
Kentville NS
B4N 1J5
Tel. 902–679–6034 (MacPhail), 902–679–6041 (Thompson), 902–679–6025 (Wilson)
Fax 902–679–6062
E-mail: bmacphail@gov.ns.ca, Cthompson@gov.ns.ca, jwilson@gov.ns.ca

Crop specialties: Exploration of all specialty vegetable and herb crops, including medicinal, for value added possibilities

Professional specialties: new marketing opportunities, all types of production research and extension (including ginseng), focus mainly on extension

Information will be provided to: mainly residents of Nova Scotia

Ms. K. Laurie Sandeson

Market Development Officer
Nova Scotia Department of Agriculture and Marketing
P.O. Box 550
Truro NS
B2N 5E3
Tel. 902–893–6387
Fax 902–895–9403
E-mail: L.Sandeson@nsar.ns.ca

Professional specialties: Identification of new product opportunities through market research; marketing & promotion

Information will be provided to: industry associations, individuals, growers, processors

NEWFOUNDLAND (& LABRADOR)

Mr. Cyril J. Hookey

Pest Management Specialist
Dept. of Forest Resources & Agrifoods
Government of Newfoundland & Labrador
Box 8700
Provincial Agriculture Building
Brookfield Rd.
Mt. Pearl NF
A1B 4J6
Tel. 709–729–6633
Fax 709–729–0205
E-mail: chookey@agric.dffa.gov.nf.ca

Crop specialties: All medicinal plants, native trees and shrubs, weeds

Professional specialties: organic gardening, pest identification and control, plant identification, landscaping, horticulture

Information will be provided to: Newfoundland & Labrador residents and medicinal herb growers

Mr. Richard Oram

Alternative Crops Coordinator
Dept. of Forest Resources & Agrifoods
Government of Newfoundland & Labrador
P.O. Box 640
Bishop's Falls NF
A0H 1C0
Tel. 709–258–5334
Fax 709–258–5873

Crop specialties: ginseng, echinacea, all new crops for Newfoundland
Professional specialties: agronomy, general information
Information will be provided to: residents of newfoundland, medicinal herb growers

Mr. Michael Stapleton

Crops Specialist
Dept. of Forest Resources & Agrifoods
Government of Newfoundland & Labrador
P.O. Box 8700
Provincial Agriculture Building
Brookfield Road
St. John's NF
A1B 4J6
Tel. 709–729–6867
Fax 709–729–2674
E-mail: mstaplet@agric.dffa.gov.nf.ca
Crop specialties: ginseng, echinacea, St. John's wort, medicinal plants, nutraceutical crops, all new crops with potential for Newfoundland & Labrador
Professional specialties: production, general information, organic production
Information will be provided to: priority levels: 1, residents of Newfoundland & Labrador; 2, Canadians; 3, everyone.

Research on medicinal plants has recently (1998) been initiated at:

Atlantic Cool Climate Crop Research Centre
Research Branch
Agriculture and Agri-Food Canada
308 Brookfield Road
P.O. Box 39088
St. John's NF
A1E 5Y7
Fax: 709 772–6064
Home Page: http://res.agr.ca/stjohns/

The following are available for possible consultation and cooperative ventures involving medicinal plants (as well as their stated areas of expertise):

Dr. David B. McKenzie

Agronomist - field crop production
Tel. 709–772–4784
E-mail: mckenziedb@em.agr.ca

Dr. Peggy L. Dixon

Entomologist - biocontrol
Tel. 709–772–4763
E-mail: dixonpl@em.agr.ca

Dr. Dean M. Spaner

Crop Physiologist;
Tel. 709–772–5278
E-mail: spanerd@em.agr.ca

Mr. Gary A. Bishop

Agricultural Engineer - drainage and equipment design
Tel. 709–772–4170
E-mail: bishopg@em.agr.ca

Mr. Boyd G. Penney

Horticulturalist - vegetables and small berries
Tel. 709–772–5277
E-mail: penneyb@em.agr.ca

Organizations

Canadian

Canadian Society for Herbal Research, P.O. Box 82, Station A, Willowdale, ON M2N 5S7, **Tel**. 705–432–2418. (Purpose: public education concerning herbalism; publication: The Herbalist (quarterly)).

Ginseng Growers Association of Canada, 395 Queensway West, 2nd Floor, Simcoe, ON, N3Y 2N4; **Tel**. 519–426–7046; **Fax** 519–426–9087.

Ontario Herbalists Association, 11 Winthrop Place, Stoney Creek, ON L8G 3M3. **Tel.** 416–536–1509. (Members include professional herbalists, naturopaths, and consumers.)

The Canadian Herb Society, c/o VanDusen Botanical Garden, 5251 Oak Street, Vancouver, BC, V6M 4H1. See: http://www.hedgerows.com/Canada/clubbrochures/CanHerbSoc.htm and http://www.herbsociety.ca/intro.html ("A society of herb enthusiasts founded to share knowledge and promote the cultivation, use and enjoyment of herbs." President: Dr. A.R. McCutcheon (see Canadian Expert Contacts, above; **E-mail**: info@herbsociety.ca; **Fax**: 604–222–7613). This is a non-profit, educational organization dedicated to providing Canadians with accurate and reliable information on herbs, their cultivation and usage. Publication: Herbal Times (quarterly). The organization is in the process of compiling a directory of herb business people, organizations, and individuals interested in herbs.)

Saskatchewan Herb and Spice Association. Contact: Connie Kehler, **Tel**. 306–694–4622, **E-mail**: g.musings@dlcwest.com; see http://paridss.usask.ca/specialcrop/c_herb_feedback.html (A producer-driven organization supporting research, development and promotion of crops.)

Saskatchewan Nutraceutical Network, #101 - 111 Research Drive, Saskatoon, SL S7N 3R4; web site: www.nutranet.org. Contact: Kelley Fitzpatrick, President, **Tel**. 306–668–2654, **Fax** 306–975–1966, **E-mail**: kfitzpatrick@innovationplace.com. ("An organization whose role is to lead the development, and support the growth of, an economically viable nutraceutical and functional foods industry in the province. Our membership is comprised of manufacturers, processors, producer organizations, research and analytical agencies, the University of Saskatchewan, and various levels of government including Federal, Provincial and Municipal." Issues a newsletter.)

American

American Botanical Council, P.O. Box 201660, Austin, TX 78720 (Nonprofit research/education. Publishes HerbalGram, distributes books).

American Herb Association, P.O. Box 1673, Nevada City, CA 95959 (Publishes quarterly newsletter).

American Herbal Products Association, P.O. Box 2410, Austin, TX 78768.

American Herbalists Guild, P.O. Box 1683, Soquel, CA 95073 (Publishes quarterly newsletter and sponsors annual symposium).

Herb Growing and Marketing Network, 3343 Nolt Road, Lancaster, PA 17601 (Publishes Herbal Green Pages).

Herb Research Foundation, 1007 Pearl Street, #200 F, Boulder, CO 80302 (Members receive the publication HerbalGram).

The Herb Society of America, Inc., 9019 Kirtland Chardon Road, Kirtland, OH 44094.

International Herb Association, 1202 Allanson Road, Mundelein, IL 60060 (Professional trade organization. Publishes Proceedings of annual meeting, listed below).

North East Herbal Association, P.O. Box 146, Marshfield, VT 05658–0146 (Publishes quarterly newsletter and sponsors yearly conference).

Resource guides

Clavio, L.Z. (*Editor*). 1994. Directory of herbal education. IntraAmerican Specialties, West Lafayette, IN. 48 pp.

Craker, L.E. 1994.The directory of specialists in herbs, spices, and medicinal plants. 2nd. ed. 101 pp. (A directory of addresses of individuals throughout the world, specializing in herbs, spices, and medicinal plants.)

McRae, B.A. 1992. Herb companion wishbook & resource guide. Interweave Press, Loveland, CO. 304 pp. (Mail-order sources for herbal products, organizations, gardens, and festivals).

Oliver, P. (*Editor*). 1996. Herb resource directory. 4th edition. Northwind Publications, Jemez Springs, NM. (Book and accompanying diskette. Sources for seeds/plants, herbs, supplies, biological pest controls. Lists books, periodicals, classes, festivals, garden tours. Available from Northwind Publications, 439 Ponderosa Way, Jemez Springs, NM 87025–8025, U.S.A.).

Rogers, M. (*Editor*). 1996. The herbal green pages: an herbal resource guide. HGMN, Lancaster, PA. 300+ pp. (Annual directory with over 3000 listings of herb businesses, publications, associations, educational programs, and suppliers. Available from Herb Growing & Marketing Network, 3343 Nolt Road, Lancaster, PA 17601, U.S.A.).

Simon, J. E. et al. (*Editors*). International Herb Association annual conference proceedings. (Annual, started 1986, numerous contributions on diverse practical subjects related to the business of growing and marketing herbs. Available from IHA, 1202Allanson Road, Mundelein, IL 60060, U.S.A.).

Trade publications, magazines and newsletters

Canadian

Canadian Journal of Herbalism (Published by the Ontario Herbalists Association. Emphasis is on medicinal herbs).

The Gilded Herb, Canada's Herbal Magazine (Published from 15 Clifford Drive, Dunsford, ON, K0M 1L0; see herbs@lindasaycomp.on.ca).

American

The American Herb Association quarterly newsletter, American Herb Association, P.O. Box 353, Rescue, CA 95672.

The business of herbs, Northwind Publications, 439 Ponderosa Way, Jemez Springs, NM 87025–8025.

Foster's botanical & herb reviews and Foster's herb business bulletin. S. Foster, P.O. Box 1343, Fayetteville, AR 72702.

Herb companion, Interweave Press, Inc., 201 E. Fourth St., Loveland, CO 80537.

The herb connection, Herb Growing and Marketing Network, 3343 Nolt Rd., Lancaster, PA 17601.

The herb, spice, and medicinal plant digest, University of Massachusetts Cooperative Extension Service, Amherst, MA 01003.

The herbal connection, P.O. Box 245, Silver Spring, PA 17575–0245.

Herbalgram, Herb Research Foundation and American Herbal Products Association, 1007 Pearl Street, #200 F, Boulder, CO 80302.

New crops news, Indiana Center for New Crops and New Products, Purdue University, 1165 Horticulture Building, West Lafayette, IN 47907.

Small farm today, Route 1, 3903 Ridgetrail Road, Clark, MO 65243.

Web guides to regional Canadian new crop information sources

British Columbia Ministry of Agriculture and Food [no information related specifically to medicinal plants; horticultural staff listed]:
http://www.agf.gov.bc.ca/index.htm

Alberta Agriculture, Food and Rural Development - special crops [provides extensive information on essential oils, ginseng, and herbs and spices]:
http://www.agric.gov.ab.ca/navigaion/crops/speciaL/index.html

Saskatchewan Agriculture and Food, Crop Development Services ["to assist the industry in identifying and capitalizing on opportunities for development and diversification of the crops sector;" an address is provided to find advice on special crops]:
http://www.agr.gov.sk.ca/saf/

Herb research and industry development in Saskatchewan [a very useful orientation document on who is doing what in Saskatchewan; has list of 20 western herbal companies]:
http://www.agric.gov.ab.ca/crops/speciaL/medconf/barl.html

Manitoba Crop Diversification Centre ["MCDC carries out the evaluation and demonstration of speciality and niche crops that offer higher value diversification opportunities, good market potential, and value adding possibilities. Program delivery mechanisms consist of field demonstrations, publication of fact sheets and bulletins, and presentation of seminars and courses for producer training in partnership with provincial, federal and industry initiatives."]:
http://www.agr.ca/pfra/mcdcgene.htm

Prairie Farm Rehabilitation Administration ["Prairie Farm Rehabilitation Administration (PFRA) works with Prairie people to develop a viable agricultural industry and sustainable rural economy in Manitoba, Saskatchewan, Alberta and the Peace River Region of British Columbia. In partnership with other levels of government, farm groups, producers and industry, PFRA develops and conserves the area's soil and water resources, and encourages diversification into new crops, value-added processing and other wealth-creating opportunities."]:
http://www.cbsc.org/fedbis/bis/1618.html

Ontario Ministry of Agriculture, Food and Rural Affairs (OMAFRA), specialty crops [information on ginseng and stevia, and non-medicinal specialty crops]:
http://www.gov.on.ca/OMAFRA/english/crops/hort/specialty.html

Ontario Ministry of Agriculture, Food and Rural Affairs - 1998 crop advisory staff [a brief but excellent guide to crop advisory personnel in Ontario; should be emulated by web sites of other provinces, for which locating this kind of essential information is often like finding a needle in a haystack]:
http://www.gov.on.ca/OMAFRA/english/crops/resource/stafcrop.htm

Quebec: Ministère de l'Agriculture, des Pêcheries et de l'Alimentation [we were unable to find information related to medicinal plants]:
http://www.cbsc.org/fedbis/bis/lnag.html

New Brunswick Agriculture and Rural Development, Horticulture Industry Development [horticulture staff are listed, and an empty category "specialty crops" is present, but information related to medicinal plants was not found]:
http://www.gov.nb.ca/agricult/horticuL/hortpage.htm

Nova Scotia Department of Agriculture and Marketing [no information related specifically to medicinal plants was found; horticultural coordinator listed]:
http://agri.gov.ns.ca/

Prince Edward Island, Agriculture and Forestry, - Agriculture [no information related specifically to medicinal plants was found; horticultural coordinator listed]:
http://www.gov.pe.ca/af/index.asp

Newfoundland and Labrador Department of Forest Resources and Agrifoods, Food and Agriculture Branch [no information on medicinal or even new crops was found]:
http://public.gov.nf.ca/forest/fra_p&s.htm#agr

Appendix 1:
A Regional Review of Medicinal Plant Research in Canada[1]

Colleen Simmons
University of Guelph

Introduction

The healthcare system in Canada is undergoing a period of dramatic change. As budgets and services continue to be eroded, consumers are taking a more active role in managing their own healthcare needs. From 1992 to 1994, herbal product sales in the US increased an average of 18% per annum, while a conservative estimate of the value of the herbal medicinal market in Canada for 1992 is $84 million (Barl and Loewen 1996). Coincidental with this increase in plant product sales are increased sales in herbal/medicinal books and magazines. Organizations such as the American Botanical Council (Austin, Texas) and the Herb Research Foundation (Boulder, Colorado) jointly publish a quarterly, *Herbalgram*, that bridges the gap between scientific research and popular understanding of medicinal herb use (Foster 1993). The new "Herbs for health" aims "to present reliable, well-researched information in plain English" (Chamberlain 1996). In Canada, a home-study course for pharmacists entitled *The power of herbs - A pharmacist's practical introduction to common herbs* (Smith 1995), has been provided to the pharmacy industry so that they may effectively council consumers on the most common herbal products. Also, a bimonthly publication, *Health Naturally*, Canada's self-health care magazine, reinforces the belief that our health and well-being depend upon taking responsibility for our actions — whether it be counseling the federal government to amend Controlled Drugs and Substances Act Bill C-7 (Rowland 1996) or by providing a professional directory of alternative health care services.

This review was initiated with a limited distribution list, and letters which were mailed to Agriculture and Agri-Food Canada and provincial research stations as well as to Canadian universities where some aspect of research on herbal/medicinal plants was known to exist. Replies usually included an ever widening list of contacts, and in all over 30 letters were sent out asking for information regarding current research in Canada on herbal/medicinal plants, incorporating such aspects as: weed control, tissue culture, active ingredient analysis, propagation/planting techniques and ethnobotanical issues. Response was good, with over half the number of people replying. The scope of the review was then extended beyond "published research" and "clinical trials" so as to provide a more comprehensive and useful survey of available information. Areas exist where much needed information has been consolidated and made more accessible, that did not fall under the original outline. These include the Saskatchewan herb database (Barl et al. 1996), reviews entitled Herbal medicine in the *Canadian Pharmaceutical Journal*, (Appendix I) and a continuing series on native medicinal plants of Canada in the *Bulletin of the Canadian Botanical Association* (Small and Catling 1994, 1995). An effort has been made to provide as broad a coverage as possible within a limited time frame, of where and by whom medicinal plant research is being done in Canada, including literature reviews and conference details.

The format of the presentation is geographic by province, with universities noted first, followed by federal and provincial research establishments, and finally private and industrial initiatives.

British Columbia

Nancy Turner, University of Victoria, has identified and documented several hundred medicinal plants native to B.C., substantially increasing our knowledge of ethnobotany in Canada (Turner 1975, 1978; Turner et al. 1980, 1990). Turner's work was used as a selection guide for plant species and type of material to be collected in several ethnopharmacological studies conducted at the University of British Columbia (McCutcheon et al. 1992, 1994, 1995; Saxena et al. 1994). Extensive work on plant photosensitizers and their therapeutic potential (Hudson and Towers 1991) has

[1] Completed in 1996.

led to such investigations as antiviral activities of *Hypericum* species (St. John's wort) (Lopez-Bazzocchi et al. 1991) and light-mediated activities of thiarubrines against HIV-1 (Hudson et al. 1993).

Allison McCutcheon is also involved in the "Western Regional Scientific Liaison Group on Traditional Medicines." This group of people from government, industry and academia, brought together by Health Canada's Health Protection Branch, are working together to resolve such pertinent issues as herbal monographs, as well as quality control and manufacturing standards for the herbal industry. Dr. McCutcheon is also co-chair of the recently formed Canadian Herb Society, whose mandate is "a society of herbal enthusiasts founded to share knowledge and promote the use and enjoyment of herbs" (CHS Newsletter 1996).

Joe Mazza of the Food Research Program at the Agriculture and Agri-Food Canada Research Center at Summerland, B.C. is currently developing high value products from Canadian crops. Examples are using isothiocyanates as natural antimicrobial agents to enhance shelf-life of foods (Delaquis and Mazza 1995), studies on essential oils (Mazza et al. 1993; Mazza and Kiehn 1992), and physiologically important constituents of flaxseed (Mazza and Oomah 1995; Oomah et al. 1995a, b). Thomas Li, also stationed at Summerland, has worked on nettle (*Urtica)* a potential organic nitrogen source fertilizer for herb growers (Li 1994a), studied ginseng replant disease (Li 1994b), and has published a review on Asian and American ginseng (Li 1995).

Pat Bowen and David Ehret at the Pacific Agricultural Research Center, Agassiz, B.C. have just begun to research medicinal plants, examining field production methods as well as plasticulture and hydroponics, focussing on commercial production aspects, while emphasizing plant quality and active agent content.

Alberta

Larry Wang of the University of Alberta's Zoology department has co-authored several pharmacological studies with members of the Institute of Pharmacology in Taiwan (Tsai et al. 1995a, b, c). As well, he has combined his expertise with colleagues Christina Benishin and Peter Pang, Physiology, and Hsin J. Liu, Chemistry, to provide a traditional Chinese medicine research program to the University of Alberta. One aspect of his current research involves the influence of herbal compounds on learning and memory, aging, exercise and obesity. A 1991 study using a saponin of ginseng, ginsenoside Rb_1, partially prevented memory deficits in rats induced by scopolamine (Benishin et al. 1991). Peter Pang, Physiology, is also conducting active ingredient analysis and efficacy studies on herbals.

At the University of Calgary, Maurice Moloney is working on a blood anti-coagulant derived from canola genetically altered with material from leeches. By using plant biotechnology techniques to alter canola with human genetic materials, Moloney also hopes to mass produce cytokinds, a family of proteins that stimulate the human immune system, and are used as an alternative to chemotherapy in treating cancer patients (Canadian Living 1995).

Alberta Agriculture, Food and Rural Development has put together a "special crops product team" of which Refe Gaudiel, a research agronomist at Crop Diversification Center South - Brooks, is a member. Herbs are included in the special crops designation and there are 1000 ha of spearmint in cultivation in Alberta, which in 1994 accounted for a major portion of the 87.5 metric tonnes of spearmint oil exported to the US, with a market value of $2.5 million. Lesser amounts of monarda, dill and peppermint are grown for their essential oil components. One of the responsibilities of the special crops team is to identify and prioritize research needs and to encourage partnerships between government and private sectors.

Terry Willard is Director and founder of the Wild Rose College of Natural Healing with offices in Calgary and Vancouver, which offer diploma courses entitled "Wholistic Therapy" and "Community Health Care Counselors." His book Edible and medicinal plants of the Rocky Mountains and neighbouring territories draws heavily from the ethnobotany of the native peoples of B.C. and Alberta.

Saskatchewan

Branka Barl of the Saskatchewan Herb Research Center, University of Saskatchewan has prepared the 1995 Saskatchewan Herb Database (Barl et al. 1996), which is a comprehensive resource of 26 economically important herbs that either grow or have growth potential in Saskatchewan as alternative cash crops. This resource includes botanical, cultural, chemical, pharmacological, processing and marketing information and was compiled in an effort to assist establishment of the herb industry in Saskatchewan. A revised and expanded edition will be available in 1996. A 1995 Herb Market Survey

(Barl and Loewen 1996) of 22 herbs was conducted at four selected health food stores in Saskatoon, with the hope that the information could be of interest to companies marketing herb-based products as well as to the general public.

Alison Stephen of Pharmacy and Nutrition, University of Saskatchewan, is studying the effects of dietary fiber on human colonic function and lipid metabolism. She stresses the importance of dietary controlled efficacy studies on nutraceuticals or medical foods, so that small differences in physiological function can be detected and the role of the active principle under study can be clarified (Stephen 1995).

Rick Kulow, President of Bioriginal Food and Sciences Corp., Saskatoon is a recognized authority on gamma-linolenic acid (GLA). GLA is presently being used in the health and nutrition industry, the pharmaceutical industry, in functional foods, in pet foods and veterinary applications, and in mass market foods (Kulow 1995). He sees increased opportunities for growing and marketing medicinal plants in the prairie provinces due to a variety of factors; a 15 - year decrease in commodities, especially wheat, as well as such factors as the aging of the "Baby Boomer" generation, economic factors such as rising health care costs, and increased research in areas such as the health effects of antioxidants (Kulow 1996).

Fytokem Products Inc., Saskatoon, has prepared a 279 page publication, describing and discussing the phytochemicals that have been identified from 1000 native and naturalized plants of the Canadian prairies (Hetherington and Steck 1997).

Manitoba

Norm Kenkel and Candace Turcott, Botany Dept., University of Manitoba are studying the biology of seneca snakeroot (*Polygala senega* L.). Field work will include habitat descriptions, summaries of species phenology (e.g. dates of shoot growth, flowering etc.), studies of population structure and propagation techniques. This information will be used to make recommendations for environmentally-friendly cultivation techniques and sustainable harvesting of wild populations in central and northern Manitoba (Kenkel and Turcotte 1996).

Robin Marles has developed an interesting and rewarding ethnobotanical program at Brandon University (Marles 1996). He has hired and trained 14 First Nation students to participate in a study, teaming them up with other botany students. It has proven to be a "win-win" situation since the native elders pass on their knowledge to younger members of the tribe, the younger natives learn of their heritage, and the botany students receive the information first hand. The students can catalogue medicinally important plant specimens, but also benefit in the cultural experience. This program funded by National Resources Canada is also exploring alternative food products, value-added processing, and increasing the tourist market as ways to supplement the economy of native people.

Jack Moes, New Crop Agronomist, Manitoba Agriculture at Brandon, has had only limited involvement in medicinal plant research, focussing on germination problems of direct seeding *Echinacea angustifolia*. Also with Manitoba Agriculture, Clayton Jackson at the Manitoba Crop Diversification Centre in Carberry, is currently working on a directory of companies involved in the purchase, manufacture and marketing of herbs.

Agriculture and Agri-Food Canada also has a research facility at Morden, where Dr. Dave Oomah is working on functional properties of flaxseed (Oomah and Mazza 1995; Mazza and Oomah 1995; Oomah et al. 1995a, b) and has explored alternative crops such as cow cockle (*Saponaria vaccaria*) for potential agronomic use (Mazza et al. 1992).

Ontario

Frank DiCosmo, Centre for Plant Biotechnology, Dept. of Botany, University of Toronto, is exploring the potential of plant cell and tissue culture of *Taxus* spp. as a potential alternative source of taxol and related taxanes used in cancer chemotherapy (Fett-Neto and DiCosmo 1992; Fett-Neto et al. 1992, 1993, 1994). At present despite increasing demand the supply of the drug is limited, since the only commercial source is the bark of the Pacific yew (*T. brevifolia* Nutt.), a slow growing gymnosperm which yields relatively low amounts of taxol per unit of biomass. DiCosmo is optimistic in his review (DiCosmo and Misawa 1995) that although plant cell culture technology for the production of valuable plant products such as taxol is not yet commercially feasible, this goal will be attained when eventually there is a more complete understanding of the factors that regulate and limit plant metabolic processes.

At the University of Guelph, Tissa Senaratna and Praveen Saxena are working on a process to develop artificial herb seeds from plant species that are normally propagated by cuttings, such as French tarragon (U. of Guelph 1995). Tissue culture techniques stimulate the formation of somatic embryos, which then undergo a developmental desiccation process that enables the "seeds" to be packaged and shipped. Also, work is underway to develop an artificial seed coat - a nutrient-rich coating that enables enhanced growth, as well as providing protection during automated planting.

Also at Guelph, John Proctor, a widely recognized authority on ginseng, is working with Marilyn Hovius, on the effects of hormone application and stratification methods to shorten the dormancy period and stimulate seed germination in ginseng (Hovius et al. 1995).

Ernie Small and Paul Catling, Agriculture and Agri-Food Canada at Ottawa are co-authors of a continuing series entitled "Poorly Known Economic Plants of Canada," published in the Bulletin of the Canadian Botanical Association, the first 10 of which are cited here with additional information. Dr. Small is also author of a book entitled Culinary herbs (NRC Research Press) which includes extensive chemical and medicinal information on 125 species of edible herbs grown in Canada. He has recently completed a review "Crop diversification in Canada with particular reference to genetic resources" (Small 1995), which examines plant classes and species with potential as alternative crops, increasing diversity with the hope of improving overall productivity and marketability. Pharmaceutical crops with diversification potential include: ginseng, borage, evening primrose, feverfew, seneca snakeroot, May-apple, fenugreek and hemp.

The Agriculture and Agri-Food Canada Research Station at Delhi, along with Ontario Ministry of Agriculture and Food (OMAF) Simcoe, have carried out feasibility studies on alternative crops for southwestern Ontario, which at one time was a prime tobacco growing region (Loughton et al. 1991). Recent field tests have included stevia (*Stevia rebaudiana* (Bertoni) Bertoni) a perennial native to Paraguay, grown as a natural sweetener with potential to replace synthetic sweetening agents (Brandle and Rosa 1992); and evening primrose (*Oenothera biennis* L.), a native weed, whose seed is high in gamma-linolenic acid (GLA), an essential fatty acid that occurs in many plant species, but is commercially extracted from only a few (Brandle et al. 1993, Court et al. 1993).

Dennis V.C. Awang, previously with Health and Welfare Canada, has established his own consulting business called MediPlant Inc. For many years the official spokesperson in herbal science for the Canadian government, he was primarily responsible for development of standards (Awang et al. 1991) for the Canadian regulation of feverfew (*Tanacetum parthenium*), which led to the subsequent granting of a Drug Identification Number (DIN) to a British feverfew product. The product was recognized as a prophylactic to migraine based on an independent certification of botanical identity and a minimum content of 0.2% parthenolide (Awang 1993). In an effort to identify and control the quality of herbs and herbal products, he helped in the development of chemotaxonomic methods of analysis for some of the prominent herbs on the Canadian market (Awang et al. 1991, 1993, 1994). Currently, he is directing an analytical survey of commercial ginseng products on behalf of the American Botanical Council and, with J.T. Arnason, University of Ottawa, is serving as chief consultant to a Canadian Industrial Development Agency (CIDA) project in Togo, West Africa, developing a crude plant drug industry.

Quebec

Vince DeLuca, University of Montreal, has worked on alkaloid biosynthesis of Madagascar periwinkle (*Catharanthus roseus* L.) (DeLuca et al. 1986; Eilert et al. 1986, 1987). The leaf alkaloids that are most important in medicine are vinblastine sulphate used in the treatment of Hodgkin's disease and vincristine sulphate used to treat acute leukemia in children. Gene technology applications to accumulate vindoline and vinblastine in periwinkle cell cultures were reviewed by Constable (1990).

The Agriculture and Agri-Food Canada research station at St.-Jean-sur-Richelieu has conducted studies into essential oil content of yarrow (*Achillea millefolium*) a widely distributed perennial whose medicinal properties have been employed since the Trojan War, when the Greek hero Achilles, is said to have used the leaves of the plant to check the flow of blood from the wounds of his soldiers — hence the generic name *Achillea* (Belanger and Dextraze 1993). Also, in two cooperative studies with Laval University, the effects of herbicide use on quality and yield of thyme, horehound and dandelion (Michaud et al. 1993) and viability of angelica root as a viable crop for Quebec soil conditions (Charbonneau et al. 1993) were researched.

The Biotechnology Research Institute, a branch of National Research Council Canada located in Montreal, has completed a study on bloodroot (*Sanguinaria canadensis* L.) comparing phenanthridine alkaloid contents of rhizomes to tissue culture cell suspensions (Rho et al. 1992). Bloodroot is a low-growing perennial plant indigenous to eastern and central US and Canada and has traditionally been used in preparing herbal cough syrups, expectorants and tinctures. Today it is most often used in dentistry, for its antiplaque and antigingivitis activity (Karlowsky 1991).

Maritime Provinces

Valuable contributions were made by Frank Chandler, Dalhousie University, Halifax NS, who wrote several papers concerning use of plants by the Maritime Indians (Chandler et al. 1979, Chandler 1983, Hooper and Chandler 1984). He has increased our knowledge of medicinal plants by contributing several articles to the Canadian Pharmaceutical Journal's Herbal Medicine column (Appendix I) and is also a member of the Government of Canada Expert Advisory Committee on Herbs and Botanical Preparations.

Canadian Conferences

Several recent conferences brought together educators, growers, health practitioners, government advisors and the interested public for information sharing sessions, promoting herbal/medicinal plants in Canada. Proceedings from the following conferences, will provide the reader with a broad representation of Canadians who have a vested interest in promoting medicinal plants:

- First Canadian Workshop/Symposium on Functional Foods, Toronto ON September 28–29, 1995.
- First Canadian International Conference on Herbal Medicine, Mohawk College, Hamilton ON, October 13–15, 1995
- Prairie Medicinal & Aromatic Plants Conference, Olds AB, March 3–5, 1996
- A commercial herb growers conference sponsored by Conrad Richter of Richter's herbs took place in October 1996 at Goodwood, ON and addressed needs unique to commercial herb growers. Dr. John Proctor, University of Guelph, spoke on ginseng production, including improved production techniques and ways to enhance seed germination.

Conclusion

The herbal/medicinal plant industry in Canada is undergoing a period of accelerated growth, stimulated by a growing concern in our population for alternatives to conventional health care methods. As the industry grows, the need for research into these phytomedicinals increases as well. There is a requirement for standardized procedures, not only in plant identification, but also in therapeutic dosages. The buyers and consumers deserve to know what they are paying for, and expect consistency in the herbal preparations that they are using. What good is it if they buy a feverfew product to treat migraine, only to be disappointed when it doesn't work, because the preparation lacks parthenolides? This paper has shown that our Canadian institutions and researchers are working together toward the vision of health by using natural remedies.

Acknowledgment

My sincere thanks to all the people who replied to my inquiries and provided assistance for this review, as it would not have been possible without your interest and input.

Appendix I

Canadian Pharmaceutical Journal 1989–1995

Medicinal Plant Reviews

Bellflower	*Platycodon grandiflorus* (Jacq.) A. DC. (as "*Platycodon grandiflorum*"); Sept. 1991, pp. 422, 423, 426; N. Hamon, University of Saskatchewan
Bloodroot	*Sanguinaria canadensis* L.; May 1991, pp. 260, 262–263, 267; J.A. Karlowsky, University of Manitoba
Boneset	*Eupatorium* spp.; May 1990, pp. 229, 231, 233; R.A. Locock, University of Alberta
Borage	*Borago officinalis* L.; March 1990, pp. 121, 123, 125–126; D.V.C. Awang, Health and Welfare Canada
Fennel	*Foeniculum vulgare* Mill.; Dec. 93/Jan. 94, pp. 503–504; R.A. Locock, University of Alberta
Feverfew	*Tanacetum parthenium* (L.) Schultz-Bip.; May 1989, pp. 266–268, 270; D.V.C. Awang, Health and Welfare Canada
Ginger	*Zingiber officinale* Roscoe; July 1992, pp. 309–311; D.V.C. Awang, Health and Welfare Canada
Goldenseal	*Hydrastis canadensis* L.; November 1990, pp. 508–510; N.W. Hamon, University of Saskatchewan
Guarana	*Paullinia cupana* Kunth ex H.B.K.; May 1992, pp. 222–224; C.J. Briggs, University of Manitoba
Horse chestnut	*Aesculus hippocastanum* L.; Jul./Aug. 1993, pp. 297, 300, 306; R.F. Chandler, Dalhousie University
Horsetail	*Equisetum arvense* L.; Sept. 1992, pp. 399–401; N.W. Hamon, University of Saskatchewan; D.V.C. Awang, Health and Welfare Canada
Leopard's bane	*Arnica montana* L.; Mar. 1992, pp. 125, 126, 128; S. MacKinnon, University of Ottawa
Lobelia	*Lobelia* spp.; Jan. 1992, pp. 33–35; R.A. Locock, University of Alberta
Milk thistle	*Silybum marianum* L.; Oct. 1993, pp. 403–404, 422; D.V.C. Awang, MediPlant Natural Products
Nutmeg, mace	*Myristica fragrans* Houtt.; July 1991, pp. 349–350, 352; C.J. Briggs, University of Manitoba
Pennyroyal	*Mentha pulegium* L.; July 1989, pp. 369, 371–372; C.J. Briggs, University of Manitoba
Peppermint	*Mentha piperita* L.; Mar. 1993, pp. 89–92; C.J. Briggs, University of Manitoba
Podophyllum	*Podophyllum peltatum* L.; July 1990, pp. 330–331, 333; R.F. Chandler, Dalhousie University; N.A. Graham, Pharmacist, Whitby
Precatory bean	*Abrus precatorius* L.; Nov. 1992, pp. 502, 504, 514; R.A. Locock, University of Alberta
St. John's wort	*Hypericum perforatum* L.; Jan. 1991, pp. 33–35; D.V.C. Awang, Health and Welfare Canada
Yams	*Dioscorea* spp.; Sept. 1990, pp. 413–415; C.J. Briggs, University of Manitoba
Yarrow	*Achillea millefolium* L.; Jan. 1989, pp. 41–43; R.F. Chandler, Dalhousie University
Yew	*Taxus* spp.; May 1993, pp. 192, 196, 199–200; N.W. Hamon, University of Saskatchewan

References

Awang, D.V.C. 1993. The North American herbal entanglement: of food, drugs and nutritional supplements. Acta-Hortic. **332**: 19–23.

Awang, D.V.C., Dawson, B.A., Kindack, D.B., Crompton, C.W., and Heptinstall, S. 1991. Parthenolide content of feverfew (*Tanacetum parthenium*) assessed by HPLC and H-NMR spectroscopy. J. Plant Prod. **54**: 1516–1521.

Awang, D.V.C., Dawson, B.A., Fillion, J., Girad, M., and Kindack, D. 1993. Echimidine content of commercial comfrey (*Symphytum* spp.: Boraginaceae). J. Herbs, Spices & Med. Plants **2**(1): 21–34.

Awang, D.V.C., Dawson, B.A., Ethier, J-C., Gentry, A.H., Girard, M., and Kindack, D. 1994. Naphthoquinone constituents of commercial Lapacho/Pau d'arco/Taheebo products. J. Herbs, Spices & Med. Plants **2**(4): 27–43.

Barl, B., and Loewen, D. 1996. Herb Market Survey 1995. Dept. of Hort. Science, University of Saskatchewan, SK. 48 pp.

Barl, B., Loewen, D., and Svendsen, E. 1996. Saskatchewan Herb Database. Dept. of Hort. Science, University of Saskatchewan, SK. 164 pp.

Belanger, A., and Dextraze, L. 1993. Variability of chamazulene within *Achillea millefolium*. Acta-Hortic. **330**: 141–144.

Benishin, C.G., Lee, R., Wang, L.C.H., and Liu, H.J. 1991. Effects of ginsenoside Rb_1 on cholinergic metabolism. Pharmacology **42**: 223–229.

Brandle, J.E., and Rosa, N. 1992. Heritability for yield, leaf:stem ratio and stevioside content estimated from a landrace cultivar of *Stevia rebaudiana*. Can. J. Plant Sci. **72**: 1263–1266.

Brandle, J.E., Court, W.A., and Roy, R.C. 1993. Heritability of seed yield, oil concentration and oil quality among wild biotypes of Ontario evening primrose. Can. J. Plant Sci. **73**: 1067–1070.

Canadian Herb Society Newsletter. 1996. **1**(1).

Canadian Living. 1995. Canola, a success story. October 1995. pp 36–39.

Chamberlain, L. 1996. Notes from the publisher. Page 4. *In* Herbs for Health. Mar. 1996.

Chandler, R.F., Freeman, L., and Hooper, S.N. 1979. Herbal remedies of the Maritime Indians. J. Ethnopharmacol. **1:** 49–68.

Chandler, R.F. 1983. Vindication of Maritime Indian herbal remedies. J. Ethnopharmacol. **9**: 323–327.

Charbonneau, J., Michaud, M.H., Gosselin, A., Martel, C., and Tremblay, N. 1993. Effect of substrate and soil type on angelica root productivity. Acta-Hortic. **331**: 331–335.

Constabel, F. 1990. Medicinal plant biotechnology. Planta Med. **56**: 421–425.

Court, W.A., Hendel, J.G., and Pocs, R. 1993. Determination of the fatty acids and oil content of evening primrose *Oenothera biennis* L. Food Res. Int. **26**: 181–186.

Delaquis, P.J., and Mazza, G. 1995. Antimicrobial properties of isothiocyanates in food preservatives. Food Technol. Nov. 1995, pp. 73–74, 79, 81, 83–84.

DeLuca, V., Balsevich, J., Tyler, R.T., Eilert U., Panchuk, B.D., and Kurz, W.G.W. 1986. Biosynthesis of indole alkaloids: developmental regulation of the biosynthetic pathway from tabersonine to vindolene in *Catharanthus roseus*. J. Plant Physiol. **125**: 147–156.

DiCosmo, F., and Misawa, M. 1995. Plant cell and tissue culture: alternatives for metabolite production. Biotech. Adv. **13**: 425–453.

Eilert, U., DeLuca, V., Constabel, F., and Kurz, W.G.W. 1986. Elicitation of indole alkaloid biosynthesis in periwinkle. *In* Recognition in microbe-plant interactions. *Edited by* B. Lugtenberg. Springer-Verlag, Berlin. pp. 363–366.

Eilert, U., DeLuca, V., Kurz, W.G.W., and Constabel, F. 1987. Alkaloid formation by habituated and tumerous cell suspension culture of *Catharanthus roseus*. Plant Cell Rep. **6**(4): 271–274.

Fett-Neto, A.G., and DiCosmo, F. 1992. Distribution and amounts of taxol in different shoot parts of *Taxus cuspidata*. Planta Med. **58**: 464–466.

Fett-Neto, A.G., F. DiCosmo, W.F. Reynold and K. Sakata. 1992. Cell Culture of *Taxus* as a source of the anti-neoplastic drug taxol and related taxanes. Biotechnology. **10**: 1572–1575.

Fett-Netto, A.G., Melanson, S.J., Sakata, K., and DiCosmo, F. 1993. Improved growth and taxol yield in developing calli of *Taxus cuspidata* by medium composition modification. Biotechology. **11**: 731–734.

Fett-Neto, A.G., Zhang, W.Y., and DiCosmo, F. 1994. Kinetics of taxol production, growth, and nutrient uptake in cell suspensions of *Taxus cuspidata*. Biotech. Bioeng. **44**(2): 205–210.

Foster, S. 1993. Herbal Renaissance. Gibbs-Smith Pub., Salt Lake City, UT. p 8.

Hetherington, M., and Steck, W. 1997. Natural chemicals from northern prairie plants. Fytokem Publications, Saskatoon, SK.

Hooper, S.N., and Chandler, R.F. 1984. Herbal remedies of the Maritime Indians: phytosterols and triterpenes of 67 plants. J. Ethnopharmacol. **10**: 181–194.

Hovius, M.H.Y., Proctor, J.T.A., and Reeleder, R. 1995. Shortening the dormancy period and improving germination in American ginseng seed. Amer. Soc. Hort. Sci. **30**: 67 (abstract).

Hudson, J.B., and Towers, G.H.N. 1991. Therapeutic potential of plant photosensitizers. Pharmacol. & Ther. **49**: 181–222.

Hudson, J.B., Balza, F., Harris, L., and Towers, G.H.N. 1993. Light-mediated activities of thiarubrines against human immunodeficiency virus. Photochem. Photobiol. **57**: 675–680.

Karlowsky, J.A. 1991. Bloodroot: *Sanguinaria canadensis* L. Can. Pharm. J., May 1991: 260–263, 267.

Kenkel, N.C., and Turcotte, C. 1996. The ethnobotany and economics of seneca snakeroot, *Polygala senega* L. Unpublished manuscript, Dept. of Botany, University of Manitoba.

Kulow, F.C. 1995. Gamma-linolenic acid - status and prospects. *In* Pre-conference proceedings of the First Canadian Workshop/Symposium on Functional Foods, Toronto, ON Sept. 28–29, 1995. pp. 32–34.

Li, T.S.C. 1994a. Use of stinging nettle as a potential organic fertilizer for herbs. J. Herbs, Spices & Med. Plants **2**(2): 93–98.

Li, T.S.C. 1994b. Evaluation of chemical and nonchemical treatments for the control of ginseng replant disease. Acta-Hortic. **363**: 141–146.

Li, T.S.C. 1995. Asian and American ginseng - a review. Hort Technol. **5**(1): 27–34.

Lopez-Bazzocchi, E., Hudson, J.B., and Towers, G.H.N. 1991. Antiviral activity of the photoactive plant pigment hypericin. Photochem. Photobiol. **54**(1): 95–98.

Loughton, A., Columbus, M.J., and Roy, R.C. 1991. The search for industrial uses of crops in the diversification of agriculture in Ontario. Alt. Crops Noteb. **5:** 20–27.

Marles, R. 1996. Perspectives on medicinal uses of native plants. *In* Proceedings, Prairies Medicinal & Aromatic Plants Conference, March 3–5, 1996, Olds College, Olds, AB. pp. 45–48.

Mazza, G., and Kiehn, F.A. 1992. Essential oil of *Agastache foeniculum*, a potential source of methyl chavicol. J. Essent. Oil Res. **4**: 295–299.

Mazza, G., Biliaderis, C.G., Przybylski, R., and Oomah, B.D. 1992. Compositional and morphological characteristics of cow cockle (*Saponaria vaccaria*) seed; a potential alternative crop. J. Agr. Food Chem. **40**: 1520–1523.

Mazza, G., Kiehn, F.A., and Marshall, H.H. 1993. Monarda: a source of geraniol, linalool, thymol and carvacrol-rich essential oils. *In* New Crops. *Edited by* J. Janick and J. E. Simon. John Wiley & Sons, New York, NY. pp. 628–631.

McCutcheon, A.R., Ellis, S.M., Hancock, R.E.W., and Towers, G.H.N. 1992. Antibiotic screening of medicinal plants of the British Columbian native peoples. J. Ethnopharmacol **37**: 213–223.

McCutcheon, A.R., Ellis, S.M., Hancock, R.E.W., and Towers, G.H.N. 1994. Antifungal screening of medicinal plants of British Columbian native peoples.

J. Ethnopharmacol **44**: 157–169.

McCutcheon, A.R., Roberts, T.E., Gibbons, E., Ellis, S.M., Babiuk, L.A., Hancock, R.E.W., and Towers, G.H.N. 1995. Antiviral Screening of British Columbian medicinal plants. J. Ethnopharmacol **49**: 101–110.

Michaud, M.H., Gosselin, A., Tremblay, N., Benoit, D.L., Belanger, A., and Desroches, B. 1993. Effect of a herbicide and two plant densities on the yield of medicinal plants grown in Quebec (Canada). Acta Hortic. **331**: 311–318.

Oomah, B.D., Kenaschuk, E.O., Cui, W., and Mazza, G. 1995a. Variation in the composition of water-soluble polysaccharides in flaxseed. J. Agric. Food Chem. **43**: 1484–1488.

Oomah, B.D., Kenaschuk, E.O., and Mazza, G. 1995b. Phenolic acids in flaxseed. J. Agric. Food Chem. **43**: 2016–2019.

Oomah, B.D., and Mazza, G. 1995. Functional properties, uses of flaxseed protein. Inform **6**: 1246–1252.

Rho, D., Chauret, N., Laberge, N., and Archambault, J. 1992. Growth characteristics of *Sanguinaria canadensis* L. cell suspensions and immobilized cultures for production of benzophenanthridine alkaloids. Appl. Microbiol. Biotechnol. **36**: 611–617.

Rowland, D. 1995. A first step for health freedom. Health Naturally **20**: 3.

Saxena, G., McCutcheon, A.R., Farmer, S., Towers, G.H.N., and Hancock, R.E.W. 1994. Antimicrobial constituents of *Rhus glabra*. J. Ethnopharmacol **42**: 95–99.

Small, E. 1995. Crop diversification in Canada with particular reference to genetic resources. Can. J. Plant Sci. **75**: 33–43.

Small, E., and Catling, P.M. 1995 – 1998. Poorly known economic plants of Canada. Bull. Can. Bot. Assoc. (Quarterly articles on Canadian medicinal plants.)

Smith, M.J. 1995. The power of herbs. A pharmacist's practical introduction to common herbs. A home study program for the pharmacy industry.

Stephen, A.M. 1995. Biologically active components in food — an overview. *In* Pre-conference proceedings of the First Canadian Workshop/Symposium of Functional Foods, Toronto, Sept. 28–29, 1995. pp. 16–17.

Tsai, T-H., Lee, T-F., Chen, C-F., and Wang, L.C.H. 1995a. Modulatory effects of magnolol on potassium-stimulated 5-hydroxytriptamine release from rat cortical and hippocampal slices. Neurosci. Lett. **186**: 49–52.

Tsai, T-H., Lee, T-F., Chen, C-F., and Wang, L.C.H. 1995b. Thermoregulatory effects of alkaloids isolated from Wu-Chu-Yu in afebrile and febrile rats. Pharm. Biochem. Behav. **50**: 293–298.

Tsai, T-H., Westly, J., Lee, T-F., Chen, C.F., and Wang, L.C.H. 1995c. Effects of honokiol and magnolol on acetylcholine release from rat hippocampal slices. Planta Med. **61**: 477–478.

Turner, N.J. 1975. Food Plants of the British Columbian Indians: Coastal Peoples. British Columbia Provincial Museum Handbook No. 34, Royal British Columbian Provincial Museum, Victoria, BC. 253 pp.

Turner, N.J. 1978. Food plants of the British Columbian Indians: Interior Peoples. British Columbia Provincial Museum Handbook No. 36, Royal British Columbian Provincial Museum, Victoria, BC. 241 pp.

Turner, N.J., Bouchard, R., and Kennedy, D.D. 1980. Ethnobotany of the Okanagan-Colville Indians of British Columbia and Washington. British Columbia Provincial Museum No. 21, Occasional Papers Series, British Columbia Provincial Museum,Victoria, B.C. 156 pp.

Turner, N.J., Thompson, L.E., Thompson, M.T., and York, A.Z. 1990. Thompson ethnobotany: knowledge and uses of plants by the Thompson Indians. British Columbia Provincial Museum, Memoir No. 25, British Columbia Provincial Museum,Victoria, BC. 321 pp.

University of Guelph. 1995. News File. Native herbs from artificial seeds. 2 pp.

General references

There are thousands of sources of general information on medicinal plants. With particular regard to the medicinal plants of Canada, much information is available in works dealing with medicinal plants that occur or are used in other countries, since the species are often the same. The following selection of information sources relevant to Canadian medicinal plants is biased towards journal articles, booklets and books, mostly in English, treating plants used medicinally in North America. Although only one publisher and city of publication is indicated for books, many are also printed in several cities, and found under the names of other publishers. Many books and booklets are available in several editions, and generally the most recent is listed. We follow the library convention of listing number of pages for one-volume books, but not where a book is published in more than one volume. The best source of bibliographic information we have found for medicinal books is the internet catalogue search menu of the Library of Congress, currently available on the World Wide Web at http://lcweb.loc.gov/catalog/browse/. The most comprehensive guide to medicinal research papers is MEDLINE (address listed below in guide to world wide web resources).

Anon. 1997. Prairie medicinal and aromatic plants conference 1997 proceedings [9–11 March 1997, Brandon, Man.]. Western Economic Diversification Canada. 132 pp. [Available on internet: http://www.agric.gov.ab.ca/crops/special/medconf/ #top]

Adams, J. 1940. Medicinal Plants and their cultivation in Canada. Farmers' Bulletin No. 4. Canada Dept. of Agriculture, Ottawa, ON. 31 pp.

Adams, J. 1916. Les plantes médicinales et leur culture au Canada. Bull. N° 23, deuxième Série. Bull. Ministère Fédéral de l'Agriculture — Canada, Fermes Expérimentales du Dominion, Service de la Botanique, Ottawa, ON. 63 pp.

Akerele, O., Heywood, V., and Synge, H. 1992. Conservation of medicinal plants [Proceedings of an international consultation, 21–27 March 1988, held at Chiang Mai, Thailand]. Cambridge University Press, Cambridge, UK. 362 pp.

Altschul, S. Von Reis. 1973. Drugs and foods from little-known plants: notes in Harvard University herbaria. Harvard Univ. Press, Cambridge, MA. 366 pp.

Andrews, T., Corya, W.L., and Stickel, D.A., Jr. 1982. A bibliography on herbs, herbal medicine, "natural" foods, and unconventional medical treatment. Libraries Unlimited, Inc., Littleton, CO. 339 pp.

Angier, B. 1978. Field guide to medicinal wild plants. Stackpole Books, Harrisburg, PA. 320 pp.

Arnason, T., Hebda, R.J., and Johns, T. 1981. Use of plants for food and medicine by Native Peoples of eastern Canada. Can. J. Bot. **59**: 2189–2325.

Arnason, J., Mata, R., and Romeo, J.T. (*Editors*). 1995. Phytochemistry of Medicinal Plants. [Proceedings of the 34th annual meeting of the Phytochemical Society of North America held in Mexico City in 1994.] Plenum, New York, NY. 372 pp.

Baba, S., Akerele, O., and Kawaguchi, Y. (*Editors*). 1992. Natural resources and human health: plants of medicinal and nutritional value. World Health Organization symposium on Plants and Health for All: Scientific Advancement (1st, Kobe-shi, Japan). Elsevier, New York, NY. 227 pp.

Baker, J.T., Borris, R.P., Carte, B., Cordell, G.A., Soejarto, D.D., Cragg, G.M., Gupta, M.P., Iwu, M.M., Madulid, D.R., and Tyler, V.E. 1995. Natural product drug discovery and development: new perspectives on international collaboration. J. Nat. Prod. **58**: 1325–1357.

Balandrin, M.F., Kinghorn, A.D., and Farnsworth, N.R. 1993. Plant-derived natural products in drug discovery and development - an overview. *In* Human medicinal agents from plants. *Edited by* A.D. Kinghorn and M.F. Balandrin. American Chemical Society, San Francisco, CA. pp. 2–12.

Balick, M.J., Elisabetsky, E., and Laird, S.A. (*Editors*). 1996. Medicinal resources of the tropical forest - biodiversity and its importance to human health. Columbia University Press, New York, NY. 464 pp.

Barrett, S., and Jarvis, W.T. (*Editors*). 1993. The health robbers: a close look at quackery in America. Prometheus Books, Buffalo, NY. 526 pp.

Becerra, A.V. 1993. Medicinal flora of Mexico. J. Herbs, Spices, Med. Plants (Recent Adv. Bot. Hortic. Pharmacol.) **2**(1): 55–91.

Bellamy, D.J., and Pfister, A. 1992. World medicine: plants, patients, and people. Blackwell, Cambridge, MA. 456 pp.

Berzins, R., and Richter, C. (*Editors*). 1997. Richters first commercial herb growing conference — transcripts. Richters, Goodwood, ON. 93 pp.

Berzins, R., Snell, H., and Richter, C. (*Editors*). 1998. Richters second commercial herb growing conference — transcripts. Richters, Goodwood, ON. 189 pp.

Bethel, M. 1968. The healing power of herbs. Thorsons Publishers, London, UK. 160 pp.

Bézanger-Beauquesne, L. 1980. Plantes médicinales des régions tempérées. Maloine, Paris, France. 439 pp.

Bianchini, F., and Corbetta, F. 1977. Health plants of the world. Atlas of medicinal plants. Newsweek Books, New York, NY. 242 pp.

Blackburn, J.L. (*Chair*). 1993. Second report of the expert advisory committee on herbs and botanical preparations. Health Canada. Irregularly paginated (ca. 80 pp.).

Blackwell, W.H. 1990. Poisonous and medicinal plants. Prentice Hall, Englewood Cliffs, NJ. 329 pp.

Blumenthal, M., Hall, T., and Rister, R. (*Editors*); Klein, S., and Rister, R. (*Translators*). 1996. The German Commission E Monographs. American Botanical Council, Austin, TX. [Over 300 monographs available.]

Boyle, W. 1991. Official herbs: botanical substances in the United States pharmacopoeias: 1820–1990. Buckeye Naturopathic Press, East Palestine, OH. 97 pp.

Bradley, P.R. (*Editor*). 1992. British herbal compendium: a handbook of scientific information on widely used plant drugs. Vol. 1. British Herbal Medical Association, Bournemouth, Dorset, UK.

British Medical Association. 1993. Complementary medicine. New approaches to good practice. Oxford University Press, Oxford, UK. 173 pp.

Brown, R. 1868. On the vegetable products used by North-west American Indians - as food and medicine, in the arts, and in superstitious rites. Trans. Bot. Soc. Edinburgh **9**: 378–396.

Bunney, S. 1992. The illustrated encyclopedia of herbs - their medicinal and culinary uses. Chancellor Press, London, UK. 320 pp.

Bruneton, J. 1995. Pharmacognosy, phytochemistry, medicinal plants. Intercept Ltd., Andover, Hampshire, England. 909 pp.

Budavari S., O'Neil, M.J., Smith, A., Heckelman, P.E., and Kinneary, J.F. (*Editors*). The Merck index: an encyclopedia of chemicals, drugs, and biologicals. 12th ed. Merck & Co., Whitehouse Station, NJ. (Pagination irregular)

Carmen-Kasparek, M. 1993. The state of herbal medicines in Canada. Drug Info. J. 27: 155–157.

Carper, J. 1997. Miracle cures: dramatic new scientific discoveries revealing the healing powers of herbs, vitamins, and other natural remedies. HarperCollins, New York, NY. 308 pp.

Castleman, M. 1991. The healing herbs. Rodale Press, Emmaus, PA. 436 pp.

Ceres. 1985 [Line illustrations by J. Renny and A. Ross, color photography by P. Turner.]. The healing power of herbal teas. Thorsons, New York, NY. 128 pp.

Chandler, R.F. 1985. Traditional remedies still valued in modern pharmacy. Can. Pharm. J. **118**: 419.

Chevallier, A. 1996. The encyclopedia of medicinal plants. The Reader's Digest Association (Canada) Ltd., Westmount, QC. 336 pp.

Childs, N.M. (*Editor*). 1997(vol. 1) and continuing. J. Nutraceuticals, Functional & Medical Foods. Howarth Medical Press.

Cox, P.A., and Balick, M.J. 1994. The ethnobotanical approach to drug discovery. Sci. Am. **270**(6): 82–87.

Cragg, G.M., Snader, K.M., Boyd, M.R., Cardellina, J.H., II, Schepartz, S.A., and Suffness, M. 1991. The search for new pharmaceutical crops: drug discovery and development at the national cancer institute. *In* New Crops. *Edited by* J. Janick and J.E. Simon. John Wiley and Sons, Inc., New York, NY. pp. 161–167.

Cragg, G.M., Newman, D.J., and Snader, K.M. 1997. Natural products in drug discovery and development. J. Nat. Prod. **60**: 52–60.

Craker, L.E. 1989. Herbs, spices, and medicinal plants gain in scientific and commercial importance. Diversity **5**(2–3): 47.

Craker, L.E. 1994. The directory of specialists in herbs, spices, and medicinal plants. 2nd edition. (Publication of The Herb, Spice, and Medicinal Plant Digest). University of Massachusetts, Amherst, MA. 101 pp.

Craker, L.E., and Simon, J. (*Editors*). 1986–1989. Herbs, spices and medicinal plants: recent advances in botany, horticulture, and pharmacology. Oryx Press, Phoenix, AZ. 4 vol.

Craker, L.E., Chadwick, A.F., and Simon, J.E. 1986. An introduction to the scientific literature on herbs, spices, and medicinal plants. J. Herbs, spices, and medicinal plants (Recent Adv. Bot. Hortic. Pharmacol.) **1**: 1–9.

Crellin, J.K., and Phillpott, J. 1997. A reference guide to medicinal plants: herbal medicine past and present: Duke University Press, Durham, NC. 551 pp.

Crop Development Centre. 1991. Special cropportunities [Proceedings Agricultural Canada Workshop on Alternative Crops, Aug. 6–7, 1991]. Crop Development Centre, Univ. Saskatchewan, SK. 157 pp.

Culhane, C. 1995. Nutraceuticals / functional Foods - an exploratory survey on Canada's potential [for Agriculture and Agri-Food Canada]. Summary report. International Food Focus Limited, Toronto, ON. ca. 9 pp. [Both summary and complete report available on world wide web at http://foodnet.fic.ca/trends/enutra.html; also see http://foodnet.fic.ca/industry/foodfax1.html]

Davidson, C.G. 1995. Canadian wild plant germplasm of economic significance. Can. J. Plant Sci. **75**: 23–32.

Deans, S.G., and Svoboda, K.P. 1990. Biotechnology and bioactivity of culinary and medicinal plants. AgBiotech News and Information **2**: 211–216.

Densmore, F. 1928. Uses of plants by the Chippewa Indians. Forty-fourth annual report of Bureau American Ethnology, Smithsonian Inst. **27**: 279–397. [Reprinted 1974 as: How Indians used wild plants for food, medicine and Crafts. Dover Publishing Co. New York, NY.]

DeSmet, P.A.G.M., Keller, K., Hansel, R., and Chandler. R.F. (*Editors*). 1992. Adverse effects of herbal drugs. Springer-Verlag, New York, NY. 2 vol. (Vol. 3 - 1996.)

Drew, A.K., and Myers, S.P. 1997. Safety issues in herbal medicine: implications for the health professions. Med. J. Aust. **166**: 538–541.

Duke, J.A. 1982. Herbs as a small farms enterprise and the value of aromatic plants as economic intercrops. U.S. Dep. Agric. Misc. Publ. **1422**: 76–83.

Duke, J.A. 1983. Medicinal plants of the Bible. Trado-Medic Books, Buffalo, NY. 233 pp.
Duke, J.A. 1985. CRC handbook of medicinal herbs. CRC Press, Boca Raton, FL. 677 pp.
Duke, J.A. 1986. Handbook of Northeastern Indian medicinal plants. Quarterman Publications, Lincoln, MA. 212 pp.
Duke, J.A. 1990. Promising phytomedicinals. *In* Advances in new crops — proceedings of the first national symposium, new crops: research, development, economics. *Edited by* J. Janick and J.E. Simon. Timber Press, Portland, OR. pp. 491–497.
Duke, J.A. 1992. Handbook of phytochemical constituents of GRAS herbs and other economic plants. CRC Press, Boca Raton, FL. 654 pp.
Duke, J.A. 1993. Medicinal plants and the pharmaceutical industry. *In* New Crops. *Edited by* J. Janick and J.E. Simon. John Wiley and Sons, Inc., New York, NY. pp. 664–669.
Duke, J.A. 1997. The green pharmacy. Rodale Press, Emmaus, PA. 507 pp.
Duke, J.A., and Ayensu, E.S. 1985. Medicinal plants of China. Reference Publications, Inc., Algonac, MI. 2 vol.
Duke, J.A., and Hurst, S.J. 1975. Ecological amplitudes of herbs, spices, and medicinal plants. Lloydia **38**: 404–410.
Duke, J.A., and Martinez, R.V. 1994. Amazonian ethnobotanical dictionary. CRC Press, Boca Raton, FL. 215 pp.
Dymock, W., and Hooper, D. 1988. Pharmacographia indica: a history of principal drugs of vegetable origin. State Mutual Book and Periodical Service, New York, NY. 3 vol.
Elisabetsky, E. 1991. Sociopolitical, economical and ethical issues in medicinal plant research. J. Ethnopharmacol **32**: 235–239.
Elliott, D. 1995. Wild roots: a forager's guide to the edible and medicinal roots, tubers, corms, and rhizomes of North America. Healing Arts Press, Rochester, VT. 128 pp.
Erichsen-Brown, C. 1979. Medicinal and other uses of North American plants: a historical survey with special reference to the eastern Indian tribes. Dover Publications Inc., New York, NY. 512 pp.
Farnsworth, N.R. 1993. Relative safety of herbal medicines. HerbalGram **29**: 36A–36H.
Farnsworth, N.R. 1993. Ethnopharmacology and future drug development: the North American experience. J. Ethnopharmacol. **38**:145–152.
Farnsworth, N.R., and Morris, R.W. 1976. Higher plants — the sleeping giant drug development. Am. J. Pharm. **148**: 46–52.
Farnsworth, N.R., and Soejarto, D.D. 1985. Potential consequence of plant extinction in the United States on the current and future availability of prescription drugs. Econ. Bot. **39**: 231–40.
Farnsworth, N.R., Akerlele, O., Bingel, A.S., Guo, Z.G., and Soejarto, D.D. 1985. Medicinal plants in therapy. World Health Organization Bull. **63**: 965–981.
Feinsilver, J.M., and Chapela, I.H. 1996. Will biodiversity prospecting for pharmaceuticals strike "green gold?" Diversity **12**(2): 20–21.
Ference, D. [D. Ference and Associates Ltd.] 1989. Economic opportunities for Canada in essential oils and medicinal crops. Agriculture Canada Policy Branch Working Paper 10/89. 287 pp.
Flannery, M.A. 1998. The medicine and medicinal plants of C.S. Rafinesque. Econ. Bot. **52**: 27–43.
Ford, R.I. 1981. Ethnobotany in North America: an historical phytogeographic perspective. Can. J. Bot. **59**: 2178–2187.
Foster, S. 1989. Phytogeographic and botanical considerations of medicinal plants in Eastern Asia and Eastern North America. J. Herbs, Spices, and Medicinal plants (Recent Adv. Bot. Hortic. Pharmacol.) **4**: 115–144.
Foster, S. 1992. Herbs of commerce. American Herbal Products Association, Austin, TX. 78 pp.
Foster, S. 1995. Medicinal plants in American forests. Forest History Society. Durham, NC. 58 pp.
Foster, S. 1995. Europe — medicinal plant use in the modern world. The Herbarist **61**: 33–39.
Foster, S. 1998. An illustrated guide - 101 medicinal herbs. Interweave Press, Loveland, CO. 240 pp.
Foster, S., and Chongxi, Y. 1992. Herbal emissaries - bringing Chinese herbs to the West. Healing Arts Press, Rochester, VT. 356 pp.
Foster, S., and Duke, J.A. 1990. A field guide to medicinal plants: Eastern and Central North America. Houghton Mifflin Co., Boston, MA. 366 pp.
Fournier, P. 1947–48. Le livre des plantes médicinales et vénéneuses de France. Paul Lechavalier, Paris, France. 3 vol.
French, D.H. 1981. Neglected aspects of North American ethnobotany. Can. J. Bot. **59**: 2326–2330.
Fuller, D.O. 1991. Medicine from the wild: an overview of the U.S. native medicinal plant trade and its conservation implications. World Wildlife Fund, Washington, D.C. 28 pp.
Garcia Rivas, H. 1982 [1983]. Enciclopedia de plantas medicinales mexicanas. 3d ed. Editorial Posada, Mexico. 655 pp.
Garland, S. 1984. The herb garden: A complete guide to growing scented, culinary and medicinal herbs. Viking Press, New York, NY. 168 pp.
Gibbons, E. 1989. Stalking the healthful herbs. A.C. Hood, Putney, VT. 301 pp. [Reprint. Originally published 1966 by D McKay, New York, NY.]
Grieve, M. 1931. A modern herbal. [Reprinted 1978.] Penguin Books, New York, NY. 912 pp.
Griggs, B. 1991. Green Pharmacy. The history and evolution of Western herbal medicine. Healing Arts Press, Rochester, VT. 379 pp.
Gruenwald, J., Brendler, T., and Jaenicke, C. (*Editors*). 1998. Physician's desk reference for herbal medicines ["PDR for herbal medicines"]. Medical Economics Co., Montvale, NJ. 1244 pp.
Gullo, V.P. 1994. The discovery of natural products with

therapeutic potential. Biotechnology Series: Vol. 26. Butterworth-Heinemann, Boston, MA.

Harding, A.R. 1972. Ginseng and other medicinal plants. Revised ed. A.R. Harding Publishing Co., Columbus, OH. 386 pp.

Henkel, A. 1904. Weeds used in medicine. Farmers' Bulletin No. 188. U.S. Dep. Agriculture, Washington, DC. 47 pp.

Henkel, A. 1906. Wild medicinal plants of the United States. Bulletin No. 89. U.S. Dep. Agriculture, Washington, DC. 76 pp.

Heinerman, J. 1992. Healing power of herbs. Globe Communications Corp., Boca Raton, FL. 95 pp.

Hoffmann, D. 1994. The Information Sourcebook of Herbal Medicine. [A comprehensive guide to information on Western herbal medicine, providing resources on all topics including on-line and database sources]. The Crossing Press, Freedom, CA. 308 pp.

Holm, W.R., and MacGregor, D. 1998. Processing guide for specialty crops. Science council of British Columbia – Okanagan. Irregularly paginated (ca. 150 pp.)

Holmes, P. 1989. The energetics of Western herbs: integrating Western and Oriental herbal medicine traditions. Artemis Press, Boulder, CO. 2 vol.

Hornock, L. (ed.). 1993. Cultivation and processing of medicinal plants. Wiley, New York, NY. 337 p.

Hostettmann, K, Marston, A., Maillard, M., and Hamburger, M. (*Editors*). 1995. Phytochemistry of plants used in traditional medicine. Proceedings of the 1993 International Symposium of the Phytochemical Society of Europe. Clarendon Press, New York, NY. 408 pp.

Houghton, P.J. 1995. The role of plants in traditional medicine and current therapy. J. Altern. Complement. Med. **1**: 131–143.

Insight Press. 1996. Nutraceuticals (papers from the May 1996 conference). Insight Press, Toronto, ON. 150 pp.

Insight Press. 1996. Nutraceuticals — a burgeoning market opportunity (papers from the October 1996 conference). Insight Press, Toronto, ON. 118 pp.

Instituto Nacional Indigenista. 1994. Flora medicinal indígena de Mexico. Instituto Nacional Indigenista, Ticopac, Mexico. 3 vol.

Israelsen, L.D. 1991. Phytomedicines as a new crop opportunity. *In* New Crops. *Edited by* J. Janick and J.E. Simon. John Wiley and Sons, Inc., New York, NY. pp. 669–671.

Jackson, S., and Prine, L. 1978. Wild plants of central North America for food and medicine. Penguis, Winnipeg, MN. 77 pp.

Joyce, C. 1994. Earthly goods: medicine hunting in the rainforest. Little Brown/Time Warner Books, New York, NY. 228 pp.

Kasparck, M., Gröger, A., and Schippmann, U. 1996. Directory for medicinal plant conservation. Networks, organizations, projects, information sources. IUCN/SSC Medicinal Plant Specialist Group, German Federal Agency for Nature Conservation, Bonn, Germany. 154 pp.

Kapadia, G.J., Rao, G.S., and Morton, J.F. 1983. Herbal tea consumption and esophageal cancer. Carcinogens and Mutagens in the Environment (Boca Raton, FL) **3**: 3–12.

Kavasch, B. 1979. Native harvests - recipes and botanicals of the American Indian. Random House, New York, NY. 202 pp.

Kaye, C., and Billington, N. 1997. Medicinal plants of the heartland. Cache River Press, Vienna, IL. 344 pp.

Kenner, D., and Requena, Y. 1996. Botanical medicine: a European professional perspective. Paradigm Publications, Brookline, MA. 393 pp.

Kindscher, K. 1992. Medicinal wild plants of the prairie — an ethnobotanical guide. University Press of Kansas, Lawrence, KS. 256 pp.

Kinghorn, A.D., and Balandrin, M.F. (*Editors*). 1993. Human medicinal agents from plants. American Chemical Society, Washington, DC. 356 pp.

Kingsbury, J.M. 1964. Poisonous plants of the United States and Canada. Prentice-Hall, Engelwood Cliffs, NJ. 626 pp.

Kloss, J. 1992. Back to Eden: a human interest story of health and restoration to be found in herb, root, and bark. 2nd ed. Back to Eden Books Loma Linda, CA. 886 pp.

Köhler, A.D. 1883–1898. Köhler's Medizinal Pflanzen. Verlag von Fr. Eugen Köhler, Germany. 3 vols.

Kozyrskyj, A. 1997. Herbal products in Canada. How safe are they? Can. Fam. Physician **43**: 697–702.

Krochmal, A., and Krochmal, C. 1984. A field guide to medicinal plants. Times Books, New York, NY. 274 pp.

Krochmal, A., Walters, R.S., and Doughty, R.M. 1971. A guide to medicinal plants of Appalachia. Agricultural Handbook No. 400. Forest Service, U.S. Dep. Agriculture. 291 pp.

Lampe, K.F., and McCann, M.A. 1985. A.M.A. Handbook of poisonous and injurious plants. American Medical Association, Chicago, IL. 432 pp.

Larkin, T. 1983. Herbs are often more toxic than magical. FDA Consumer **17**(8): 5–10.

Lanthier, A. 1977. Les plantes médicinales canadiennes. Editions Paulines, Montréal, QC. 92 pp.

Launert, E. 1981. The Hamlyn guide to edible and medicinal plants of Britain and Northern Europe. Hamlyn, London, UK. 288 pp.

Leung, A.Y., and Foster, S. 1996. Encyclopedia of common natural ingredients used in foods, drugs and cosmetics, 2nd edition. J. Wiley & Sons, New York, NY. 649 pp.

Lewington, A. 1993. A review of the importation of medicinal plants and plant extracts into Europe. TRAFFIC International, Cambridge, UK. 37 pp.

Lewis, W.H., and Elvin-Lewis, M.P.F. 1977. Medical Botany: plants affecting man's health. John Wiley & Sons, New York, NY. 515 pp.

Lust, J.B. 1974. The herb book. Bantam Books, New York, NY. 660 pp.

Mann, C., and Staba, E.J. 1986. Herbs, Spices, and

Medicinal Plants (Recent Adv. Bot. Hortic. Pharmacol.) **1**: 235–280.

Marles, R.J. 1997. Registering a herbal remedy as a “traditional medicine” under Health Canada regulations. *In* Prairie medicinal and aromatic plants conference 1997 proceedings [9–11 March 1997, Brandon, Manitoba]. *Edited by* Anonymous. Western Economic Diversification Canada. pp. 79–84.

Martin, C. 1990. Earthmagic: using New England medicinal herbs. Dirigo Books, Inc., North Woodstock, ME. 250 pp.

Martin, C. 1991. Earthmagic: finding and using medicinal herbs. Countryman Press, Woodstock, VT. 228 pp.

Mathieu, G. 1977. La santé par les plantes. Presses Médicales Européennes, Europe. 441 pp.

McCarthy, S. 1992. Ethnobotany and medicinal plants: July 1991-July 1992. Quick Bibliography Series, QB 93–02. United States Department of Agriculture, National Agricultural Library, Beltsville, MA. 134 pp.

McGuffin, M., Hobbs, C., Upton, R., and Goldberg, A. (*Editors*). 1997. The botanical safety handbook. CRC Press, Boca Raton, FL. 231 pp. [Published for The American Herbal Products Association.]

Meares, P. 1987. The economic significance of herbs. HerbalGram **13**: 1, 6–8.

Mendelsohn, R., and Balick, M.J. 1995. The value of undiscovered pharmaceuticals in tropical forests. Econ. Bot. **49**: 223–228.

Micozzi, M.S. 1996. Fundamentals of complementary and alternative medicine. Churchill Livingstone, New York, NY. 303 pp.

Mills, S.Y. 1993. The essential book of herbal medicine. Arkana, New York, NY. 677 pp.

Millspaugh, C.F. 1974. American medicinal plants. Dover Publications, New York, NY. 801 pp. [Reprint, first published 1884–1887 as two volumes.]

Ministry of Agriculture and Fisheries. 1941. Medicinal herbs and their cultivation. His Majesty’s Stationary Office. Bulletin No. 121. London, UK. 22 pp.

Moerman, D.E. 1982. Geraniums for the Iroquois. Reference Publications, Inc., Algonac, MI. 242 pp.

Moerman, D.E. 1986. Medicinal plants of native America. Univ. Mich. Mus. Anthropol. Tech. Rep. 19. 2 vols.

Moerman, D.E. 1991. The medicinal flora of native North America: an analysis. J. Ethnopharmacol. **31**: 1–42.

Moore, M. 1979. Medicinal plants of the mountain West. A guide to the identification, preparation and uses of traditional medicinal plants found in the mountains, foothills and upland areas of the American West. Museum of New Mexico Press, Santa Fe, NM. 200 pp.

Moore, M. 1990. Medicinal plants of the desert and canyon west. A guide to identifying, preparing, and using traditional medicinal plants found in the deserts and canyons of the American West and Southwest. Museum of New Mexico Press, Santa Fe, NM. 184 pp.

Moore, M. 1993. Medicinal plants of the Pacific West. Red Crane Books, Santa Fe, NM. 359 pp.

Morton, J. 1976. Herbs and spices. Golden Press, NY. 160 pp.

Morton, J.F. 1977. Major medicinal plants, botany, culture and uses. Charles C. Thomas, Springfield, IL. 431 pp.

Morton, J.F. 1981. Atlas of medicinal plants of Middle America. Charles C. Thomas, Springfield, IL. 1420 pp.

Moulds, R.F.W., and McNeil, J.J. 1988. Herbal preparations - to regulate or not to regulate. Med. J. Aust. **149**: 572–574.

Mowrey, D.B. 1986. The scientific validation of herbal medicine. Cormorant Books, Keats Pub., New Canaan, CT. 316 pp.

Murray, M.T. 1995. The healing power of herbs: the enlightened person’s guide to the wonders of medicinal plants. 2nd ed. Prima Pub., Rocklin, CA. 410 pp.

Murray, M.T., and Pizzorno, J. 1998. Encyclopedia of natural medicine. 2nd ed. Prima Pub., Rocklin, CA. 946 pp.

Naegele, T.A. 1996. Edible and medicinal plants of the Great Lakes Region. Revised edition. Wilderness Adventure Books, Davisburg, MI. 200 pp.

Newall, C.A., Anderson, L.A., and Phillipson, J.D. 1996. Herbal medicines — a guide for health-care professionals. The Pharmaceutical Press, London, UK. 296 pp.

Nigg, H., and Seigler, D. (*Editors*). 1992. Phytochemical resources for medicine and agriculture. [Proceedings based on American Chemical Society Symposium, Plant Chemicals Useful to Humans, Sept. 10–15, 1989, Miami, FL]. Plenum Press, New York, NY. 445 pp.

Ody, P. 1993. The complete medicinal herbal. Dorling Kindersley, New York, NY. 192 pp.

Office of Alternative Medicine. 1995. Alternative medicine: expanding medical horizons. A report to the National Institutes of Health on alternative medical systems and practices in the United States. Office of Alternative Medicine, National Institutes of Health Publ. No. 94–066. US Government Printing Office, Washington, DC. 372 pp.

Osol, A., Farrar, G.E., Beyer, K.H., Detweiler, D.K., Brown, J.H., Pratt, R., and Youngken, W.H. 1955. The dispensatory of the United States of America, 25th Edition. J.B. Lippincott Company. Philadelphia, PA. 2130 pp.

Ott, J. 1996. Pharmacotheon: entheogenic drugs, their plant sources and history. 2nd ed. Natural Products Co., Kennewick, WA. 639 pp.

Pammel, L.H. 1911. A manual of poisonous plants: chiefly of eastern North America, with brief notes on economic and medicinal plants and numerous illustrations. Torch Press, Cedar Rapids, IA. 977 pp.

Perkin, J. 1993. Herbal extracts: health aid or marketing aid? Food Processing (Tonbridge) **62**(10): 17, 20, 23, 24.

Perry, L.M., and Metzger, J. 1980. Medicinal plants of East and Southeast Asia. MIT Press, Cambridge, MA. 620 pp.

Pettit, G.R., Pierson, F.H., and Herald, C.L. 1994. Anticancer drugs from animals, plants, and microorganisms. Wiley, New York, NY. 670 pp.

Pizzorno, J., and Murray, M. (*Editors*). 1991. A textbook of natural medicine. Prima Pub., Rocklin, CA.

Prescott-Allen, C., and Prescott-Allen, R. 1986. The first resource — wild species in the North American economy. Yale University Press. New Haven, CT. 507 pp.

Rajak, R.C., and Rai, M.K. 1996. Herbal medicines, bio-diversity, and conservation strategies: proceedings of the national seminar, sponsored by All India Association for Christian Higher Education, New Delhi, India. International Book Distributors, Dehra Dun, U.P., India. 292 pp.

Randal, J. 1995. Experts warn of health risks from loss of biodiversity. J. Natl. Cancer Inst. **87**: 714–717.

Richters 1998. Richters herb catalogue. Otto Richter and Sons Ltd., Goodwood, ON. 103 pp.

Reid, D.P. 1986 [Reissued 1992]. Chinese herbal medicine. Shambhala Publications, Inc., Boston, MA. 174 pp.

Reis, S. Von, and Lipp, F.J. 1982. New plant sources for drugs and foods from the New York Botanical Garden herbarium. Harvard University Press, Cambridge, MA. 363 pp.

Reynolds, J.E.F., Parfitt, K., Parsons, A.V., and Sweetman, S.C. (*Editors*). 1989. Martindale: the extra pharmacopoiea, 29th edition. Council of the Royal Pharmaceutical Society of Great Britain, London, UK. 1896 pp. [30th ed. published in 1993 by The Pharmaceutical Press, London.]

Riddle, J.M. 1997. Eve's herbs: a history of contraception and abortion in the West. Harvard University Press, Cambridge, MA. 341 pp.

Robbers, J.E., Speedie, M.K., and Tyler, V.E. 1996. Pharmacognosy and pharmacobiotechnology. Williams & Wilkins, Baltimore, MD. 337 pp. [Revised edition of 9th (1988) edition of Pharmacognosy by V.E. Tyler, L.R. Brady, and J.E. Robbers.]

Robinson, T. 1988. An introduction to the chemistry of herbs, spices, and medicinal plants. The Herb, Spice, and Medicinal Plant Digest **6**(3): 2–4, 10.

Roecklein, J.C., and Leung, P. 1987. A profile of economic plants. Transaction, Inc., New Brunswick, NJ. 623 pp.

Saxe, T.G. 1987. Toxicity of medicinal herbal preparations. Am. Fam. Physician **35**(5): 135–142.

Schar, D. 1992. Thirty plants that can save your life! Elliott and Clark, Washington, DC. 134 pp.

Schulz, V., Hansel, R., and Tyler, V.E. 1998. Rational phytotherapy: a physician's guide to herbal medicine. 3rd ed. Springer, New York, NY. 306 pp.

Shaw, D., House, I., Kolev, S., and Murray, V. 1995. Should herbal medicines be licensed? Br. Med. J. **311**: 451–452.

Sheldon, J.W., Balick, M.J., and Laird, S. 1997. Medicinal plants: can utilization and conservation coexist? The New York Botanical Garden, Bronx, NY. 104 pp.

Sievers, A.F. 1930. American medicinal plants of commercial importance. Misc. Publ. No. 77. U.S. Dep. Agriculture, Washington, DC. 74 pp.

Sievers, A.F. 1948. Production of drug and condiment plants. Farmers' Bulletin No. 1999. U.S. Dep. Agriculture, Washington, DC. 99 pp.

Small, E. 1995. Crop diversification in Canada with particular reference to genetic resources. Can. J. Plant Sci. 75: 33–43.

Small, E. 1997. Biodiversity priorities from the perspective of Canadian agriculture: ten commandments. Can. Field-Nat. 111: 487–505.

Small, E. 1997. Culinary herbs. NRC Research Press, Ottawa, ON. 710 pp. [Advertisement: http://www.nrc.ca/cisti/journals/40393/tc40393.html]

Small, E. 1999. Why is crop diversification inportant for Canada? *In* Special Crops Conference "Opportunities and Profits into the 21st Century" (Edmonton, Nov. 1–3, 1998). *Edited by* S. Blade. Alberta Agriculture, Food and Rural Development, Edmonton, AB. pp. 9-18

Small, E. 1999. New crops for Canada. Proceedings of the Fourth National New Crops Symposium (Phoenix, AZ, Nov. 8–11, 1998). (In press).

Snider, S. 1991. Herbal teas and toxicity. FDA Consumer **25**(4): 30–33.

Soejarto, D.D. 1996. Biodiversity prospecting and benefit-sharing: perspectives from the field. J. Ethnopharmacol. **51**: 1–15.

Solecki, R.S., and Shanidar, I.V. 1975. A Neanderthal flower burial in northern Iraq. Science **190**: 880–881.

Spak, S. 1998. Nutraceuticals. Bi-weekly Bull. [Agriculture and Agri-Food Canada] **11**(1). [available online at http://www.agr.ca/policy/winn/biweekly/English/biweekly/volume11/v11n01e.htm]

Speck, F.G., and Dexter, R.W. 1951. Utilization of animals and plants by the Micmac Indians of New Brunswick. J. Wash. Acad. Sci. **41**(8): 250–259.

Spoerke, D. G., Jr. 1990. Herbal medications. Woodbridge Press, Santa Barbara, CA. 192 pp.

Stockberger, W.W. 1915. Drug plants under cultivation. Farmers' Bulletin No. 663. U.S. Dep. Agriculture, Washington, DC. 39 pp.

Stuart, M. (*Editor*). 1979. The encyclopedia of herbs and herbalism. Grosset and Dunlap, New York, NY. 304 pp.

Stuart, M. (*Editor*). 1981. Encyclopédie des herbes. Éditions Atlas, Paris, France. 303 pp.

Swanson, T. (*Editor*). 1995. Intellectual property rights and biodiversity conservation: an interdisciplinary analysis of the values of medicinal plants. Cambridge Univ. Press, Cambridge, UK. 285 pp.

Taylor, R.L. 1981. Plants and the indigenous peoples of North America. Can. J. Bot. **59**: 2175–2177.

Thomson, W.A.R. (*Editor*). 1978. Medicines from the earth. A guide to healing plants. McGraw-Hill, New York, NY. 208 pp.

Tilford, G.L. 1993. The ecoherbalist's fieldbook: wildcrafting in the mountain west. Mountain Weed Pub., Conner, MT. 295 pp.

Tilford, G.L. 1997. Edible and medicinal plants of the West. Mountain Press, Missoula, MO. 256 pp.

Torkelson, A.R. 1996. The cross name index to medicinal plants. CRC Press, Boca Raton, FL. 3 vol.

Torres, E. 1983. Green medicine: traditional Mexican-American herbal remedies. Nieves Press, Kingsville, TX. 62 pp.

Tucker, A.O., Duke, J.A., and Foster, S. 1989. Botanical nomenclature of medicinal plants. J. Herbs, Spices, and Medicinal plants (Recent Adv. Bot. Hortic. Pharmacol.) **4**: 169–242.

Turner, N.J. 1981. A gift for the taking: the untapped potential of some food plants of North American Native Peoples. Can. J. Bot. **59**: 2331–2357.

Tyler, V.E. 1979. Plight of plant-drug research in the United States today. Econ. Bot. **33**: 377–83.

Tyler, V.E. 1984. Hazardous herbs — a brief review. The Herb, Spice and Medicinal Plant Digest **2**(1): 1, 2, 4, 6.

Tyler, V.E. 1985. Hoosier home remedies. Purdue University Press, West Lafayette, IN, U.S.A. 212 pp.

Tyler, V.E. 1986. Plant drugs in the 21st century. Econ. Bot. **40**: 279–88.

Tyler, V.E. 1993. The honest herbal. 3rd ed. Pharmaceutical Products Press (Haworth Press) Binghamton, NY. 375 pp.

Tyler, V.E. 1993. Phytomedicines in western Europe — potential impact on herbal medicine in the United States. *In* Human medicinal agents from plants. *Edited by* A.D. Kinghorn and M.F. Balandrin. American Chemical Society, San Francisco, Washington, DC. pp. 25–37.

Tyler, V.E. 1994. Herbs of choice — the therapeutic use of phytomedicinals. Pharmaceutical Press, New York, NY. 209 pp.

Tyler, V.E. 1996. What pharmacists should know about herbal remedies. J. Am. Pharm. Assoc. (Wash.) NS **36**: 29–37.

Tyler, V.E., and Foster, S. 1996. Herbs and phytomedicinal products. *In* Handbook of nonprescription drugs, ed. 11 (2 vol.). *Edited by* T.R. Covington. American Pharmaceutical Association, Washington, DC. pp. 695–713.

United States Pharmacopeial Convention Inc. 1995. The United States pharmacopeia - the national formulary. 23rd ed. (including 9 supplements). Pharmacopeial Convention Inc., Rockville, MD. 2391 pp.

Uphof, J.C. 1968. Dictionary of economic plants. 2nd. ed. Verlag Von J. Cramer, Lehre, Germany. 591 pp.

Vanier, P., and Lefrançois, P. 1994. Les produits à base de plantes: des règles simples pour choisir des plantes médicinales de qualité. Série Produits naturels, n° 6. Le Guide Ressources, Montréal, QC. 26 pp.

Verlet, N. 1990. New markets for herbs in France and Europe. The Herb, Spice, and Med. Plant Digest **8**(2): 1–5.

Viereck, E. 1987. Alaska's wilderness medicines: healthful plants of the Far North. Alaska Northwest Books, Edmonds, WA. 107 pp.

Watt, J.M., and Breyer-Brandwijk, M.G. 1962. The medicinal and poisonous plants of southern and eastern Africa. 2nd edition. E. & S. Livingstone, London, U.K. 1457 pp.

Weiner, M.A. 1991. Earth medicine, earth foods: plant remedies, drugs, and natural foods of the North American Indians. Ballantine Books, New York, NY. 230 pp.

Weiss, G., and Weiss, S. 1985. Growing & using the healing herbs. Rodale Press, Emmaus, PA. 360 pp.

Weiss, R.F. 1988. Herbal medicine. Beaconsfield Publishers Ltd., Beaconsfield, UK. 362 pp.

Werbach, M.R., and Murray, M.T. 1994. Botanical influences on illness: a sourcebook of clinical research. Third Line Press, Tarzana, CA. 344 pp.

WHO 1996. [World Health Organization Expert Committee on Specifications for Pharmaceutical Preparations.] WHO (Geneva) Thirty-fourth report, 1996 (WHO Technical Report Series No. 863), pp.178–184. [Available on World Wide Web at: http://itnet.rsu.ac.th/pharmacy/WHO_GUIDE.html]

Wichtl, M. [*Edited by* N.G. Bisset]. 1994. Herbal drugs and phytopharmaceuticals. A handbook for practice on a scientific basis. CRC Press, Boca Raton, FL. 568 pp.

Willard, T. 1992. Edible and medicinal plants of the Rocky Mountains and neighbouring territories. Wild Rose College of Natural Healing, Ltd., Calgary, AB. 277 pp.

Williams, L.O. 1960. Drug and condiment plants. U.S. Department of Agriculture, Agricultural Research Service, Agriculture Handook No. 172. Washington, DC. 37 pp.

Williams, T.I. 1947. Drugs from plants. Sigma, London, UK. 119 pp.

Wren, R.C. (Re-edited and enlarged by). 1956. Potter's new cyclopedia of botanical drugs and preparations. 7th ed. Pitman for Potter & Clarke, London, UK. 400 pp. [First - 6th ed. published under title: Potter's cyclopaedia of botanical drugs and preparations.]

Young, J.H. 1992. American health quackery: collected essays. Princeton University Press, Princeton, NJ. 299 pp.

Youngken, H.W., Jr. 1983. On herbal medicines — the last half century. The Herbarist **49**: 92–98.

World Wide Web sites of general interest on medicinal herbs

The internet is a valuable adjunct to information concerning medicinal herbs obtained from printed sources, especially of very recently generated information. However, the quality and reliability of the information is extremely variable and often erroneous. Information should be examined in the light of the biases of individuals and organizations sponsoring web sites. Private companies marketing herbal products are of course pimarily interested in convincing consumers of the merits of the herbs. Sites that are maintained by governments, universities and other major research organizations tend to produce relatively reliable information. Internet links change, move, and often rapidly become defunct. As the internet is a transient, volatile medium, addresses may not continue to be accurate, and use of search engines and web crawlers is recommended. Many of the sites noted below are interlinked, and have numerous related links. Even if many of the following URLs become dated, their titles furnish key words for a search.

Health Canada (a very large and useful web site; particularly note options of "Therapeutics Products Programme," which deals extensively with aspects of both herbal and pharmaceutical drugs in Canada):
http://www.hc-sc.gc.ca/links/english.html

The alternative medicine homepage:
http://www.pitt.edu/~cbw/altm.html

American Association of Colleges of Pharmacy basic booklist for pharmaceutical education preliminary 1996 edition [a huge list of references on drugs]:
http://www.aacp.org/aacp/info/booklist.html

Anancyweb Medical links, resources, sites, pages [not specifically for herbs, but provides comprehensive list of medical links, including a valuable set of links to medical libraries]:
http://www.anancyweb.com/medical.html

EthnoMedicinals home page:
http://walden.mo.net/~tonytork/

Facts about the importation and sale of herbal products [relevant to Canada; Health Canada's position as of May 1997]:
http://www.hc-sc.gc.ca/main/hc/web/datapcb/communc/home/news/herbsale.htm

Getting wired: the medical writer's guide to online resources [not specifically for herbs, but has an extremely impressive set of links to health databases]:
http://www.journalism.iupui.edu/caj/jrodden/HealthOnline.html#anchor1717333

HealthWWWeb [for an excellent set of links, see Herbs and botanical medicine links - http://www.HealthWWWeb.com/LinksHerbs.html]:
http://www.healthwwweb.com/

Henriette's herbal homepage:
http://sunsite.unc.edu/herbmed/

Herbal applications to conditions [extensive list of herbs said to be useful for treating various medical ailments]:
http://www.bryrus.net/essence/education/herbal_applications_to_condition.htm

Herbal bookworm:
http://www.teleport.com/~jonno/index.shtml

Herbal Hall:
http://www.herb.com/herbal.htm

Herbfaqs:
http://sunsite.unc.edu/herbmed/

The Herb Research Foundation: herb information greenpaper: useful herb references [extensive guide to library resources, databases, scientific journals, magazines; has useful links]:
http://ddmi.he.net/~herbs/greenpapers/resource.html

HerbNET:
http://www.herbnet.com

Herb Research Foundation:
http://www.herbs.org and http://sunsite.unc.edu/herbs/

Herb Society (UK) home page:
http://sunsite.unc.edu/herbmed/HerbSociety/

Herb world news online (from the Herb Research Foundation):
http://www.herbs.org/current/topnews.html

Holistic Healing [provides a very large number of links to a very large number of medical topics, several related to medicinal herbs]:
http://www.inforamp.net/~marcotte/healing.htm

Howie Brounstein's home page:
http://www.teleport.com/~howieb/howie.html

IBIDS database [The International Bibliographic Information on Dietary Supplements (IBIDS) database is produced by the Office of Dietary Supplements of the National Institutes of Health, in conjunction with the Food and Nutrition Information Center of the National Agricultural Library, US Department of Agriculture. IBIDS contains bibliographic records, including abstracts published in international scientific journals on the topic of dietary supplements, including vitamins, minerals, herbal and botanical supplements. The general public, scientists, researchers, and other interested parties will be able to search the database using keywords to obtain the citations of research journal articles. A limited group of 85 important medicinal plant species has been included initially. This is a key guide to the literature on medicinal plants.]:
http://www.nal.usda.gov/fnic/IBIDS/index.html

Internet resource guide for botany:
http://herb.biol.uregina.ca/liu/bio/subject/botecon.html

Lloyd Library:
http://www.libraries.uc.edu/lloyd/

Medical botany introduction by James A Duke [leads to phytochemical and ethnobotanical databases - http://sun.ars-grin.gov/~ngrlsb/]:
http://www.inform.umd.edu:808

MEDLINE [extraordinarily valuable resource, provides free access to database of more than 9 million references to articles published in 3800 biomedical journals]:
http://www.ncbi.nlm.nih.gov/PubMed/

USDA herbal info (James Duke):
http://www.ars-grin.gov/~ngrlsb/

MedNets [not specifically for medicinal herbs; provides huge collection of direct links to on-line medical search engines and databases]:
http://www.internets.com/mednets/medgovt.htm

Medicinal plants of Native America database (MPNADB), USDA:
http://probe.nalusda.gov:8300/cgi-bin/browse/mpnadb

Michael Moore's medical herbal glossary:
http://chili.rt66.com/hbmoore/ManualsMM/MedHerbGloss2.txt0/PBIO/MEDICAL_BOTANY/index.html

Missouri Botanical Garden research databases:
http://www.mobot.org/MOBOT/database.htm

Medical herbalism: a clinical newsletter for the herbal practitioner:
http://www.concentric.net/~bergner/MHHOME.HTM

NAPRALERT, DEREP and MEDFLOR [see below for information on NAPRALERT]:
http://pcog8.pmmp.uic.edu/mcp/MCP.html

Natural Products Alert [NAPRALERTSM, an acronym for NAtural PRoducts ALERT, is the largest relational database of world literature describing the ethnomedical or traditional uses, chemistry, and pharmacology of plant, microbial and animal (primarily marine) extracts. In addition, NAPRALERTSM contains considerable data on the chemistry and pharmacology (including human studies) of secondary metabolites of known structure, derived from natural sources. The database was specifically designed to be of value in the drug discovery and development processes. NAPRALERTSM currently contains the extracted information from over 150,000 scientific research articles, and is increasing at a rate of 600 articles per month. The database has: specific information on the use of herbal medicines and dosages; drug interactions and contraindications of botanicals with prescription drugs or food; and information concerning the pharmacokinetics and toxicity of botanical products. The NAPRALERT database is maintained by the Program for Collaborative Research in the Pharmacological Sciences (PCRPS), College of Pharmacy, University of Illinois at Chicago, 833 South Wood Street (M/C 877), Chicago, IL 60612, U.S.A. Access is on a fee basis.]:
http://www.ag.uiuc.edu/~ffh/napra.html

National Agricultural Library:
http://probe.nalusda.gov/

New Crop Compendium. Contents. Policy. International development. Regional & state development. Research & development. Germplasm & breeding. New crops. [published papers from the Advances in new crops symposia; has a section on medicinal plants; a valuable resource]:
http://www.hort.purdue.edu/newcrop/compendium/comp-toc.html

Phytochemical and ethnobotanical databases [leads to EthnobotDB, USDA: http://probe.nalusda.gov:8000/related/aboutethnobotdb.html]:
http://www.ars-grin.gov/~ngrlsb/

Michael Tierra's Planet Herbs site:
http//:www.planetherbs.com

Rainfo - the regulatory affairs home page [for many Canadian links, see - http://www.medmarket.com/tenants/rainfo/canada.htm]:
http://www.medmarket.com/tenants/rainfo/rainfo.htm

Prairie medicinal and aromatic plants conference:
http://www.agric.gov.ab.ca/crops/specia L/medconf/#top

The medicinal plants of Native America database:
http://probe.nalusda.gov:8000/related/aboutmpnadb.html

PubMed - public access to Medline:
http://www.ncbi.nlm.nih.gov/PubMed/

Richters herb specialist home page [an extremely important Canadian herbal firm]:
http://www.richters.com/

Soaring Bear's herb resources [extensive list of herbal and ethnobotanical sites]:
http://ellington.pharm.arizona.edu/~bear/herb.html

Southwest School of Botanical Medicine and Michael Moore's Herbal Treasurehouse:
http://www.rt66.com/hbmoore/HOMEPAGE/HomePage.html

Top herbal products encountered in drug information requests (Part 1) by J.L. Muller and K.A. Clauson [requires registration (free) with Medscape; also available in Drug Benefit Trends 10(5): 43–50]:
http://www.medscape.com/SCP/DBT/1998/v10.n05/d3287.mulL/d3287.mull-01.html

Top herbal products encountered in drug information requests (Part 2) by J.L. Muller and K.A. Clauson [requires registration (free) with Medscape; also available in Drug Benefit Trends 10(6):21–23, 31]:
http://www.medscape.com/SCP/DBT/1998/v10.n06/d5131.mulL/d5131.mull-01.html

University of Washington medicinal herb garden:
http://www.nnlm.nlm.nih.gov/pnr/uwmhg/

Wild Rose College of Natural Healing:
http://www.nq.com/netquest/wildrose/

The Fire Effects Information System (FEIS) [a site developed at the USDA Forest Service Intermountain Research Station's Fire Sciences Laboratory (IFSL) in Missoula, Montana; not a site dedicated to herbs, but does contain extremely competently researched information on the biology of certain herbs]:
http://svinet2.fs.fed.us/database/feis/plants/

Research on crop production, genetics, germplasm improvement, micropropagation, and protection of medicinal herbs [outlines research studies on medicinal herbs carried out at the Southern Crop Protection and Food Research Centre of Agriculture and Agri-Food Canada]:
http://res.agr.ca/lond/pmrc/study/newcrops/ncweb.html

Glossary of pharmacological and medical terms relevant to medicinal plants

Pharmacological and medicinal terminology are generally not needed in this work, as we have avoided technical medical words. Nevertheless, technical medical terms are used so frequently by professionals dealing with medicinal plants that access to a comprehensible dictionary of these terms is advantageous. In addition, technical terms are often employed in the literature references and World Wide Web links provided. Those exploring the subject of medicinal crops often must be able to decipher specialized medical terminology. For these reasons, this extensive glossary is provided.

Herbs have been used to treat virtually every affliction known to mankind, and so all of the terminology of medicine and related disciplines could be used with specific reference to medicinal plants. The present selective list of terms was prepared by inspecting the vocabulary used in recent major surveys of herbal medicine such as Duke (1985) and Wichtl (1994), which are typical of such works in not providing the user with definitions of unfamiliar terminology. If a term is not present here, see the list of internet and hardcopy dictionaries at the end of the glossary. In most cases the shortest spelling of a word is given, with diphthongs reduced (e.g., ae to e, as in hemoglobin, not haemoglobin) and American spellings preferred (e.g., tumor, not tumour). In many cases alternative spellings and synonyms are given.

Abortifacient Induces expulsion of an embryo or foetus.

Acetylcholine A neurotransmitter that slows heart rate; controlled by the parasympathetic nervous system (which see).

Achlorhydria The absence or insufficiency of hydrochloric acid from the gastric juice; this results in insufficient digestion.

Acidosis Excess acidity in the body, as may result for example from diabetes and kidney disease. This usually causes the pH of the blood to drop.

Acrid Sharp and harsh, or bitter and unpleasant tasting; causing heat and irritation; corrosive, caustic.

Adaptogen A substance that is "innocuous, causing minimal physiological disorder; non-specific in action, increasing resistance to the adverse influences of a wide range of physical, chemical and biological factors; and capable of a normalizing action irrespective of the direction of the pathological change." Adaptogens are said to help adapt people to stress, thereby preventing many chronic degenerative diseases. The validity of the concept is debatable.

Addison's disease A destructive disease characterized by deficient secretion of the adrenal hormones aldosterone and cortisol, and weakness, loss of weight, low blood pressure, gastrointestinal disturbances and brownish pigmentation of the skin and mucous membranes.

Adenitis Inflammation of a gland. Often specifically the inflammation of lymph nodes or tissues.

***Adenoid* (usually plural, adenoids)** An enlarged body of lymphoid tissue at the back of the pharynx, which may obstruct breathing.

Adenopathy Glandular disease (a general term).

Adrenalin See *Epinephrine.*

Adrenaline A trademark for a preparation of adrenalin.

Adrenergic See *Sympathomimetic*.

Aerophagia The swallowing of air, particularly in hysteria.

Aesthesia Perception and feeling, the opposite of anaesthesia.

***Agalactia* (agalactosis)** Absence or failure of the secretion of milk.

Ague A fever accompanied by chills and sweating that recurs at regular intervals (as with malaria); a fit of shivering.

Albuminuria Presence of protein in the urine, principally albumin, generally indicating disease.

Alexiteric Antidotal, especially to snakebites.

Alkaloid One of a large group of nitrogenous substances found naturally in plants. These are alkaline, and react with acids to form salts. They are usually very bitter and although often poisonous, may have pharmacological value. Examples: atropine, berberine, caffeine, coniine, morphine, nicotine, quinine, strychnine. The term is also applied to similar synthetic substances such as procaine.

Allergen A substance that can cause an allergic reaction (e.g., pollen, dander, mold spores).

Allergy Hypersensitivity of the immune system to exposure to given substances (antigens) such as pollen, bee stings, poison ivy, drugs, or foods. In response, the body summons antibodies (immunoglobulin E) to fight the allergens. Special cells called mast cells are frequently inadvertently injured, releasing a variety of chemicals including histamine (which see) into the tissues and blood that frequently cause allergic reactions. These irritating chemicals may produce itching, swelling, muscle spasm, and lung and throat tightening as is found in asthma. Anaphylactic shock is a severe form of allergy response (symptoms include dizziness, loss of consciousness, labored breathing, swelling of the tongue and breathing tubes, blueness of the skin, low blood pressure, and death).

Allopathy 1. A system of medicine combatting disease by remedies producing effects different from those produced by the disease treated; 2. A system of medicine using all measures proven of value in treating disease. A term invented by S. Hahnemann to designate the ordinary practice of medicine, as opposed to homeopathy (which see).

Alopecia Baldness.

Alterative Gradually restores healthy functions; a very general and rather vague term.

Amaurosis Blindness, often the result of a cortical lesion or from no change in the eye itself.

Amblyopia Impairment of vision without detectable organic lesion of the eye.

Amblyopy Weakness of sight (without opacity of the cornea or of the interior of the eye).

***Amebiasis* (amoebiasis)** A protozoan infection, especially by *Entamoeba histolytica*, which occurs worldwide in areas of poor sanitation. Typically such infections occur in the large intestine, rarely in the liver.

Amebicidal Kills amebae.

Amenorrhea Absence, discontinuation or abnormal stoppage of the menstrual periods. In primary amenorrhea menses have not begun by age 16, while secondary amenorrhea is slow reinitiation of menstruation from such causes as pregnancy, illness or dieting.

Analeptic A drug that acts as a restorative, such as caffeine and amphetamine.

Analgesic Alleviates pain without causing loss of consciousness.

Anaphrodisiac Inhibits or decreases sexual desire.

Anaphylactic shock A serious, often life-threatening allergic reaction that is characterised by low blood pressure, shock, and difficult breathing (see *Allergy*).

Anaphylaxis Hypersensitivity from sensitization following prior contact with a causative agent. In immunology, an inflammatory reaction involving release of histamine and histamine-like substances, causing immune responses, and resulting in an acute allergic reaction with shortness of breath, rash, wheezing, and hypotension (also see *Allergy*).

Anasarca Generalized massive edema; subcutaneous accumulation of serum, causing a soft, pale, inelastic swelling of the skin.

Androgens Primarily male steroid hormones, produced in the adrenal glands and (in women) the ovaries.

***Anemia* (anaemia)** Abnormally low number of red blood cells or amount of hemoglobin in the blood. The lowered oxygen-transporting capacity results in fatigue, paleness, palpitations, and shortness of breath. Causes are varied (e.g., bleeding, sickle cell anemia, iron deficiency, bone marrow diseases).

***Anesthetic* (anaesthetic)** Causes unconsciousness or a loss of general or local sensation.

Angina Chest pain due to inadequate oxygen supply to heart muscle.

Angina pectoris Sudden pain in the chest or arms brought on by lack of oxygen supply to the heart muscle. Generally refers to a chronic heart condition of this nature.

Anodyne Externally-applied material that relieves pain or discomfort.

Anorexia Uncontrolled lack or loss of appetite for food.

Anorexia nervosa An eating disorder; individuals with this disease often believe they are overweight even when grossly underweight.

***Anorexiant* (anorectic, anorexiac)** A drug or substance that leads to anorexia or diminished appetite; an appetite suppressant.

Antacid Counters or neutralises excess acidity, usually of the stomach.

Antarthritic Counteracts or alleviates gout (cf. *Antiarthritic*; gout is (like arthritis) an inflammation of joints, but is specifically caused by deposits of urates in and around the joint).

***Anthelmintic* (anthelminthic)** Destroys and (or) expels internal parasitic worms from gut of animals and humans (see *Vermifuge*).

Anthrax An infectious highly communicable disease caused by the bacterium *Bacillus anthracis*; symptoms include external ulcerating nodules and lesions in the lungs.

Antiacid Counteracts or neutralises acidity.

Antiandrogen An agent that interferes with the activity of male sex hormones.

Antiaphonic Alleviates voice problems such as laryngitis.

Antiarrhythmic An agent that prevents or alleviates cardiac arrhythmia (see *Arrhythmia*).

Antiarthritic Alleviates arthritis (cf. *Antarthritic*).

Antiasthmatic Alleviates spasm of asthma.

Antiatherogenic Counteracts atherogenesis (the formation of lipid deposits in the arteries, which leads to atherosclerosis; see *Antiatherosclerotic*).

Antiatherosclerotic Counteracts atherosclerosis (progressive narrowing and hardening of the arteries).

Antibacterial Destroys bacteria or suppresses their growth or reproduction.

Antibilious Alleviates or prevents excess bile.

Antibiotic Chemical substance produced by a microorganism which inhibits growth of or kills other microorganisms.

***Antibiotic resistance* (drug resistance)** Acquired ability of bacteria and other microorganisms to withstand an antibiotic to which they were once susceptible.

Antibodies Specialized proteins produced by white blood cells that circulate in the blood, attaching to and neutralizing foreign proteins, microorganisms or toxins.

Anticarcinogen A chemical that counteracts the effect of a cancer-causing agent.

Anticatarrhal Efficacious against catarrh (which see); reduces or stops production of mucus.

Anticholesterolemic Reduces or prevents the build up of cholesterol in arteries.

Anticholinergic An agent that blocks the parasympathetic nerves. This affects induction of spasms and cramps of the intestinal tracts and related ducts (see *Parasympathetic nervous system*).
Anticoagulant Prevents or slows blood clotting, or removes blood clots.
Anticonvulsant An agent that prevents or relieves convulsions.
Antidepressant Alleviates depression.
Antidermatosic Treats skin problems.
Antidiabetic Prevents or alleviates diabetes.
Antidiuretic Suppresses rate of urine formation.
Antidote Counteracts poison.
Antiedemic Counteracts edema (which see).
***Antiemetic* (antinauseant)** Prevents or alleviates nausea and vomiting, for example during treatment with chemotherapy or radiotherapy.
Antifebrile Alleviates fever (see *Febrifuge*).
Antiflatulent Reduces intestinal gas (= carminative).
Antifungal Effective against fungal infections.
Antigen A substance, generally a protein, that induces counteractive antibodies.
Antigonadotropic Inhibits functions of reproductive organs.
***Antihelmintic* (antihelminthic)** Counteracts parasitic worms, especially of the intestinal tract.
***Antihemorrhagic* (antihaemorrhagic)** Reduces or stops bleeding.
***Antihemorrhoidal* (antihaemorrhoidal)** Reduces inflammation of hemorrhoids.
Antihepatotoxic Counteracting liver toxins.
***Antihistamines* (synonym of antihistamine as adjective & noun: antihistaminic)** Drugs that counteract the actions of histamine released during an allergic reaction. Antihistamines do not stop the formation of histamine nor stop the allergic reaction, but protect tissues from some of its effects, although often causing side effects (e.g., mouth dryness, sleepiness, occasionally urine retention in males and fast heart rate); (see *Allergy*).
Antihydropic Effective against dropsy (which see).
Antihydrotic Reduces perspiration.
Antihypertensive Reduces high blood pressure.
Antihypertonic Counteracts hypertonia (which see).
Antihysteric Alleviates hysteria.
Anti-infective Inhibits spread of or kills infective agents.
Anti-inflammatory Alleviates inflammation.
Antilithic An agent preventing the formation of calculi (stones), especially uric-acid calculi, in the kidneys and other parts of the urinary system.
Antimicrobial Kills microorganisms or suppresses their growth.
Antimitogenic The reverse of mitogenic (which see).
Antimitotic A drug for inhibiting or preventing mitosis (cell division). The term is often used for compounds such as colchicine that cause metaphase arrest. Many antitumor drugs are antimitotic, blocking cell proliferation.
Antimycotic Suppresses the growth of fungi.
Antineoplastic Inhibits tumors.
Antineuralgic Counters neuralgia (which see).
Antiodontalgic Counters toothache.
Antioxidant Prevents or delays deterioration by action of oxygen in the air. Chemically, oxidation consists of an increase of positive charges on an atom or the loss of negative charges. Most biological oxidations are accomplished by the removal of a pair of hydrogen atoms (dehydrogenation) from a molecule. At the cellular level, oxidative reactions produce energy; however, free radicals and other oxidizing agents can damage membranes and other cell components and interfere with regulatory systems. Antioxidants such as vitamins C and E are able to counteract the damage from oxidation from oxygen-based free radicals. Beta carotene (the precursor of vitamin A) is an antioxidant said to protect cells against oxidation damage that can lead to cancer.
Antipellagral Counters pellagra (which see).
Antiperiodic Prevents return of recurring disease, for example of intermittent fever.
Antiphlogistic Counteracts inflammation and fever.
Antiplaque Counteracts buildup of plaque on teeth.
Antipruritic Alleviates or prevents itching of the skin.
Antipyretic Alleviates fever (see *Febrifuge*).
Antirheumatic Reduces rheumatism.
Antiscorbutic Counteracts scurvy by providing vitamin C.
***Antiscrophulatic* (antiscrofulatic)** Treats scrofula (which see).
Antiseptic Inhibits growth and development of microorganisms.
Antispasmodic Relieves muscular spasms and cramps.
Antisyphilitic Efficacious against syphilis; a medicine used against syphilis.
Antithyrotropic Inhibits thyroid function.
***Antitumor* (antitumour)** Counteracts tumor formation.
Antitussive Relieves or suppresses cough (compare *Expectorant*).
Antiveratrinic Acts against veratrine [a poisonous alkaloid mixture obtained from the root of hellebore (*Veratrum*) and from sabadilla (*Schoenocaulon officinalis*) seeds, sometimes used externally as a counterirritant to treat neuralgia and rheumatism, and also as an insecticide].
Antiviral Alleviates or prevents virus diseases.
Anuria The complete suppression of urinary secretion by the kidneys.
Aperient A gentle laxative.
Aperitive Stimulates the appetite.
Aphonia Inability to produce speech sounds, often caused by disease of the vocal structures.
Aphrodisiac Increases libido, i.e., desire for sex.
Aphtha See *Thrush*.
Apoplexy Sudden loss or reduction of consciousness and voluntary motion, due to rupture or obstruction of an artery in the brain.

Aposteme An abscess; a swelling filled with pus. (Corrupted as "imposthume.")

Appetizer Improves appetite.

Aromatherapy Utilizes the presumed medicinal properties of essential oils extracted from plants. Treatments may be administered through inhalation, external application (e.g., bath, massage, compress, or topically), or (untypically) by ingestion. Essential oils may be antibacterial, antiviral, antispasmodic, and may have such varied pharmacological actions as widening or narrowing blood vessels, or acting on the adrenals, ovaries, and the thyroid. The potential for toxicity exists.

Aromatic 1. Chemically, molecules with one or more benzene rings. 2. Plant constituents which vaporize and can be smelled. 3. An agent added to a medicine to give it aroma or flavor.

Arrhythmia Abnormal rhythm, generally referring to heart. The heartbeats may be too slow, too rapid, irregular, or too early. Rapid arrhythmias (>100 beats/minute) are tachycardias. Slow arrhythmias (<60 beats/minute) are bradycardias. Irregular heart rhythms are called fibrillations (e.g. atrial fibrillation).

***Arteriosclerosis* ("hardening of the arteries")** A chronic disease marked by abnormal thickening and hardening of the walls of the arteries.

Arthralgia Pain in joints.

Arthritis Inflammation of a joint (symptoms may include stiffness, warmth, swelling, redness, and pain). There are over 100 types of arthritis. Osteoarthritis (degenerative arthritis) is a type caused by inflammation, breakdown, and eventual loss of the cartilage of the joints.

Arthrogryposis A congenital disease in which those affected are born with stiff joints and weak muscles.

Arthrosis A disease of a joint.

Ascariasis Infection by the nematode *Ascaris lumbricoides* (the "common roundworm"). Adult worms are 15–40 cm in length and live in the small intestine. Infection occurs after ingesting eggs in contaminated food or more commonly, by transmission to the mouth by the hands after contact with contaminated soil.

Ascaridial Removes roundworms (nematode worms of the family Ascaridae), especially the common roundworm, *Acaris lumbricoides*.

Ascites Abnormal accumulation of serous fluid in the abdominal cavity. This most commonly results from trauma.

Aseptic Refers to absence of microorganisms.

Astringent Causes contraction, usually locally after topical application. By reducing the size of liquid-carrying channels, or reducing the area of injuries, astringents can reduce the flow of secretions and discharges such as blood, mucus and diarrhea.

Ataractic An agent that can produce sedation without profound drowsiness; tranquillizer.

Ataxia Muscular incoordination, irregularity of muscular action.

Atheroma 1. An encysted tumor containing curdy matter. 2. A disease characterised by thickening and fatty degeneration of the inner coat of the arteries.

Atonic 1. Characterized by atony, or want of vital energy or strength. 2. A remedy capable of alleviating excitement or irritation.

Autonomic nervous system A component of the nervous system regulating key functions including the activity of the cardiac (heart), muscle, smooth muscles (e.g., of the gut), and glands. It has two divisions: the sympathetic nervous system that accelerates the heart rate, constricts blood vessels, and raises blood pressure; and the parasympathetic nervous system which slows the heart rate, increases intestinal and gland activity, and relaxes sphincter muscles.

Avicidal Kills birds.

Azotemia Abnormally high level of nitrogen-type wastes in the bloodstream, associated with kidney dysfunction (compare *Uremia*).

Bactericidal Capable of killing bacteria.

Bacteriostatic Inhibits growth or multiplication of bacteria.

Balsamic Healing and soothing tree resins or other plant materials useful as dressings, expectorants and diuretics.

***Bell's palsy* (Bell's facial nerve palsy)** A paralysis of one side of the face, involving dysfunction of the facial nerve (7th cranial nerve).

Benign Not cancerous; not malignant.

Benign prostatic hyperplasia A noncancerous enlargement of the prostate that can interfere with urination.

***Beriberi* (beri-beri)** A disease caused by lack of or inability to utilize thiamine (vitamin B_1). Symptoms include marked inflammatory or degenerative changes of the nerves, digestive system and heart.

Bile Fluid secreted by the liver that helps break down fats in the small intestine.

Bile acids Cholesterol-rich gallbladder secretions that emulsify fatty foods in the intestine and help in digestion.

Bilious Suffering from disordered liver function, often manifested by jaundice.

Bitters Bitter-tasting substances that increase appetite and stimulate digestion; may increase flow of bile, stimulate repair of gut wall lining, and regulate secretion of insulin and glucogen.

Blennorrhea Excessive secretion and discharge of mucus.

Blepharitis Inflammation of the eyelids.

Blood poisoning See *Septicemia*.

***Blood pressure, high* (hypertension, which see)** A repeatedly elevated blood pressure exceeding 140 over 90 mm Hg.

***Blood purifier* (blood cleanser)** An old fashioned term (rarely used today) for agents that were supposed to remove toxins from the blood. Alterative (which see) herbs were once considered to be blood purifiers.

Blood-thinner See *Anticoagulant*.

Borborygmus A rumbling, bubbling or gurgling noise from propulsion of gas through the intestines.

Bradycardia See *Arrhythmia*.

Bright's disease Any of several kidney diseases characterized by albumin in the urine.

Bronchitis Inflammation of the bronchi (the trachea divides into the left and right bronchus, each leading to a lung).

Bronchodilator Dilates (enlarges) the airway, which can ease breathing difficulties resulting from inflammation. Such medications are commonly given to those with asthma who manifest wheezing; e.g., adrenaline.

***Bronchorrhea* (bronchorrhoea)** Excess secretion of mucus by the bronchial mucosa.

***Bulla* (plural bullae)** A large vesicle or blister. (Anatomically, the ovoid prominence below the opening of the ear in the skull; e.g., the auditory bulla.)

Bunion A swelling or deformity from chronic inflammation of the small sac on the first joint (the metatarsal) of the big toe.

Bursa A tissue space (fibrous sac) lined with synovial membrane (joint tissue) and containing joint fluid. Bursas are found between tendon and bone, skin and bone, and muscles, and facilitate movement.

Bursitis Inflammation of a bursa (which see), especially of the shoulder, elbow or knee.

Cachexia A profound state of constitutional disorder, general ill health and malnutrition.

Cacoethes A habitual, uncontrollable desire, such as a mania.

Calcium channel blocker Drugs, useful for angina and blood-pressure reduction, that dilate the coronary arteries and increase blood flow through them.

Calculus A concretion, typically of mineral salts around organic material, especially found in hollow organs or ducts of the body.

Candicidal Having the ability to kill *Candida* (see *Candidiasis*).

***Candidiasis* (moniliasis, candidosis, oidiomycosis, blastodendriosis).** Infection by the fungus genus *Candida*. *Candida* (*Monilia*) *albicans* commonly infects moist skin, oral mucous membranes, the respiratory tract, and the vagina. It is commonly responsible for thrush (which see) in infants and vaginal yeast infections.

Carbuncle A skin abscess marked by a collection of pus formed inside the body.

Carcinogen An agent that causes cancer.

Carcinoma Cancer that begins in the tissues lining or covering an organ.

Carcinostatic Weakens or halts progression of cancer.

Cardialgia 1. Pain in the heart. 2 Heartburn (which see).

Cardiotonic Having a tonic, strengthening or regulating effect on the heart.

Cardiotoxic Having a poisonous or deleterious effect upon the heart.

Caries 1. Tooth decay (dental caries). 2. Progressive destruction of bone or tooth.

Carminative Expels gas from the intestines and therefore reduces flatulence and colic (= antiflatulent).

Cataplasm A poultice.

Catalepsy A state of diminished responsiveness characterized by a trance-like state and constantly maintained immobility usually with rigidity.

Cataract A clouding or fogging of the crystalline lens of the eye.

Catarrh Inflammation of a mucous membrane, with a free discharge or congestion, especially in air passages of the head and throat. A rather old term.

Cathartic A strong laxative (which see).

Caustic A substance that burns organic tissue by chemical action.

***Celiac* (coeliac)** Referring to abdomen.

Celiac disease A disease in which sensitivity to gluten, a protein found in wheat, rye, and barley results in defective absorption of nutrients; symptoms include diarrhea, weight loss, and malnutrition.

Cellulitis An acute spreading, inflammation of the deep subcutaneous tissues and sometimes also muscle.

Cephalgia Headache.

Cephalic Pertaining to the head.

Chancre In general, a primary sore or ulcer at the site of infection of a pathogen; specifically, the initial lesion of syphilis, caused by the spirochete *Treponema pallidum*.

Chancroid A sexually transmitted disease caused by the bacterium *Haemophilus ducreyi*. Symptoms may include multiple painful ulcers on the penis or the vulva, and tender and enlarged lymph nodes of the groin.

Cheilitis Inflammation and cracking of the lips.

Chemotherapy Treatment with anticancer drugs.

Chilblain A redness and swelling of toes, fingers, nose, ears or cheeks in cold weather, with itching, burning, and (or) skin cracking and ulceration, thought to be due to insufficiency of local circulation.

Chlorosis Iron deficiency anemia.

Cholagogue An agent that increases the flow of bile (into the intestines).

Cholangitis Inflammation of a bile duct.

Cholecystalgia Cramps of gall bladder or associated ducts.

Cholecystitis Inflammation of the gallbladder and associated ducts.

Cholelithiasis 1. Production of gallstones. 2. Abnormal presence of gallstones.

***Cholemia* (cholaemia)** Excess bile in the blood, usually the result of liver disease.

Choleretic An agent inducing the flow of bile from

the liver (compare *Cholagogue*).

Cholesterol The most common steroid in the body, produced in the liver and carried in the bloodstream by lipoproteins. Low-density lipoprotein cholesterol ("bad" cholesterol) is associated with an increased risk of coronary artery disease. High-density lipoprotein cholesterol is "good" cholesterol. This fatlike substance is found in all food from animal sources, and is an essential component of body cells and a precursor of bile acids and certain hormones.

Cholinergic Refers to functions controlled by parasympathetic nervous system, a portion of the involuntary (autonomic) nervous system; controls many digestive, reproductive, vascular, secretory, and cardiopulmonary functions, for examples salivary glands, thoracic and abdominal viscera, bladder and genitalia. The system operates by releasing acetylcholine, a neurotransmitter. Cholinergic agents resemble acetylcholine in action (see *Parasympa-thetic nervous system*).

***Chorea* (Saint Vitus's dance)** A nervous disorder characterized by spasmodic movements and incoordination.

Chronotropic Affecting rate, especially the rate of contraction of the heart.

Cicatrizant Promoting the healing of a wound or the formation of a scar.

Ciguatera A form of fish poisoning that results from eating reef fish that are normally nontoxic. Thought to be caused by seasonal accumulation of toxins in the tissues of the fish.

Cirrhosis Fibrosis especially of the liver with hardening caused by excessive formation of connective tissue.

Clonic Pertaining to spasmic movements marked by contractions and relaxations of a muscle, perhaps occurring in rapid succession but with intermittent rest periods (e.g., birthing contractions).

Colchicine Chemical from the plant *Colchicium* (autumn crocus) used to treat gouty arthritis and in the laboratory to arrest cells during cell division (by disrupting the mitotic spindle) so their chromosomes can be more easily observed.

Colic 1. A syndrome in early infancy (in about 1 in 10 babies, lasting from 2–3 weeks of age to 4 months) characterised by episodic loud crying, apparent abdominal pain (legs drawn up and rigid abdomen) and irritability (see *Carminative*). 2. Spasms or cramping of smooth muscle tube, such as of stomach or uterus.

Colitis Inflammation of the colon (large intestine). There are many forms of colitis.

Collagen A structural protein that gives strength and tone to skin and other tissues.

Collyrium A lotion or wash for the eyes.

Colpitis Vaginitis.

***Condyloma* (plural: condylomata)** A warty papilloma (benign tumor) with a central core of connective tissue in a treelike configuration, usually occurring on the mucous membrane or skin of the external genitals or in the perianal region.

Conjunctiva The clear mucous membrane that coats the inner aspect of the eyelids and is continued over the forepart of the eyeball.

Conjunctivitis Inflammation of the conjunctiva.

Consumption A progressive wasting away of the body, especially from tuberculosis of the lungs (see *Phthisis*).

Contraceptive Diminishes the likelihood of or prevents conception.

Contusion A bruise, with skin intact but underlying blood vessels injured.

Cordial An invigorating medicine.

Corn A hardening or thickening of the epidermis, most familiarly on the toe.

Cornea The curved, transparent tissue making up the front of the eye, through which light first enters.

Corroborant Invigorating.

Coryza An acute inflammatory contagious disease of the upper respiratory tract, especially the common cold.

Costive Constipated, or causing constipation.

Counterirritant An agent applied locally to produce superficial inflammation in order to reduce inflammation in deeper adjacent structures (e.g., application of a mustard plaster to the chest to treat bronchitis).

Coxalgia Disease of the hip joint, with pain.

Craw-craw An itching skin disease caused by the filarial worm *Onchocerca volvulus*.

Crohn's disease A poorly understood chronic inflammatory disease of the small or large intestine; symptoms include recurrent abdominal pains, fever, nausea, vomiting, weight loss and diarrhea.

Croup An infection of the larynx, trachea, and the bronchial tubes, mainly in children.

Cyanogenetic Capable of producing cyanide (as hydrogen cyanide); therefore, cyanogenetic plants are potentially (but not necessarily) dangerous.

Cyanosis A bluish discoloration, especially of skin and mucous membranes, due to lack of oxygen in the blood.

Cynanche Diseases of the throat.

Cystitis Inflammation of the urinary bladder.

Cystorrhea Abnormally elevated level of mucus in the urine.

Cytostatic Suppresses cell growth and multiplication. Useful in slowing or controlling the growth of tumors.

Cytotoxic Chemicals that are directly toxic to cells, preventing their reproduction or growth. Useful in the treatment of diseases such as cancer although, as a side effect, healthy, noncancerous tissues or organs might be damaged.

Decoction An extract obtained by boiling in water (the strained liquor is called the decoction). In pharmacy, a decoction may be contrasted with an infusion (which see), where there is merely steeping.

Decoctions are made using roots and bark, etc., whose medicinal components are resistant to boiling.

Decongestant Removes mucus and phlegm, especially from the respiratory system. Decongestants make it easier to breath, often by reducing congestion or swelling by constricting blood vessels in the nasal passages.

Delirium tremens Hallucinations associated with alcoholism.

Dementia A mental disorder characterised by a general loss of intellectual abilities as well as changes in personality.

Dementia praecox Schizophrenia (mental disorder characterized by loss of contact with environment and disintegration of personality).

Demulcent Soothes irritation of inflamed or abraded surfaces, especially mucous membranes. Sometimes the word is restricted to soothing of internal membranes, while emollient (which see) is used for external applications.

Dengue An acute, infectious, mosquito-transmitted, tropical, viral disease characterized by headache, severe joint pain and a rash.

Deobstruent Removes obstructions from the natural ducts of the body (e.g., see *Aperient*).

Depilatory An agent that removes hair.

Depurative Purifies or eliminates toxins from the blood or other fluid systems of the body.

Dermatitis Inflammation of the skin.

Dermatitogenic Causing dermatitis.

Detersive Cleansing.

Diabetes mellitus A chronic condition characterized by abnormally high levels of sugar (glucose) in the blood. Failure of the pancreas to produce sufficient insulin results in lower blood glucose and causes diabetes. Insulin dependent (juvenile diabetes) is type I and non-insulin dependent (adult-onset diabetes or insulin-resistant diabetes) is type II. Symptoms of diabetes include increased urine output and appetite as well as fatigue.

Diaphoresis Perspiration, especially profuse perspiration (= sudoresis).

Diaphoretic Promotes perspiration. (Diaphoretics have been said to cause evident increase in sweat while sudorifics merely increase perspiration but not the accumulation of sweat.) Examples: ginger, hot peppers.

Diaphragm A thin muscle below the lungs and heart, separating the chest from the abdomen.

Diaphragmitis Inflammation of the diaphragm.

Diathesis A state of the body that makes the tissues react in special ways to certain stimuli and thus predisposes the person to certain diseases.

Digestant See *Digestive*.

Digestive Promotes digestion (see, for example, *Bitters*).

Diplopia Double vision.

Diphtheria An acute contagious disease of the throat caused by a bacterium that produces a toxin inflaming the heart and nervous system.

Discutient An agent that causes dispersal of a tumor or other lesion.

Disinfectant Disinfects, particularly wounds.

Diuretic Promotes the excretion of urine. Used inappropriately, diuretics may cause dehydration.

Diverticulitis Inflammation of congenitally-inherited pouches (diverticula), particularly of the colon.

Dropsy Edema (which see), often caused by kidney or heart disease.

***Drug, over-the-counter* (OTC)** A drug for which a prescription is not needed.

Drug, prescription A drug requiring a physician's prescription.

Drug resistance See *Antibiotic resistance*.

Dysentery Inflammation of intestines, especially of the colon, with pain in the abdomen and severe diarrhea.

Dyskinesia Impaired voluntary movement, resulting in incomplete movements.

***Dysmenorrhea* (dysmenorrhoea)** Painful, often incapacitating menstruation, with cramps and other pains.

Dyspepsia Impaired digestion, usually with discomfort following meals, often characterised by heartburn, nausea, vomiting, and flatulence.

***Dyspnea* (dyspnoea)** Breathing distress, usually result of serious disease of the heart, lungs, or airways.

Dystrophy Inadequate nutrition, especially any of several neuromuscular disorders.

Dysuria Painful or difficult urination.

Ecbolic An agent that excites uterine contractions and thereby promotes the expulsion of the contents of the uterus.

Ecchymosis A small hemorrhagic spot in the skin or mucous membrane forming a blue or purplish patch.

***Eccoprotic* (eccoproticophoric)** Laxative, often a mild laxative.

Ectoparasiticidal Kills external parasites.

Eczema An inflammation of the skin characterized by redness, itching and oozing vesicular lesions that become scaly, crusted or hardened.

***Edema* (oedema)** Presence of abnormally large amounts of fluid in the intercellular tissue spaces of the body (may be localized, due to venous or lymphatic obstruction or to increased vascular permeability; or it may be systemic, due to heart failure or renal disease).

Edemagenic Giving rise to edema (which see).

Electuary A confection (a general term, often used to refer to confections with medicinal materials, prepared with tasty materials such as honey for palatability).

Elephantiasis Enlargement and thickening of tissues, especially the enormous enlargement of a limb or the scrotum caused by filarial worms obstructing lymphatic system.

Elixir A general term, referring to a syrup or other liquid, often sweetened to be palatable, often containing alcohol or another substance to emulsify and suspend a medicine (like an extract).

Emesis Vomiting.

Emetic Induces vomiting.

Emmenagogue Restores regular menstrual discharge.

Emollient Softens (also often soothes) the skin (= malactic; compare *Demulcent*).

Emphysema A pathological accumulation of air in tissues or organs, especially of the lungs (frequently associated with impairment of heart action).

Encephalitis Inflammation of the brain.

Endometriosis Occurrence of uterine mucous membrane (the endometrium) aberrantly (in places where it normally does not develop) in various locations in the pelvic cavity.

Endometritis Inflammation of the endometrium (the lining of the uterus).

Endorphins Morphine-like substances produced in the body.

Enteralgia Pain in the intestines; colic (which see).

Enteritis Inflammation of the intestines.

Enterorrhagia Bleeding from or of the intestine.

Enuresis Involuntary discharge of urine after the age at which urinary control should have been achieved; bed wetting.

Epididymitis Inflammation of the epididymis (where sperm are matured and stored).

Epigastralgia Pain in the upper central region of the abdomen.

Epinephrine A chemical that acts as a neurotransmitter or a hormone; it constricts blood vessels and increases heart rate (= adrenaline).

Epistaxis Nosebleed.

Epithelioma A tumor (benign or malignant) derived from epithelial tissue.

Ergotism Poisoning from eating grain or grain products (such as rye bread) infected by ergot fungus (compare Saint Anthony's fire).

Erotomania Excessive (especially pathological) sexual desire.

Errhine Causes sneezing, and possibly also secretion of mucus.

Eructation Belching.

Erysipelas A severe inflammation of the skin caused by a (group A) hemolytic streptococcal bacterium. (One of the conditions developed is called St. Anthony's fire; ergotism, caused by eating grain infected with ergot fungus, has also been called St. Anthony's fire.)

Erythema Redness of skin produced by congestion of the capillaries, which may result from a variety of causes, for example by sunburn.

Erythrocytes Red blood cells, that carry oxygen to all parts of the body.

***Escharotic* (escarotic)** Caustic. A substance that produces a scar, especially a mild caustic.

***Estrogen* (oestrogen)** The female, steroid, sex hormones. In humans, estrogen is formed particularly in the ovary, but has various functions in both sexes. It is responsible for the development of the female secondary sex characteristics, and during the menstrual cycle it acts on the female genitalia to produce an environment suitable for the fertilization, implantation, and nutrition of the early embryo. Estrogen is used in oral contraceptives and in the relief of the discomforts of menopause. Estrogen deficiency in women can lead to osteoporosis.

***Estrogenic* (oestrogenic)** Promoting estrus or estrus cycle (period during which reproductive conception can occur).

Euphoriant An agent that produces a feeling of well-being. The term can include addictive, dangerous, and illegal substances (compare *Psychotropic*).

Exanthematic Ameliorates skin eruptions.

Excema An inflammation of the skin, with redness, itching and oozing vesicular lesions that become scaly, crusted, or hardened.

Exophthalmia An abnormal protrusion of the eye from the orbit.

Exostosis A non-cancerous growth on the surface of a bone, usually with a cartilage cap, due to long-term irritation resulting from osteoarthritis, infections, or trauma.

Expectorant Promotes ejection of mucus or exudate from the lungs, bronchi and trachea; clears phlegm from the chest by inducing coughing. (Sometimes equated with *Antitussive*, which see.)

Extract A concentrated preparation of a vegetable or animal drug, often standardized in concentration. Extracts are prepared in three forms: semiliquid or syrupy, pilular or solid, and as dry powder.

Extrasystole A premature contraction of the heart independent of the normal rhythm.

Fats In chemistry, fats are compounds formed from chemicals called fatty acids (see *Lipid*).

Fats, saturated Fatty acids, abundant in red meat, lard, butter, hard cheeses, and some vegetable oils (particularly tropical oils such as palm, coconut, and cocoa butter, and partially hydrogenated oils), in which each molecule has the maximum amount of hydrogen atoms.

Fats, unsaturated Fatty acids in which some of the hydrogen atoms in each molecule have been replaced by double bonds.

Fauces The narrow passage from the mouth to the pharynx, between the soft palate and the base of the tongue. On either side, two membranous folds inclose the tonsils.

Favus A contagious fungal disease of the skin, especially the scalp.

Febrifuge Reduces fever.

Febrile Pertaining to or characterized by fever.

Felon A soft tissue infection of the finger tip.

***Fibromyalgia* (fibrositis)** Pain, stiffness, tenderness of muscles, tendons, and joints, and fatigue, all without detectable inflammation. A disorder of unknown cause affecting about 1% of North Americans.

Fibrositis See *Fibromyalgia*.

Filaricidal Kills species of *Filaria*, a genus of slender, nematode worms, parasitic in the blood of various mammals.

Fistula An abnormal passage, usually between two internal organs or leading from an internal organ to the surface of the body. Fistulas are frequently created experimentally to obtain body secretions for study.

Flatulence Excessive intestinal gas, a concept difficult to quantify. The average person passes about 2 litres of gas each day through burping and farting.

Flavoring Any substance employed to give a particular flavor to a medicine.

Flesh-eating bacteria A strain of group A streptococcus (bacteria) which, in severe cases, can destroy tissues rapidly.

Flu See *Influenza*.

Flux A flow of fluid from the body, especially excessive, abnormal discharge from the bowels.

Folliculitis Inflammation of a follicle or follicles, usually of the hair.

Fomentation 1. Application of hot moist poultice to the body to ease pain. 2. The materials so applied.

Free radical An unstable oxygen molecule that can damage tissue.

Frigidity Lack of sexual desire, especially in a woman; abnormally adverse to sexual activity.

Furuncle A localized inflammatory swelling of the skin and nearby tissues, caused by a bacterial infection in a hair follicle or skin gland. A boil.

Galactofuge Reduces or stops flow of milk in nursing mother.

Galactogogue Increases flow of milk in nursing mother.

***Galactorrhea* (galactorrhoea)** Excessive or spontaneous flow of milk; persistent secretion of milk irrespective of nursing.

Galenical A medicinal preparation, usually prepared as an extract from plant material. The word is a reference to the Greek physician Galen (ca. 130–200), and can be interpreted as referring to non-surgical medicine.

Gangrene Localized death of soft tissues, caused by lack of blood supply.

Gastralgia Gastric colic, pain and cramping in the stomach.

Gastritis Inflammation of the stomach.

Gastroenteritis Acute inflammation of the lining of the stomach and intestines, characterized by anorexia, nausea, diarrhea, abdominal pain, and weakness.

Genotoxic A poisonous substance that harms an organism by damaging its DNA.

Giardiasis An infection of the gastrointestinal tract caused by the flagellated protozoan *Giardia lamblia*. This disease is often called "beaver fever" because beaver droppings containing the protozoan are thought to contaminate streams furnishing water drunk by humans.

Gingivitis A gum disease marked by inflammation.

Glaucoma An eye disease, marked by excessive pressure within the eyeball, damage to the optic disk, and gradual loss of vision.

Gleet A chronic inflammation of a body orifice, usually accompanied by an abnormal discharge; also, the discharge itself. Specifically, a transparent mucous discharge from the membrane of the urethra, commonly an effect of gonorrhea.

Glossitis Inflammation of the tongue.

Glycoside Compound that contains a carbohydrate molecule (sugar), particularly any such natural product in plants, convertible, by hydrolytic cleavage, into sugar and a nonsugar component (aglycone), and named specifically for the sugar contained, as glucoside (glucose), pentoside (pentose), fructoside (fructose).

Glycosuria The excretion of an abnormally large amount of glucose in the urine.

Goiter Enlargement of the thyroid gland (usually visible as a swelling in the anterior portion of the neck). Can be associated with normal, elevated (hyperthyroidism) or decreased (hypothyroidism) thyroid hormone levels in the blood. Both deficiency and excessive intake of iodine can lead to inadequate production of thyroid hormone from the thyroid gland (hypothyroidism). Graves' disease, the most common cause of hyperthyroidism, is the result of too much thyroid hormone. The seaweeds (kelp) discussed in this work have been used in the past to alleviate goiter caused by insufficient availability of iodine.

Gonadotrophic Stimulating the gonads; generally applied to hormones of the anterior pituitary which influence the gonads.

Gout A condition characterized by abnormally elevated levels of uric acid in the blood, recurring attacks of joint inflammation (arthritis), deposits of hard lumps of uric acid in and around the joints, and decreased kidney function and kidney stones.

Gram's stain In 1884 Danish physician Hans Christian Joachim Gram devised a method for distinguishing two classes of bacteria, that remains today extremely important for identifying bacterial species and the diseases they cause. His method consisted of pouring a solution of gentian violet stain over a dried fluid smear of bacteria on a glass slide, heating, washing with water, adding potassium triiodide solution, then rinsing with ethanol. Bacteria that remain purple were "gram-positive" while those that do not retain the color are "gram-negative." A later worker added a final staining with safranine, and today bacteria are judged gram-negative if they retain the red color of safranine but not the first purple color of gentian violet. Gram-positive bacteria have thick cell walls of cross-linked polysaccharide that absorb gentian violet well. Gram-negative bacteria have thin polysaccharide cell walls overlaid by lipid layers that resist staining by gentian violet but that can be stained by safranine. The penetrability of different antibacterial drugs can be very different depending on the wall type. Some drugs are effective only against gram-positive bacteria, others only against gram-negative bacteria, while those that act against both types are called wide-spectrum.

Granulation The process of forming small, red, grain-like prominences on wounds, which bring about healing.

GRAS Abbreviation for "Generally Recognized As Safe," a status given to consumed materials by the American Food and Drug Administration and others.

Gravel 1. A deposit of small calculi (see *Calculus*) in the kidneys and the urinary or gall bladder. 2. The disease of which the stones are a symptom.

Gripe Pinching and spasmodic pain in the bowels. (Not to be confused with grippe, which see.)

Grippe An acute febrile virus disease, resembling or identical with influenza.

Guinea worm The nematode *Dracunculus medinensis*; it grows to several feet in length, and occurs in subcutaneous tissues of man in warm countries.

Gynecomastia Excessive development of the breast in the male.

Hallucinogenic Causes hallucinations: sense perception in the absence of an external stimuluting object.

Heartburn An uncomfortable feeling of burning and warmth occurring in waves rising up behind the breastbone toward the neck, usually due to the return of stomach acid back up into the esophagus (see *Pyrosis*). Heartburn has nothing to do with the heart.

Hematochezia **(haematochezia)** The passage of bright red blood from the rectum.

Hematoma **(haematoma)** A localized swelling (such as within an organ or a soft tissue space) filled with blood that is usually clotted or partially clotted.

Hematuria **(haematuria)** Blood in the urine.

Hemicrania A pain affecting only one side of the head.

Hemiplegia Paralysis of one side of the body.

Hemoglobin **(haemoglobin)** The oxygen-carrying pigment of red blood cells.

Hemolytic **(haemolytic)** Destroys red blood cells, often by disruption of the integrity of the cell membrane, causing release of haemoglobin.

Hemopoiesis **(haemopoiesis)** Formation and development of blood cells from stem cells, usually in bone marrow.

Hemoptysis **(haemoptysis)** Coughing up of blood or of blood-stained sputum.

Hemorrhoids **(haemorrhoids, piles)** Small, troublesome tumors or swellings associated with varicosely dilated veins about the anus and lower part of the rectum. Generally results from persistent increase in venous pressure.

Hemostatic **(haemostatic)** Stops bleeding.

Hepatic 1. Referring to the liver and its function. 2. Improves the liver and aids in ridding the body of toxins.

Hepatomegaly Enlarged liver.

Hepatosplenomegaly Enlarged liver and spleen.

Hepatotoxic Toxic to liver cells.

Herbalist A person with expertise in herbs; in regard to therapy, a herb doctor. The therapeutic use of herbs can fall within the scope of a number of alternative modalities (e.g., depending on jurisdiction, Oriental medicine, acupuncturism, naturopathy, homeopathy).

Herpes Inflammatory skin disease caused by a herpes virus and characterised by the formation of clusters of small vesicles. The word may refer to herpes simplex or herpes zoster.

Herpes simplex A virus responsible for several different infections in humans: gingivostomatitis (inflammation of mouth and gums; in children), pharyngitis, oral and lip lesions (recurrent Herpes simplex type 1), proctitis (inflammation of the rectum) (type 2), and genital herpes (type 2). Type 1 herpes is commonly said to occur mostly above the waist, while type 2 is usually sexually transmitted and the symptoms are mostly below the waist.

Herpes zoster The virus responsible for chicken pox. In the disease called shingles, the virus is reactivated, resulting in a painful blistery red rash confined to one side of the body. Facial rash can lead to optic nerve involvement and blindness.

Hirudicidal Kills leeches.

Histamine A substance that plays a major role in many allergic reactions. It dilates blood vessels and makes their walls abnormally permeable (see *Allergy, Antihistamine*).

Hives **(urticaria)** Raised, itching areas of skin, often indicative of an allergic reaction. Also called "welts" or "nettle rash" (more information at *Urticaria*).

Hodgkin's disease A cancer of lymph tissue (lymphoma) that appears to originate in a particular lymph node and later spreads to the spleen, liver and bone marrow. It occurs mostly in persons 15 - 35 years of age, and has a high remission rate when detected early.

Homeopathic Referring to homeopathy, a system of medicine founded by the German physician Samuel Hahnemann (1755–1843) about 2 centuries ago. It is based on providing usually minute doses of alleged remedies which produce in a healthy person effects similar to the symptoms of the complaint of the patient. Substances employed may be of animal, vegetable or mineral origin. In some jurisdictions of the world, homeopaths are empowered to employ as adjunctive therapies a variety of other treatments, such as acupuncture, neuromuscular integration, nutrition, pharmaceutical medicine, and minor surgery. (For the viewpoint that classical homeopathy is without scientific merit, see "Homeopathy: a position statement by the National Council Against Health Fraud (from Skeptic vol. 3, no. 1, 1994, pp. 50–57.): http://www.skeptic.com/03.1.jarvis-homeo.html) (compare *Allopathy*).

Hormone A chemical released from a gland into the bloodstream and affecting organs or tissues elsewhere in the body.

Hydrocele Accumulation of serous fluid in a sac-like cavity (most commonly in the scrotum).

***Hydrocephaly* (hydrocephalus)** Abnormal increase in the volume of cerebrospinal fluid in the brain, that can increase pressure, enlarge the skull and atrophy the brain.

Hydrogogue Causes body to lose water.

Hydrogogue cathartic A cathartic that causes abundant watery evacuation.

Hydrophobia 1. Rabies. 2. A morbid fear of water.

Hydrotic Causing a discharge of water or phlegm.

Hyperaldosteronism Overproduction of aldosterone, the hormone produced by the adrenal glands that causes the kidneys to retain sodium and water; can lead to hypertension.

Hypercholesteremia = Hypercholesterolemia (which see).

***Hypercholesterolemia* (hypercholesterolaemia)** Presence of abnormally high concentrations of cholesterol in the bloodstream; can lead to heart disease, hardening of the arteries, heart attacks, and strokes.

Hyperemesis Excessive vomiting.

***Hyperemia* (hyperaemia)** Excessive blood in a body part; engorgement (cf. *Plethora, Anemia*).

Hyperemic An agent used to reduce vomiting. [Note possible confusion with above definition of hyperaemia.]

***Hyperglycemia* (hyperglycaemia)** Abnormally increased content of glucose in the blood.

Hyperhidrosis (hyperidrosis, polyhidrosis, polyidrosis) Excessive perspiration.

Hyperkinesia Abnormally increased motor function or activity; hyperactivity.

Hyperlipidemia High levels of blood lipids.

Hyperpigmentation Abnormal darkening of the skin.

Hyperplasia Abnormal multiplication or increase in the number of normal cells in normal arrangement in a tissue.

Hyperpyrexia Exceptionally high fever.

Hypersensitivity State of altered reactivity, the body responding with an exaggerated immune response to a foreign substance.

Hypertension Persistently high arterial blood pressure; may be associated with other primary diseases, or may have no known cause (essential or idiopathic hypertension). Constitutes a major risk factor for stroke, because it puts excess stress on the walls of blood vessels and damages their delicate inner lining.

Hyperthyroidism Excessive functional activity of the thyroid gland, or the resulting abnormal condition marked by increased metabolic rate, enlargement of the thyroid gland, rapid heart rate, high blood pressure, and other symptoms (compare *Hypothyroidism*).

***Hypertonia* (hypertony)** Pathological rigidity, tension and spasticity of the muscles.

Hypnotic Induces sleep (cf. *Soporific*).

***Hypoglycemia* (hypoglycaemia)** Abnormally diminished concentration of glucose in the blood.

Hypoglycemic Reduces sugar level in the blood by increasing the production of insulin in the pancreas.

Hypotension Abnormally low blood pressure.

Hypotensive 1. Characterized by or causing diminished tension or pressure, as abnormally low blood pressure. 2. An agent that reduces high blood pressure.

Hypothyroidism Deficient activity of thyroid gland; also, the resulting syndrome characterized by lowered metabolic rate and general loss of vigor (compare *Hyperthyroidism*).

Hypoxia Inadequate supply of oxygen to tissues.

Iatrogenic Induced by a physician; particularly refers to ailments unintentionally induced in a patient, susceptible to suggestion, by a doctor's words or behavior. The term is now applied to any adverse condition in a patient resulting from treatment by a physician or surgeon, especially to infections acquired during treatment.

Ichthyosis A group of mostly genetically determined skin disorders characterised by noninflammatory scaling of the skin; these may be known be such terms as alligator or fish skin.

Icterus Jaundice in the eye's sclera (tough white outer coat of the eyeball).

Idiopathic Without a known cause.

Ileus An obstruction of the intestines. Specifically, a condition characterized by painful distended abdomen, vomiting, toxemia and dehydration, and due to failure of peristalsis.

Immunogen A substance capable of inducing an immune response.

Immunostimulant Improving the action of the immune system.

Immunosuppressant Suppressing or lowering the body's immune response.

Impetigo An acute, contagious, bacterial skin disease marked by vesicles, pustules and yellowish crusts. It is most commonly seen in children, usually on the face, especially about the nose and mouth.

Imposthume A collection of pus in any part of the body; an abscess (see *Aposteme*).

Impotence Inability to have or maintain an erection.

Indolent Causing little or no pain or annoyance (e.g., an indolent tumor).

Induration An abnormally hard area, particularly of the skin.

Inflammation Localized redness, warmth, swelling

and pain as a result of infection, irritation or injury.

***Influenza* (flu)** An acute highly contagious disease, marked by rapid onset, fever, weakness, severe aches and pains, and inflammation of the respiratory mucous membranes. The flu is caused by viruses that infect the respiratory tract, and are divided into types A, B, and C.

Infusion An extract prepared by soaking or steeping, usually in water; sometimes called a tea (compare *Decoction*). An infusion is usually made by pouring boiling water over the materials/herbs and letting it steep, and is usually stronger than a tea. It can also be made by adding concentrated extracts to water.

Insomnia Inability to sleep.

Insulin A hormone made by the islet cells of the pancreas, which controls the level of glucose sugar in the blood. A lack of or resistance to insulin causes diabetes mellitus.

Irritant Causes inflammation, irritation or sensitivity following contact with skin or mucous membranes.

Ischemia A hazardous decrease in the supply of blood to tissue, generally caused by atherosclerotic narrowing of the vessel.

Jaundice A disease (frequently because of a liver problem) marked by yellowish pigmentation of the skin (and whites of the eyes), tissues, and body fluids resulting from deposition of bile pigments, especially bilirubin.

Keratitis Inflammation of the cornea.

Keratoconjunctivitis Inflammation of the cornea and conjunctiva.

Kino 1. Any of several dark red to black tannin-containing dried juices or extracts from various tropical trees (e.g., from the East Indian tree *Pterocarpus marsupium Roxb.*, which is used as an astringent in diarrhea). 2. A tree species that produces kino, especially *Pterocarpus marsupium*.

Lachrymal gland A gland that produces tears, located in the upper, outer section of the eye's orbit (the bony socket that surrounds the eyeball).

***Lachrymitis* (lacrimitis)** Inflammation of the tear ducts.

Lacrimation Secretion of tears, especially excessively.

Lacteal Enhances lactation (see *Galactogogue*). (A different meaning anatomically: lacteals are specialized lymphatic vessels in the small intestinal mucosa.)

Larvicidal Kills parasitic larvae.

Laxative Promotes evacuation of the bowel. (The terms purgative, cathartic, and laxative are often used to indicate progressively gentler action. Laxatives usually cause a more or less normal evacuation of the bowel, usually without griping or irritation.)

Leishmaniasis An infectious disease caused by protozoan parasites of the genus *Leishmania*.

***Leucorrhea* (leucorrhoea)** A discharge of mucus, resulting from inflammation or irritation of the membrane lining the vagina.

Lenitive 1. Eases pain or protects against irritants. 2. Soothes; allays passion, excitement, or pain; a palliative. 3. A laxative.

Libido Sexual desire.

Lipid Any of a heterogeneous group of fats and fatlike substances characterized by being water-insoluble, extractable by nonpolar (or fat) solvents such as alcohol, ether, chloroform, benzene, etc. All contain aliphatic hydrocarbons as a major constituent. They include fatty acids, neutral fats, waxes, and steroids. Compound lipids include the glycolipids, lipoproteins, and phospholipids. Lipids are easily stored in the body and serve many functions, e.g., they are a source of fuel, and an important constituent of cell structure.

Lithiasis Formation of stony concretions or calculi in any part of the body, especially in the bladder and urinary passages.

Lithontriptic Dissolving or destroying stones in the bladder or kidneys.

Lumbago A usually painful muscular rheumatism in the lumbar region.

Lumbar The loins [part of back between thorax (area between neck and abdomen) and pelvis]; the vertebrae between the thoracic vertebrae and sacrum (five united vertebrae at the base of the spine, in the pelvic region).

Lupus Any of several diseases characterized by skin lesions. A systemic disease resulting from an autoimmune problem (individuals with lupus will produce antibodies to their own body tissues; the resultant inflammation can cause kidney damage, arthritis, and heart problems).

Lupus erythematosus Skin disease in which there are red scaly patches, especially over the nose and cheeks.

Lymph An almost colorless fluid in the lymphatic system which carries cells that help fight infection and disease.

***Lymph nodes* (lymph glands)** Small, bean-shaped organs located along the channels of the lymphatic system.

Lymphadenitis Inflammation or swelling of lymph nodes.

Lymphangitis Inflammation of lymph nodes and (or) vessels.

Lymphatic system The channels that carry lymph, and the tissues and organs (the bone marrow, spleen, thymus, and lymph nodes) that produce and store cells that fight infection and disease.

Malignant Cancerous.

Mange Persistent contagious skin diseases of man and other animals, marked by eczematous inflammation and loss of hair, especially such a disease caused by a minute parasitic mite.

Marasmus A progressive emaciation, especially in the young, usually associated with faulty assimilation and utilization of food.

Mastitis Inflammation of the breast (or udder of cattle), usually caused by infection.

Melanoma A tumor (malignant unless otherwise qualified) arising from the melanocytic system of the skin and other organs.
Menagogue Promotes menstruation.
Meningitis Inflammation of the meninges, the membranes surrounding the brain and spinal cord.
Menorrhagia Excessive uterine bleeding occurring at the regular intervals of menstruation.
Menometorrhagia Excessive uterine bleeding both at the usual time of menstrual periods and at other irregular intervals.
Metastasis Transfer of a disease-producing agent (infective agent or cells capable of infecting other cells) from the site of disease to another site in the body; specifically, growth of a tumor established by transfer from another site. The ability to metastasize is a characteristic of malignant tumors.
***Meteorism* (tympanites)** The presence of gas in the abdomen or intestine.
Metritis Inflammation of the womb, may be associated with purulent vaginal discharge.
Metrorrhagia Non-menstrual uterine bleeding.
Miasma Vaporous exhalation, formerly believed to cause disease.
Micturition Urination.
Migraine A condition marked by recurrent severe headache, often with nausea, irritability, constipation or diarrhea, and vomiting. Attacks are preceded by constriction of the cranial arteries, usually with sensory (especially visual) symptoms.
Miotic In ophthalmology, pertains to miosis, the contraction of the pupil. In pharmacology, a miotic is an agent (typically an eye-drop) that causes the pupil to constrict. Should not be confused with meiotic, pertaining to cell division. Miotics can be used to treat glaucoma.
Mitogen A substance that can induce mitosis (cell division) of certain cells. This might be desirable, for example, to stimulate the production of lymphocytes in the blood.
Mitogenic Acting as a mitogen (which see).
Molluscicidal Having the properties of a chemical (or pesticide) used to kill molluscs or mussels.
Mucolytic Destroying mucus.
Mucositis Inflammation of a mucous membrane.
Mutagenic Induces genetic mutation.
Myalgia Muscle pain.
Myasthenia gravis A neurological disorder, characterized by easy fatigue of certain voluntary muscle groups on repeated use, especially of the face and upper trunk.
Mydriatic Dilates the pupil of the eye.
Myelitis 1. An inflammation or infection of the spinal cord. 2 (less common than 1) An infection of the bone marrow.
Myocarditis Inflammation of the myocardium (the muscular walls of the heart).
Myosis Prolonged contraction of the pupil of the eye.
***Myxedema* (myxoedema)** Severe hypothyroidism. Symptoms may include dry skin, puffiness, coarse hair, intolerance to cold, and learning impairment.

Narcolepsy A common disorder of sleep associated with excessive daytime sleepiness, involuntary daytime sleep episodes, and disturbed nocturnal sleep.
Narcotic Produces drowsiness, insensibility or stupor, and a sense of well-being; applied especially to the opioids. The word is often used pejoratively, and narcotics have often been defined simply as being certain materials that are listed in legislation as illegal (irrespective of their properties).
***Naturopathy* (natureopathy)** A term first used at the turn of the century, and variously defined, for examples: 1. A system of treatment of diseases emphasizing assistance to nature and sometimes including the use of medicinal substances (such as herbs) and physical means (as manipulation). 2. a distinct system of healing - a philosophy, science, art and practice that seeks to promote health by stimulating and supporting the body's inherent power to regain harmony and balance. 3. A drugless system of therapy based on the use of physical forces such as heat, water, light, air and massage. Naturopathy is often interpreted relatively comprehensively. The emphasis is usually on the use of "natural" forms of health care treatment, for example "naturally occurring substances" or "natural medicines," and may include nutritional supplements, counselling, and education. Sometimes the scope of practice for naturopathy includes such alternative modalities as acupuncture, biofeedback, homeopathy, hypnotherapy or massage. Obviously most if not all of the activities listed above are practised in conventional medicine. In some jurisdictions of the world, naturopaths are permitted to perform complex medical activities, such as minor surgery, while in other jurisdictions naturopathy is illegal.
Necrosis Death, usually localized, of tissues.
Nephritic A medicine adapted to relieve or cure diseases of the kidney.
Nephritis Kidney inflammation or infection.
Nephrolithiasis The presence of calculi (stones) in the kidney or urinary collecting system. The stones are usually 2–12 mm in diameter, and composed of calcium, phosphate or uric acid.
Nephrosis A degeneration of the kidneys, principally affecting the renal tubules.
Nervine Acts upon the nerves, calming nervous excitement, relieving anxiety, and preventing over-anxiety.
Neuralgia Pain that extends along the course of one or more nerves. Varieties of neuralgia are distinguished according to the part affected (e.g., facial) or cause (e.g., anemic, diabetic, gouty, malarial).
Neurasthenia A condition of nervous debility thought to result from impairment of the spinal cord.
Neuritis Inflammation or degeneration of nerves, usually with pain.

Neuroleptic A tranquillizer or ataractic (which see).
Nutritive Nourishing, nutritious.
Nyctalopia Night blindness or difficulty in seeing at night; may result from deficiency of vitamin A.

Odontalgia Toothache.
***Odontaligic* (odontalgic)** A toothache remedy (also see *Antiodontalgic*).
Oedema See *Edema*.
Ointment A salve for application to the skin. Because the word ointment comes from the Latin *ungere*, meaning anoint with oil, some have restricted the term to oil-based salves, while contrasting "cream," which is said to be water-soluble. However, any medicinal preparation that has the consistency of butter, petroleum jelly, lip balm, etc. can be called an ointment (= unguent).
Oliguria Diminished urination in relation to the fluid intake.
Oncogenic Cancer-causing.
Opisthotonus A spasm, with the head and heels bent backward and the body bowed forward.
Opthalmalgia Eye pain.
Opthalmia An inflammation of the conjunctiva of the eyeball.
Opthalmic 1. Of or relating to the eye. 2. For use on or in the eye.
Orchitis Inflammation of a testis, evidenced by swelling and tenderness.
Orexigenic Appetite-stimulating.
Osteoporosis A reduction in the amount of bone mass, with susceptibility to fractures.
Otitis Inflammation of the ear.
Oxytocic Speeds up parturition (delivery of a baby from the uterus).
***Ozena* (ozaena, ozoena)** A chronic nose disease, with atrophy of the nasal structures and a fetid discharge.

Palsy 1. Paralysis. 2. Uncontrollable tremor of the body or a part of the body.
Panaritium 1. Paronychia (which see). 2. Foot rot.
Pancytopenia Deficiency of all cell elements of the blood; aplastic anemia.
Parasiticide Kills parasites.
Parasympathetic nervous system A part of the nervous system that serves to slow the heart rate, increase intestinal and gland activity, and relax the sphincter muscles. The autonomic nervous system includes this and the sympathetic nervous system, which accelerates the heart rate, constricts blood vessels, and raises blood pressure.
Parenteral Referring to introducing material into the body by a non-oral route (e.g., by injection).
Paresthesia A feeling of prickling, tingling, or creeping on the skin without an evident external agent.
Parkinsonism Any of several neurological disorders marked by slow, trembling, or rigid movements.
Parkinson's disease A progressive neurological disease described first in 1817 by James Parkinson. Symptoms include shuffling gait, stooped posture, tremor, speech impediments, difficulties of movement, mental deterioration and dementia.
Paronchia Inflammation of tissues beside the nail of a finger or toe (cf. *Felon*).
Parotitis Inflammation of the parotid gland, a salivary gland in front of and below the ear.
Paroxysm A convulsion.
Parturifacient A substance causing parturition.
Parturition Giving birth to a child.
Pathogen A micro-organism, (e.g., bacterium, virus) that has the ability to cause disease in a host organism.
Pathogenic Possessing the ability to cause illness.
Pectoral 1. Pertaining to the breast or chest. 2. Good for diseases of the chest or lungs, especially respiratory diseases.
Pediculoside An agent that destroys lice.
Pediculosis Infestation with lice.
Pellagra A niacin deficiency disease caused by improper diet and characterised by skin lesions, gastrointestinal disturbances and nervousness.
Pemphigus Several chronic, relapsing, sometimes fatal skin diseases characterised by successive development of vesicles and bullae.
Pericarditis Inflammation of the heart.
Peridontitis See *Pyorrhea*.
Periosteum The membrane of fibrous connective tissue covering all bones except at their articular (joining) surfaces.
Periostitis Inflammation of the periosteum (which see).
Peristalsis A rippling motion of muscles in the digestive tract. In the stomach this mixes food with gastric juices.
Peritonitis Inflammation of the membrane lining the abdominal cavity.
Pertussis Whooping cough.
Pharmacology The study of drugs, their sources, nature, and properties.
***Pharmacopeia* (pharmacopoeia)** An official authoritative listing of drugs. Various countries have their own pharmacopeia.
Pharyngitis Inflammation of the pharynx, the part of the throat between the cavity of the mouth and the esophagus. Loosely, a sore throat.
Phlebitis Inflammation of a vein, marked by the formation of a thrombus (blood clot), and characterized by edema, stiffness and pain in the affected part.
Phosphaturia Excessive discharge of phosphates in the urine. This may be associated with cloudy urine and kidney stones.
Photophobia Abnormal visual intolerance of light.
Photosensitization Development of abnormally heightened reactivity of the skin to sunlight.
Phrenitis Inflammation of the brain, or of the meninges of the brain, accompanied by acute fever and delirium.

Phthiriasis **(morbus pediculous)** An infestation of lice on the human body.

Phthisis Wasting or consumption of tissues. A term formerly applied to many wasting diseases, but now usually restricted to pulmonary phthisis, or consumption (which see).

Phyma **(plural phymas or phymata)** An external nodule or swelling; a skin tumor.

Piles See *Hemorrhoids*.

Pinworm **(threadworm)** A nematode, *Oxyurus vermicularis*, parasitic chiefly in the intestines and rectum of man. The most frequent temperate-region parasitic "worm" of man, it is most common in children and the aged.

Piscicide Kills fish.

Pityriasis A group of skin diseases characterised by the formation of fine, branny scales.

Placebo effect Improvement resulting from having been treated that cannot be attributed to the treatment used.

Plaque A fatty buildup of cholesterol, calcium, and other substances inside a blood vessel. Also, a buildup of material on teeth.

Plethora Overfullness, especially of the blood vessels; hyperemia (opposed to anemia).

Pleura The thin covering that protects and cushions the lungs.

Pleurisy Inflammation of the pleural membrane that envelops the lungs, usually accompanied by fever, painful and labored respiration, and cough.

Pleuritis Inflammation of the pleura.

Pleurodynia Pain of the side, simulating pleurisy, usually due to rheumatism.

Pleuropneumonia Combined inflammation of the pleura and lungs.

Pneumonia Inflammation of the lungs with consolidation (i.e., conversion of lung tissue to unaerated, solid state).

Pneumonitis Inflammation of the lungs (secondary to viral or bacterial infection).

Polydipsia Chronic excessive thirst, as in diabetes.

Polyuria Passage of considerable urine in a given period (a characteristic of diabetes).

Porphyria A partly genetically determined pathological condition resulting from abnormal porphyrin metabolism, characterized by excretion of excess porphyrins in the urine and extreme sensitivity to light. Porphyrins are pigments found in animals and plants, and include such important molecules as haemoglobin, chlorophyll, and cytochromes.

Priapism Persistent abnormal erection of the penis, usually without sexual desire and accompanied by pain and tenderness. This may result from diseases and injuries of the spinal cord or injuries to the penis.

Proctitis **(rectitis)** Inflammation of the rectum.

Proliferant A cell proliferant causes cell division; this can promote healing.

Prophylactic 1. Preventing disease. 2 A device for preventing venereal infection; a contraceptive.

Prostaglandins A group of compounds (originally purified from prostate tissue, hence the name) that are ubiquitous in tissues and play a variety of important roles in regulating cellular activities, especially the inflammatory response, where they can dilate (or constrict) veins and (in the bronchi) relieve breathing difficulty.

Prostate A walnut-sized gland at the base of the male bladder, which produces a fluid that forms part of semen.

Prostatism A syndrome of symptoms associated with *Benign prostatic hyperplasia* (which see), including feeling the need to urinate right away yet having to strain to do so, having a weak urinary stream, dribbling after urinating, feeling as though the bladder has not been emptied completely, needing to urinate frequently, and (or) experiencing urinary incontinence.

Prostatitis A bacterial infection of the prostate, that causes an inflammation and may result in painful or difficult urination.

Protein Any of a group of high molecular weight, complex, organic compounds that contain carbon, hydrogen, oxygen, nitrogen, and usually sulphur. These are the principal constituents of the protoplasm of all cells, consisting mostly of combinations of 20 different linked amino acids. Proteins serve as enzymes, structural elements, hormones, etc.

Proteolytic Hydrolyses proteins or peptides, producing simpler soluble products (and so can aid digestion or, depending on circumstances, harm tissues).

Protisticidal Kills protists. (Protists are usually interpreted as members of Kingdom Protista, including bacteria, protozoans, various algae, fungi, and sometimes viruses.)

Prurigo A skin disease marked by intense itching of papules (superficial solid elevations of the skin) that scarcely differ from the healthy skin in color.

Pruritus Itching. Possible cause includes drug reaction, food allergy, kidney or liver disease, cancers, parasites, aging or dry skin, contact skin reaction such as poison ivy.

Psoriasis A chronic skin disease characterized by red patches covered by white scales.

Psychedelic **(psychodelic)** Relating to psychedelic drugs, i.e., those causing abnormal mental and sensory effects.

Psychotropic Affecting the mind or mood, capable of modifying mental activity, usually applied to drugs that effect the mental state.

Ptosis Falling down of an organ or part; specifically, a drooping of the eyelid attributed to flaccid muscles.

Puerperal Pertaining to childbirth.

Puerperium The condition of a woman immediately following childbirth.

Pulicide An agent that destroys fleas.

Purgative A drastic laxative (which see).

Purpura Hemorrhagic state characterized by blood in the skin and (or) mucous membranes with appearance of patches of purplish discoloration. May refer specifically to a small hemorrhage (up to about 1 cm in diameter), or to a group of hemorrhagic diseases characterised by the presence of purpuric lesions, ecchymoses (see *Ecchymosis*) and a tendency to bruise easily.

Pyelitis Inflammation of pelvic portion of kidney (located between inner urine-secreting surface and the ureter draining into the bladder); symptoms may include pain and tenderness in the loins, irritability of the bladder, fever, bloody or purulent urine, diarrhea, and vomiting.

Pyoderma A bacterial skin inflammation with pussy lesions.

Pyorrhea Pussy inflammation of the sockets of the teeth usually leading to loosening of the teeth.

Pyrogenic Inducing fever.

Pyrosis Heartburn (which see); a burning sensation in the epigastric region of the abdomen.

Quincy An inflammation of the throat or adjacent region, especially of the fauces or tonsils, with considerable swelling, painful and impeded ability to swallow, inflammatory fever, and sometimes a danger of suffocation. (Synonyms: squinancy, squinzey.)

Rectitis Proctitis (which see).

Rectocele A bulging of the rectum into the vagina.

Refrigerant A medicine that cools the body, allays fever.

Relaxant An agent that lessens tension.

Repellent A remedy to repel from a tumefied part the fluids which render it tumid (swollen, inflated).

Repercussive Reduces swelling or tumor.

Resolvent Disperses inflammatory or other tumors, or dead or dying tissues.

Restorative Restores normal function.

Rheumatic Relating to rheumatism, any of several diseases characterized by inflammation and pain in muscles or joints.

Rheumatic fever A disease involving inflammation of joints and damage to heart valves that follows infection by *Streptococcus* (which see).

Rheumatoid arthritis A chronic inflammatory disease marked by destruction of joints.

Rickets A childhood bone formation disease resulting from inadequate calcium and phosphorus due to insufficient sunlight or vitamin D.

Rinderpest An acute infectious viral disease of cattle marked by diphtheritic inflammation of mucous membranes.

Ringworm A fungal skin infection, typically a scaly, ring-shaped reddening on the skin, and commonly seen in children. May be treated with an antifungal cream such as clotrimazole or miconazole (also see *Tinea*).

Roborant A strengthening medicine or tonic.

Rubefacient An external stimulant which when applied to the skin produces mild irritation accompanied by reddening. This may be desired to counter irritation (see *Counterirritant*).

Saint Anthony's fire Gangrenous inflammations of the skin, such as ergotism (which see; compare *Erysipelas*).

Saint Vitus's dance Chorea (which see).

Salicylism Toxic effects of excessive consumption of salicylic acid or its salts, usually marked by tinnitus, nausea and vomiting.

Salt rheum Eczema (which see).

Saluretic Promoting saluresis, the excretion of sodium and chloride ions in the urine.

Satyriasis Excessive craving for sex in males.

Scabies Itch or mange (which see), especially with exudative crusts.

Scarlatina Scarlet fever (a severe illness characterised by a reddish skin rash, caused by the bacterium *Streptococcus pyogenes*).

Schistosomiasis A disease caused by schistosomes (elongated trematode worms parasitizing blood vessels); this causes severe blood loss and tissue damage, and is endemic in much of Asia, Africa and South America.

Sciatica Pain along a sciatic nerve (the two largest nerves of the body, from the buttock, down the back and side of the leg, and into the foot and toes), often due to a herniated disk.

***Scirrhus* (plural: scirrhi)** An indurated (hardened) organ or part, especially, an indurated gland.

Scleroma Indurated (hardened) tissues.

***Sclerosis* (plural: scleroses)** An induration or hardening, especially from inflammation and disease. The term is often used for hardening of the nervous system and blood vessels.

Scrofula Tuberculosis of lymph glands, especially of neck.

Scurf Thin dry scales or scabs, especially scales exfoliated from the cuticle, particularly of the scalp; dandruff.

Scurvy A disease caused by lack of vitamin C (ascorbic acid), with development of weakness, spongy gums, loosening of teeth, and bleeding under skin.

***Seborrhea* (seborrhoea, hypersteatosis)** Abnormally large secretion and discharge of sebum, the fatty lubricant secreted by the skin's sebaceous glands.

Secretagogue A substance that induces secretion from cells (originally applied to peptides that induce gastric and pancreatic secretion).

Sedative Tranquillises and usually promotes sleep.

Sepsis An infection deep into the body. Septicemia (which see) can be considered a form of sepsis.

***Septicemia* (septicaemia, "blood poisoning")** Systemic disease associated with pathogenic microorganisms or their toxins in the blood. This may be serious, depending on the bacteria, since toxins are quickly distributed throughout the body. A possible outcome is shock.

Shigella A flagellated bacterium of the Eschericiae group, responsible for dysentery in humans, and commonly acquired by drinking contaminated water or food, or through contact with an infected individual.
Shigellosis An intestinal infection by *Shigella* (which see), marked by diarrhea, fever, and abdominal pain.
Shingles See *Herpes zoster.*
Sialagogue Stimulates secretion of saliva.
Sinapism A plaster or poultice composed mainly of powdered mustard seed or the volatile oil of mustard seed. The term is applied sometimes to preparations with herbs other than mustard.
Sinusitis Inflammation of the lining membrane of any of the hollow areas (sinuses) of the skull around the nose.
Solvent A fluid capable of dissolving substances.
Soporific Induces sleep (cf. *Hypnotic*).
Spasm 1. A sudden, violent, involuntary contraction of muscles, with pain and interference of function. 2. A sudden but transitory constriction of a passage, canal or orifice.
Spasmodic Having the nature of spasms.
Spasmolytic Checks spasms; antispasmodic.
Spermatorrhea **(spermatorrhoea)** Abnormally frequent involuntary emission of semen without copulation.
Splenalgia Pain over the region of the spleen.
Splenitis Inflammation of the spleen.
Splenomegaly Enlargement of the spleen (generally the result of splenitis).
Spondylitis Inflammation of the vertebrae.
Sprue An inherited chronic disease characterized by fatty diarrhea and deficiency symptoms. The intestinal lining is inflamed in response to eating the protein gluten, which occurs in many grains including rye, oats, barley and triticale. The disease is treated by avoiding gluten.
Steatorrhea **(steatorrhoea)** Production of fecal matter that is frothy, foul-smelling and floats because of a high fat content, typical of malabsorption syndromes.
Sternutatory Promotes sneezing and nasal discharge.
Steroid A group name for lipids that contain a hydrogenated cyclopentanoperhydrophenanthrene ring system. Examples: progesterone, adrenocortical hormones, the gonadal hormones, cardiac aglycones, bile acids, sterols (such as cholesterol), toad poisons, saponins, prednisone (a drug used to relieve swelling and inflammation), and vitamin D.
Stimulant An agent that increases functional activity. Stimulates, especially by causing tension on muscle fiber through the nervous tissue.
Stomachic Strengthens and improves stomach action. Stimulates appetite and gastric secretion.
Stomatitis Any of many inflammatory diseases of the mouth.
Stitch A local, sharp, sudden pain, especially in the side.
Strabismus **(squint)** Inability of one eye to focus with the other because of imbalance of the muscles of the eyeball.
Strangury Painful discharge of urine, drop by drop, produced by spasmodic muscular contraction.
Streptococcus A genus of bacteria responsible for numerous infections, including scarlet fever, tonsillitis, erysipelas (which see), rheumatic fever (which see), pneumonia (which see), bronchopneumonia, meningitis, pharyngitis (which see), and wound infections (also see *Flesh-eating bacteria*).
Struma An archaic term, that may refer to scrofula (which see) or (more probably) goiter (which see).
Stye An abscess in the follicle of an eyelash.
Styptic An externally applied astringent (which see) that contracts tissue and blood vessels, and thereby stops bleeding.
Sudorific Causes sweat (compare *Diaphoretic*).
Suppurative Promoting suppuration, i.e., the discharge of pus.
Sycosis A pustular eruption on the scalp or bearded part of the face; can be due to ringworm, acne, or impetigo.
Sympathetic nervous system A part of the nervous system that accelerates the heart rate, constricts blood vessels, and raises blood pressure (see *Autonomic nervous system*).
Sympathomimetic **(adrenergic)** An agent that produces effects similar to those of impulses conveyed by adrenergic postganglionic fibers of the sympathetic nervous system (which see).
Syncope A faint or swoon.

Tachycardia See *Arrhythmia.*
Taenia A genus of intestinal worms which includes the common tapeworms of man.
Taeniacide Kills or removes tapeworms (see *Taenia*).
Taenifuge Repulses or removes tapeworms (compare *Taenicide*).
Tendon A tough cord or band of dense white fibrous connective tissue joining a muscle with some other part and transmitting the force of the muscle.
Tendonitis **(tendinitis)** Inflammation of tendons and of tendon muscle attachments.
Tenesmus Straining, especially ineffective and painful straining in attempting a bowel movement or urination.
Tenia See *Taenia.*
Tenifuge See *Taenifuge.*
Teratogenic Tending to produce anomalies of formation or development; tending to produce birth defects.
Tetanus 1. A spastic paralysis induced by neurotoxin from the anaerobic spore-forming bacterium *Clostridium tetani*, which generally enters the body through a wound. The muscles of the jaw are frequently involved (hence the condition is often

called "lockjaw"). 2. A state of sustained muscular contraction (also called tonic spasm and tetany).

Tetter A vesicular disease of the skin, specifically herpes (which see) or psoriasis (which see).

Thrombogenic Causing thrombosis with clotting or coagulation of the blood.

Thrombolytic Agent used to dissolve blood clots.

Thrombosis Formation or presence of a thrombus (which see).

Thrombus A clot of blood, formed within a blood vessel, that remains at one place.

***Thrush* (aphtha)** A fungal disease, especially in infants, marked by white patches in the mouth (also see *Candidiasis*).

Thymoleptic Mood stabilizing.

Thymus A lymphoid organ located in the upper chest, and decreasing with adulthood. T. lymphocytes are matured and multiply in the thymus. T. lymphocytes are important in immune reactions and in the body's defence against cancer; they also aid B lymphocytes in making antibodies, and helping in the recognition and rejection of foreign tissues.

***Thyrotoxicosis* (exophthalmic goiter)** Pathological thyroid over-function.

Tincture A solution of a medicinal material and a carrier or extractive liquid such as alcohol, water, glycerin, or vinegar.

***Tinea* (plural: tineas)** Any of various skin diseases, but especially ringworm (which see).

Tinea capitis A fungal scalp infection. Symptoms include crusting and scaly lesion of the scalp and possibly localised hair loss.

Tisane A herbal tea, usually not as strong as an infusion, often made with flowers.

Tocolytic agent A medication that can slow down or halt the contractions of the uterus and thereby delay delivery of a baby.

Tonic An agent that stimulates the restoration of physical or mental tone. Improves or promotes general health in a slow, continuous manner.

Topical Applied to a certain area of the skin and affecting only that area.

***Toxemia* (toxaemia)** An abnormal condition with toxic substances in the blood.

Trachoma A chronic infectious disease of the conjunctiva and cornea, producing photophobia, pain, lacrimation and blindness. The disease is caused by *Chlamydia trachomatis*, a microorganism which resembles both bacteria and viruses. The disease is present in some underdeveloped countries of the world, and causes at least 15% of the world's blindness. It spreads by contact with eye discharge from an infected person (e.g., on towels or handkerchiefs) and through transmission by eye-seeking flies.

Trichomoniasis Infection with or disease caused by trichomonads (parasitic protozoans); specifically, vaginitis with a persistent discharge. An infection is usually obtained from semen at intercourse, but the organism can survive 24 hours in tap water, in hot tubs, in urine, on toilet seats and in swimming pools.

Trypanocide Kills *Trypanosoma* (which see).

Trypanosoma A genus of protozoans that cause serious infections in humans and domestic animals.

***Tumor* (tumour)** An abnormal mass of tissue. Tumors can be benign or malignant.

***Typhoid* (typhoid fever)** A communicable illness caused by the bacterium *Salmonella typhi*, which is transmitted in contaminated food, water and sewage, and indirectly by flies or by faulty personal hygiene. Symptoms include fever, diarrhea, prostration, headache, and intestinal inflammation. Asymptomatic carriers (such as the legendary "typhoid Mary") harbor the organism in their gallbladder and excrete it in their stools for years.

Typhus An acute infective illness caused by the bacterium *Rickettsia prowazekii*, which is transmitted by the human body louse, or sometimes also by ticks and fleas. Symptoms include high fever, stupor alternating with delirium, intense headache, and a dark red rash.

Unguent An ointment, or soothing or healing salve.

Uremia Accumulation in the blood of toxic constituents normally excreted in the urine, producing a severe toxic condition; usually the result of kidney disease.

Ureter Duct taking urine from the kidneys to the bladder.

Ureteralgia Pain or spasm of the ureters.

Ureteritis Inflammation of the ureter, the duct that carries urine from the kidney to the bladder.

Urethra The duct that carries urine from the bladder and semen from the prostate and seminal vesicles out through the penis.

Urethritis Inflammation of the urethra.

***Urticaria* (hives)** A condition of the skin, usually caused by an allergic reaction, characterised by pale or reddened irregular, elevated patches and severe itching (also see *Hives*).

Uteritis Inflammation of the uterus.

Uvula The central fleshy lobe hanging down from the edge of the soft palate.

Vaginismus A painful spasmodic contraction of the vagina, often preventing copulation.

Vaginitis Any inflammation of the vagina.

Vaginosis Vaginal infection with presence of anaerobic bacteria such as *Gardnerella* and *Mycoplasma*, and a discharge with a disagreeable odor.

Varicose Abnormally swollen or dilated, especially veins.

Vasculitis Inflammation of one or several blood vessels.

Vasoconstrictor Narrows blood vessels so that less blood is able to flow through, and blood pressure is increased.

Vasodilator Dilates or relaxes the blood vessels, reducing blood pressure.

Venery Pursuit of sexual pleasure.

Vermifuge Expels parasitic worms from animal bodies. An anthelmintic (which see).

Vesicant Causes blisters.
Vesicle A blister (small elevation of the cuticle of the skin, containing a clear watery fluid).
Viristatic Inhibits viruses.
Vitiligo A skin disorder characterized by smooth white spots on various parts of the body.
Vulnerary Heals external wounds.

Wheal 1. A suddenly formed elevation of the skin surface, for example a welt. 2. A flat burning or itching eminence of the skin.
White blood cells These are found in the lymph glands and spleen, and circulate in the blood and lymphatic system. They are part of the immune system attacking foreign invaders of the body.
Whitlow A painful blistery eruption of a finger or toe, resulting from a herpes viral infection.

Yaws An infectious contagious tropical disease caused by a spirochete.

Dictionaries

Dorland's Illustrated Medical Dictionary, 28th edition, Saunders [130,000 terms]:
http://www.emery-pratt.com/dorlands.htm [advertisement for book]
www.sppbooks.com.au/sppmailing/dorlands.html [advertisement for CD-ROM]

Hocking, G.M. 1997. A dictionary of natural products: terms in the field of pharmacognosy relating to natural medicinal and pharmaceutical materials and the plants, animals, and minerals from which they are derived. 2nd ed. Plexus, Medford, NJ. 994 pp.

Merriam-Webster Medical Dictionary [advertisement for CD-ROM, this must be purchased; available to subscribers of America Online; also available as book; "Webster's medical dictionary" is a paperback with ca. 35,000 entries; "Merriam-Webster's medical desk dictionary" is a hardcover book (1996, 894 pp.) with ca. 55,000 entries]: http://www.opengroup.com/open/cdbooks/087/0877794618.shtml

Stedman's electronic medical dictionary, 100,000 medical words with definitions [advertisement for CD-ROM; this must be purchased; also available as book (1995, 26th edition, 2030 pp.)]:
http://www.medproda.demon.co.uk/steddict.html

Internet sources for definitions of medicinal and pharmacological terms

Medline plus dictionaries [links to numerous medical dictionaries]:
http://medlineplus.nlm.nih.gov/medlineplus/dictionaries.html

Cancer web - the on-line medical dictionary (45787 words as of Mar 20 1998) [impressively huge, but most of the terms are not commonly encountered in herbal medicine]:
http://www.graylab.ac.uk/omd/

English dictionary of medical terms (1828 words as of Mar 22 1998) [a very useful database]:
http://allserv.rug.ac.be/~rvdstich/eugloss/DIC/dictio01.html

MedicaL/herbal glossary, 2nd edition, by Michael Moore [prepared by an expert herbalist]:
http://chili.rt66.com/hbmoore/ManualsMM/MedHerbGloss2.txt

MedicineNet medical dictionary (4195 words as of Apr 4 1998) [written in comprehensible language for laypeople]:
http://www.MedicineNet.com/ALPHAIDX.asp?li = MNI&ag = Y&Param = ADICT

OneLook dictionaries database [lists links to numerous medical dictionaries, as well as a wide variety of other dictionaries]:
http://www.onelook.com/browse.shtml

Plants for a future: species database - medical uses [prepared from a botanical viewpoint]:
http://www.scs.leeds.ac.uk/pfaf/D_med.html

Glossary of medical terms [a basic glossary of terms for introductory medical students]:
http://www.countway.med.harvard.edu/publications/Health_Publications/gloss.html#anchor5250835

American Medical Association medical glossary:
http://www.ama-assn.org/insight/gen_hlth/glossary/glos_ag.htm

Pharmaceutical Information Network glossary:
http://pharminfo.com/pia_glos.html

Merriam-Webster medical dictionary:
http://www.intelihealth.com/IH/ihtIH?t=9276&p=~br,IHWI~st,408I~r,WSIHW000I~b,*I